Nurturing the Older Brain and Mind

Pamela M. Greenwood and Raja Parasuraman

The MIT Press
Cambridge, Massachusetts
London, England

Excerpt from "Provide, Provide" (from *The Poetry of Robert Frost*, ed. Edward Connery Lathem, © 1969 Henry Holt and Company, © 1936 Robert Frost, © 1964 Lesley Frost Ballantine) reprinted by arrangement with Henry Holt and Company, LLC.

Excerpt from "In View of the Fact" (from A. R. Ammons, *Bosh and Flapdoodle*, © 2005 John R. Ammons) used by permission of W. W. Norton & Company, Inc.

For information about quantity discounts, email special_sales@mitpress.mit.edu.

Set in Stone Sans and Stone Serif by Toppan Best-set Premedia Limited. Printed and bound in the United States of America.

Library of Congress Cataloging-in-Publication Data

Greenwood, Pamela M., 1947–
Nurturing the older brain and mind / Pamela M. Greenwood and Raja Parasuraman.
  p. ; cm.
Includes bibliographical references and index.
ISBN 978-0-262-01714-5 (hardcover : alk. paper)
I. Parasuraman, R. II. Title.
[DNLM: 1. Brain—physiology. 2. Aged. 3. Aging—physiology. 4. Cognition—physiology. 5. Neuronal plasticity. WL   300]
LC classification not assigned
612.8'20846—dc23

2011030565

10 9 8 7 6 5 4 3 2 1

Man makes himself through enlightened choices that enhance his humanness.
—Rene Dubos, *So Human an Animal*

Contents

Preface

The population of older adults continues to expand, both in the United States and worldwide. Whereas physical abilities clearly decline as we age, the most recent scientific research does not support the long-held view that our minds must also inevitably grow dim. The popular notion of inexorable, progressive cognitive decline raises the prospect of nations increasingly burdened by their growing numbers of older citizens. Certainly, relative to the young, older people, on average, perform at a lower level on many (but not all) perceptual, cognitive, and motor tasks. But with the exception of disorders affecting older adults, such as Alzheimer's disease, cognitive decline in healthy old age is neither universal nor inevitable.

Some older adults maintain their level of cognitive functioning even in very old age, and a few even outperform their younger counterparts. Is there something special about these individuals? Yes and no. No, because various experiential and lifestyle factors—including education (both early and late in life), exercise, diet and nutrition, opportunities for new learning, and, more controversially, use of estrogen or cognition-enhancing drugs—may contribute to maintenance of cognitive function in late life. Yes, since certain genetic factors may help some individuals to avoid the extreme effects of cognitive aging or of dementing disease. But, as we document in this book, irrespective of genes, all older individuals can benefit cognitively from exposure to these experiences at any stage of life.

We begin the book with a reappraisal of the scientific evidence on aging, cognitive function, and the brain. Across a range of species—rats, monkeys, humans—we show that a substantial subset of older individuals do not succumb to cognitive decline and do not show age-related neuronal changes. Moreover, the neural substrate of cognitive aging has not been identified. The best-documented brain change over the course of healthy aging—regional brain shrinkage—is not reliably linked to cognitive decline. The effects of aging on biophysical properties of neurons are selective and subtle, seen only in specific brain regions and cell types. Nor do age-related synaptic changes necessarily lead to corresponding changes in neuronal transmission and thereby to cognitive decline. Finally, age-related synaptic changes are plastic and reversible.

We describe the scientific evidence supporting these statements regarding cognitive and brain aging in chapters 1–3. In chapter 4 we offer a theoretical framework for why certain experiential and lifestyle factors may promote and maintain cognitive vitality in older adults. Recent evidence shows that even the aged brain remains capable of plasticity (the ability to adapt to and benefit from experience). Such plasticity may explain why certain experiences can protect against cognitive decline late in life. These experiential factors, described in chapters 5–10, include childhood and adult education, aerobic exercise, diet and nutrition, estrogen and cognition-enhancing drugs, cognitive training, and learning. Consistent with the view that plasticity is not lost in old age, some of these behaviors can confer benefits even if initiated late in life. Other factors that appear to exert ill effects on cognitive functioning, such as a high-fat diet, can be avoided. Of course, if such lifestyle changes were easily made, more of us would be trim and fit non-smokers. Research into ways to motivate people to avoid the truly dire consequences of poor diet and lack of exercise is ongoing. Some factors moderating cognitive aging remain controversial and will have to be better assessed for risks and benefits before their use can be recommended for all. These include dietary restriction and the use of estrogen and cognition-enhancing drugs. Nevertheless, on balance, there is good evidence that there are many self-initiated behaviors and experiences that older individuals can engage in to exploit the power of brain plasticity and promote cognitive vitality. Because individual experiential factors, such as exercise or diet, yield benefits, combining them may provide even greater benefits than either alone. We discuss the combined effects of such preventative actions in chapter 10.

Complementary to the approach of adapting the older individual is the approach of adapting the environment (e.g., at work and at the home) to support optimal cognitive functioning in older adults. That approach—based on principles drawn from human factors and from cognitive engineering—involves changing the environment and changing the interfaces between people and the devices and systems they use. It is described in chapter 11. Cognitive engineering seeks to design systems that take human capabilities into account while minimizing the effects of the physical and cognitive limitations of older adults. A number of demonstration projects have been undertaken, most of them aimed at helping older people to "age in place"—that is, to remain in their own homes despite physical health problems. One example is the use of computerized "virtual assistants."

Genetic factors may temper the influence on cognitive aging of the various experiential and environmental factors we discuss in the earlier chapters. Advances in molecular genetics have increased our knowledge of how genes contribute to individual differences in various cognitive functions, of the brain networks that control these functions, and of the neurotransmitters that innervate these networks. As gene-environment interactions become better understood, we may be in a better position to understand both the potential and the limits of efforts to promote cognitive vitality in older individuals. We discuss these issues in chapter 12.

In chapter 13 we summarize several ways in which the effects of cognitive aging can be ameliorated. These methods can make a measurable difference to cognitive and brain integrity late in life. Yet it is prudent to consider not only what *can* be done but also what *should* be done. Dietary restriction is not for everybody and may pose risks that aren't known yet. As to the use of estrogen by peri- or post-menopausal women, the jury is still out. The use of cognition-enhancing drugs raises ethical issues and concerns about addiction. The use of computer-based cognitive assistants in the home raises issues of complacency, dependency, and the possibility of social neglect. But many of the other ameliorative factors we discuss in this book—educational experiences, physical exercise, diet and nutrition, opportunities for new learning—pose minimal risks to older adults.

The first wave of the post-World War II "baby boomers" is now reaching the point at which age-related cognitive decline can be detected, on the average. But nothing forces any individual to follow the mean course. In the past 20 years, scientific research on cognitive and brain aging has given us a sound knowledge base that we can exploit to point to how older adults can lead rich and cognitively vital lives.

Preparation of this book was aided by grants we received for our research from the National Institute on Aging, the Virginia Council on Aging, and the Mason-INOVA Life Sciences Research Collaboration Fund. The interpretations of data and the opinions we offer in this book are our own and should not be considered to reflect the views of those agencies. We also thank our colleagues at George Mason University and the many postdoctoral fellows, graduate, and undergraduate students who have worked with us for fruitful discussions on the effects of experiential factors on brain plasticity and cognitive aging.

1 Global Aging and Cognitive Functioning

. . . Make the whole stock exchange your own!
If need be occupy a throne,
Where nobody can call you crone . . .

No memory of having starred
Atones for later disregard,
Or keeps the end from being hard.

Better to go down dignified
With boughten friendship at your side
Than none at all. Provide, provide!

—Robert Frost, "Provide, Provide"

1.1 The Demographics of Aging

When we were young children, people of our parents' generation were generally expected to live about 60 years. Our respective grandparents, born around the end of the nineteenth century, could expect to live only to their early fifties or so. If one goes back even further in time, estimates of life expectancy become progressively lower—perhaps 40 in the Middle Ages and 30 in the Bronze Age. Although life-expectancy estimates can be fraught with error (for example, because of variation in infant mortality rates), the general trend is nevertheless clear: people are living longer and longer, especially in the developed countries. Of course, the larger hope—one that we would wish for our own children—is that people will not only live longer but will lead better, more fulfilling lives as they age. What can be done to enable older people to lead cognitively vital lives?

In the United States, as in the rest of the world, the population is growing older. In 1900, about 4 percent of the population of the US was 65 or older. By 2004, that percentage had increased to 12. By 2050, the percentage is expected to be nearly 25.[1] Figure 1.1, which is based on projections by the US government's Administration on Aging, illustrates these trends clearly. Also evident in figure 1.1 is the striking increase in the proportion of people 85 and older, the fastest-growing segment of the US population. Because similar changes are occurring in many other countries, including China and India, the term *global aging* has come into use. Some see global aging as an issue that matches global warming in importance.

In view of these population trends, it is not surprising that scholars in several disciplines are paying increasing attention to the health, social, and economic implications of global aging,[2] including the effects of longer life expectancy on overall health, disability, and susceptibility to age-related disease. Among the social consequences of global population aging are altered intergenerational relationships, changed familial organization, and lack of personal support systems. The economic issues include, in addition to rising health-care costs, the effects of aging on the labor market and the quality of pension plans. As a result of global aging, many countries are

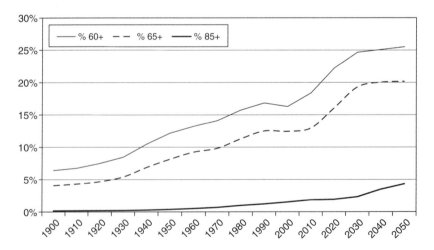

Figure 1.1
Past and projected percentages of older people in three age groups: over 60, over 75, and over 85. Source: US Administration on Aging.

now faced with the prospect of large numbers of their workers retiring in the next few years. High retirement numbers will simultaneously increase pension costs and reduce the numbers of skilled workers. A good example is Finland, where, as in a number of other European countries, the government long encouraged early retirement. The Finnish government has now reversed course and has implemented programs to help private Finnish companies retain their older skilled workers.[3]

These issues are not limited to the developed countries. China has a rapidly aging population, and the one-child policy has left many people with only one child to assist them in old age. The retirement age in China is also very low, currently 55 for women and 60 for men. However, only about 20 percent of the Chinese have pensions. The government is attempting to reform the pension system, especially in rural areas, with help from the Asian Development Bank.[4] However, because of the weak pension system, many Chinese will have to work during their official retirement years.

Other countries are raising the retirement age, forcing their older workers to stay in their jobs longer. In the United States, the retirement age for those wanting to obtain full Social Security benefits will be 67 for people born after 1960. In Italy, the retirement age was changed from 57 to 60 for women and to 65 for men in 2008. In France, on October 22, 2010, after weeks of demonstrations by workers, the parliament narrowly approved raising the retirement age from 60 to 62. Britain plans to increase the retirement age from 65 to 68 by 2044. Germany is increasing it to 67 by 2029.

As a result of these policies, there will be greater numbers of older adults in the workforce. But even during retirement, many older people will have economic incentives to continue to work. In a recent report, McKinsey & Company determined that 85 percent of American baby boomers plan to work in retirement,[5] in part because they have not saved for retirement to the same extent as previous generations.

Thus, although there are some differences between countries, a clear pattern can be seen worldwide: older people are going to be staying in the workforce longer. In order to be productive and to be financially able to support themselves late in life, those older workers will have to pay attention to their physical health. Equally important, they will have to avoid the decline in cognitive functioning that accompanies aging in the older adult on average. Only some of them will succeed at that. What can be done to help the majority of older adults enjoy cognitive vitality for a longer period of time?

1.2 Cognitive Aging and Brain Aging

The psychological ramifications of growing old—particularly as they affect independent living, mobility, mental health, and well-being—have received much attention in the past three decades from researchers and from the National Institutes of Health and other federal agencies. The psychological effects of adult aging, including the effects on sensory, cognitive, and motor functions, are collectively referred to as *cognitive aging*.

Researchers in the field of cognitive aging investigate such representative mental functions as attention, memory, perception, speech and language, decision making, and problem solving.[6] Most often, a *cross-sectional* design is used to examine age-related changes. In the typical study, groups of young adults (e.g., people in their twenties) and older adults (e.g., people in their sixties) are compared for performance on a cognitive task.[7] Occasionally, however, researchers are able to conduct a more costly and complex *longitudinal* study in which a cohort of similarly aged adults is followed periodically for several years so that changes within individuals can be assessed. Two well-known longitudinal studies are the Seattle Longitudinal Study[8] and the Berlin Aging Study.[9]

The received wisdom characterizes cognitive aging as inexorable decline. Psychologists have shown that, on average, many (though certainly not all) cognitive functions in older adults decline in efficiency.[10] Although there is agreement that the mean cognitive performance of adults declines with age, researchers disagree on when that decline begins. Cross-sectional studies find that cognitive decline begins as early as the thirties; longitudinal studies find that it doesn't begin until the sixties.[11]

Brain researchers have long sought to identify a neural basis for age-related decline, irrespective of the precise age at which it may begin. Using non-invasive brain imaging techniques, such as magnetic-resonance imaging, not only can researchers peer into the living brain; they can obtain quantitative estimates of the volumes of specific regions of the brain. The general conclusion from such studies is that some regions of the brain shrink with age.[12]

Popular wisdom also alludes to the inevitability of cognitive and brain aging: as we age, we struggle to see and hear, our minds get clouded, and we are slower to act, all because we reach, as Shakespeare wrote, "second childishness and mere oblivion / Sans teeth, sans eyes, sans taste, sans everything."[13]

If the received wisdom on cognitive and brain aging, whether popular or scientific, is true, the prospects are less than pleasant. Nations will be increasingly burdened economically with large numbers of older adults who are less capable of independent living and of enjoying meaningful, cognitively vital lives. But cognitive decline in healthy old age is neither universal nor uniform. Contrary to the popular view, there is little evidence of loss of neurons with age (except in dementing illnesses such as Alzheimer's disease), and there is increasing evidence that the aging brain retains some degree of plasticity (changed innervation in response to environmental stimulation). Nor is it in the best interests of a nation or of its aging citizens to accept the inevitability of cognitive decline. The recent McKinsey report[14] estimated that the ability of baby boomers to work during retirement would enable the US economy to generate an additional $12.9 trillion in gross domestic product by the year 2025. Thus, retention of good cognitive skills will be important not only for the financial well-being of older individuals, but also for the economic performance of the countries in which they live.

With increasing numbers of older people in the workforce worldwide, the moment has arrived for the scientific community to address the question of how to best retain good cognitive functioning late in life. Fortunately, from slow beginnings, there has been steady growth in the volume of scientific research on this topic. In the subsequent chapters, we discuss the roles of various experiential and lifestyle factors in the maintenance of cognitive integrity throughout healthy old age.

1.3 Healthy Aging and Disorders of Aging

Our focus in this book is on the cognitive features of so-called healthy aging—that is, aging in the absence of clinically evident disease. Yet many are aware that the later years of life also bring with them the threat of various forms of disease. Foremost among these are neurodegenerative disorders (including Alzheimer's disease and Parkinson's disease), heart disease, hypertension, stroke, and related vascular illnesses. Each disease can have significant and often devastating effects on cognitive function, sometimes (as in the case of Alzheimer's) years before the disease is clinically diagnosed.

Alzheimer's disease is of particular interest for several reasons. First, because its most common form is diagnosed in older adults (usually over 60), and because the risk of developing it increases with age, the phenomenon of global aging means that more and more people will be affected. Currently there are about 5 million cases of Alzheimer's disease in the United States, but the number of cases is expected to surpass 7 million by 2030 and to be between 11 million and 16 million by 2050.[15] Second, in its "pre-clinical" stage (that is, before it is diagnosed), which may last many years, the cognitive changes it causes overlap those of healthy aging. Consequently, there is much interest in understanding how the neural and cognitive changes associated with this degenerative disease differ from those associated with healthy aging.

The cognitive deficits of Alzheimer's disease have been extensively studied and described.[16] Initially, there is only a loss in memory.[17] The memory deficit, considerably greater than the occasional memory lapses that healthy older adults experience, typically gets progressively worse with time. Clinical diagnosis of Alzheimer's disease therefore requires not only that a patient score poorly on a standardized test of memory, such as the Wechsler Memory Scale, but also that the patient's memory performance be even poorer on a test conducted 6 months or a year later. (In practice, clinical diagnosis of Alzheimer's disease involves several stages and various other inclusionary and exclusionary factors.[18])

In addition to a deficit in memory, Alzheimer's disease results in deficits in attention and in visuospatial functioning.[19] Two types of attention that are especially vulnerable are *covert* attention and *executive* attention. The first involves the ability to focus attention independently of the eyes. (Think of being at a cocktail party and keeping your eyes on the face of the boring person in front of you while you attend to interesting gossip from a nearby group.) The ability to shift and focus covert visuospatial attention appears to be unchanged in healthy older adults[20] and in the early, mild stages of Alzheimer's disease.[21] In contrast, redirecting (disengaging) attention from a spatial location[22]—for example, shifting your attention from that interesting gossip back to the boring person whose face your eyes never left—is moderately affected by normal aging[23] and markedly impaired by Alzheimer's disease.[24] The deficit in covert attention in Alzheimer's disease is associated with a reduction in glucose metabolism in the right parietal lobe of the brain, as measured by positron-emission tomography.[25] The second type of attention that is affected in Alzheimer's disease is executive attention—the ability to identify goals and means of achieving them while suppressing competing tendencies. One way to measure executive attention is by measuring the ability to name the color of the word "blue" when it is printed in red ink (the Stroop test). It is difficult to suppress reading the word itself in order to name the color of the word ("red"). Such a deficit in executive control is also seen in early Alzheimer's disease.[26] These attentional problems may contribute to and exacerbate the memory disorder seen in patients with Alzheimer's.[27]

Although Alzheimer's is an age-related disease, it is not an inevitable consequence of aging. Most old people will not develop it. Even though the risk of developing it is greater for people in their eighties than for people in their sixties, many very old individuals will never develop the disease, as an increasing number of healthy centenarians testify.[28] Hence, in a sense, the study of Alzheimer's disease is secondary to an understanding of the cognitive changes that accompany healthy aging. But investigations of neuropsychological disorders can also inform understanding of normal behavior. We shall therefore make occasional reference to evidence from studies of disorders of aging, particularly Alzheimer's disease, throughout this book. But in the main we shall restrict our review to cognitive and brain changes associated with healthy aging, and to ways of ameliorating these changes and promoting optimal cognitive functioning in everyday life.

One caveat is necessary before we proceed. We discuss cognitive aging as though it were completely dissociable from Alzheimer's disease. For obvious reasons, most studies of cognitive aging attempt (with varying degrees of rigor) to exclude participants who have Alzheimer's disease and other disorders of aging that can compromise cognition. A study comparing the average performance of a group of older adults against that of a group of younger adults might wrongly conclude that healthy aging leads to cognitive decline if only a few members of the older group (those with a disorder) showed extremely poor performance, thereby bringing down the average score of their group. To guard against that possibility, most studies of cognitive aging use some simple screening test for dementia, such as the Mini-Mental State Exam (a 30-item test of orientation and memory[29]), and exclude those who score below a certain threshold.

Screening for clear signs of dementia is obviously important in a study of healthy aging. However, screening for all possible disorders is, to an extent, a fool's errand. It is easy to exclude from experimental participation those individuals who clearly meet clinical criteria for dementia. But what about people who are apparently normal at the time of testing but will develop Alzheimer's disease in the future? Although its natural history isn't fully understood, Alzheimer's disease is not generally thought to be an inevitable consequence of aging. To the extent that studies of cognitive aging include people with "pre-clinical" Alzheimer's, such studies cannot be viewed as accurately reflecting healthy cognitive aging. Martin Sliwinski and colleagues demonstrated that when people who were later found to have developed Alzheimer's disease at a 4-year follow-up were removed from group cognitive performance data collected earlier, the effect of age on cognition was lessened and the variability decreased.[30] In contrast, Lars Bäckman and colleagues found the effects of removing people who had become demented or died by the time of a 3-year follow-up to be quite weak.[31] This question, therefore, remains unresolved. Nevertheless, it is acknowledged that impending but undiagnosed Alzheimer's disease can contribute to variability in cognitive performance among older people.

In the subsequent chapters, we describe how brain plasticity can be exploited to enable the large majority of the older population to maintain cognitive vitality.

Summary

The worldwide aging of the adult population, both in the developed countries and in the developing countries, poses significant economic and social problems that must be addressed. An inexorable decline in cognitive functioning as people age was once thought to be the root of the problem. However, recent evidence has overturned that popular view. Many older adults not affected by dementia or by other diseases of aging are able to preserve effective cognitive functioning as they age. Furthermore, the human brain retains a degree of plasticity to environmental stimulation even in old age.

2 Cognitive Aging: Neither Universal nor Inevitable

Anyone who stops learning is old, whether at twenty or eighty.
—Henry Ford

2.1 Variability in Cognitive Aging

Is age-related cognitive decline inevitable? Some years ago, many scientists would have answered this question in the affirmative. The views of most laypersons also echo the older scientific account. But such views are misleading and incomplete. Not all individuals, not even most, show a decline in cognitive functioning in old age. This is true whether one considers rats, monkeys, or humans. Several research groups have reported cognitive deficits in only about half of the aged laboratory animals in their studies, the other half performing as well as younger subjects on a range of tasks.[1] Such variability has also been seen in monkeys. One study found that three out of ten aged monkeys performed as well as young monkeys on a recognition memory task.[2] Consistent with that finding, another study found that three out of five middle-aged monkeys and six out of ten old monkeys showed levels of acetylcholine (a neurotransmitter thought to be important for memory and attention) comparable to the levels in young adult monkeys.[3]

A skeptic might argue that the samples in the aforementioned studies were relatively small. The reason for the small numbers in primate aging studies is that monkeys live relatively long. Keeping and nurturing them in a controlled environment until they are old requires complex and expensive facilities. Despite the small samples, the results show clearly that a substantial minority of older monkeys do not exhibit age-related cognitive decline. Nor do all humans show cognitive decline in old age. In a large and well-respected longitudinal study of 229 older adults, Willis and Schaie observed no cognitive decline over 14 years in 46 percent of their older participants.[4] In another large sample of older adults, Salthouse observed that 68–87 percent of the variance in cognitive test scores was unrelated to age.[5] Moreover, although many cross-sectional studies report that older individuals perform more poorly than young ones, buried within the average scores reported in any given study are older adults who perform as well as their younger counterparts or even better.

That many individuals do not show age-related cognitive decline indicates that cognitive aging is not inevitable. Why do some individuals undergo cognitive decline late in life whereas others do not? Genetically or environmentally driven variations in brain plasticity probably play an important role. In chapter 4 we outline a neurocognitive framework for cognitive aging. A major element of our framework is plasticity—the brain's capacity to implement neuronal changes, including synaptic structures and even neurogenesis (the growth of new neurons), and the mind's capacity to adapt to change and benefit from experience. We argue that both cognitive plasticity and brain plasticity are present in old age and can be exploited in a number of ways to help adults maintain optimal cognitive functioning in their later years.

2.2   Brain Plasticity

That older brains retain a degree of plasticity goes against many long-held and widely accepted beliefs. One popular notion is that cognitive aging is inevitable and is associated with gradual degradation in the brain's structural integrity. All of us have heard tiresome jokes about losing brain cells as we age. There is little truth in them. Moreover, an irreversibly regressive substrate of cognitive decline in the *healthy* aging brain has not yet been identified—that is, we cannot point to a specific brain change that is reliably associated with cognitive aging.

The best-documented evidence of age-related brain change involves regional shrinkage of brain volume, especially in the prefrontal cortex.[6] (For an illustration of the regions of the human brain, see figure 2.1.) Yet such reductions in brain volume are not consistently related to cognitive decline. Longitudinal studies using magnetic-resonance imaging (MRI) to measure regional brain shrinkage and neuropsychological tests to assess

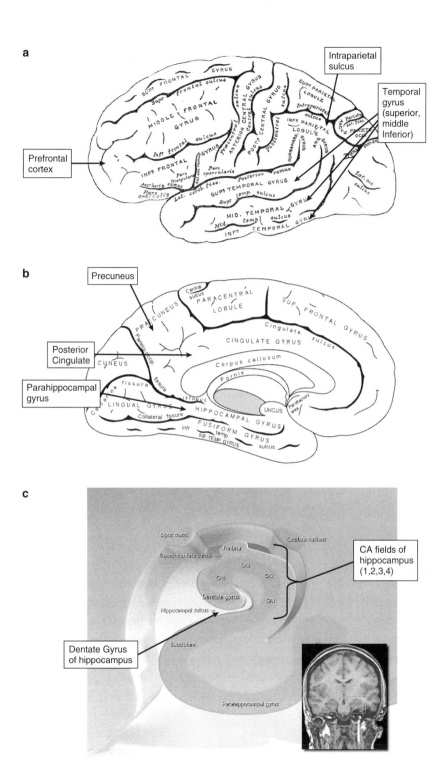

Figure 2.1
(a) Lateral view of cerebral hemispheres. (b) Medial (interior) view of cerebral hemispheres. (c) Dentate gyrus and fields CA1–CA4 of hippocampus.

age-related cognitive change in the same individuals over time have found either an inverse relation[7] or no significant association between the two.[8] These findings raise questions about the functional significance of changes in brain volume as assessed by MRI. But what is the anatomical significance of regional brain shrinkage in aging? An obvious possibility is loss of neurons, but this has not been demonstrated. Careful post-mortem neuroanatomical studies have shown that neuron loss generally doesn't occur in healthy old brains.[9] Consequently, whatever regional MRI volume measures signify, they don't appear to index neuron number.

Not only do old brains not lose neurons; old brains can generate new neurons, even in people with terminal cancer or neurological disorders. Reports of neurogenesis (the birth of new neurons) have overturned the decades-old dogma that brain neurons cannot be replaced in adulthood. There was marked skepticism[10] toward the early reports of neurogenesis in the brains of adult animals.[11] Joseph Altman was the first scientist to report adult neurogenesis in mammals (cats). He published his study in the prestigious journal *Science* in 1971.[12] The scientific community largely ignored that report, and Altman was unable to obtain funding to follow up his findings. Fernando Nottebohm reported that adult songbirds grow new neurons each season as they learn new songs.[13] However, because the work was done in birds, it was not viewed as relevant to mammals. The next study to report adult neurogenesis in mammals (rats) was by Michael Kaplan.[14] Kaplan's work too was published in *Science*, so certainly it was available to a wide audience. A critique of the notion of adult neurogenesis published by Pasko Rakic of Yale University, a well-known authority on primate brains who had not seen evidence of adult CNS neurogenesis in his own extensive work, has been cited widely.[15] Rakic is quoted as having remarked to Kaplan that "those may look like neurons in New Mexico, but they don't in New Haven."[16] Nottebohm has stated that Rakic "singlehandedly held the field of neurogenesis back by at least a decade."[17] It was not until the 1990s, when researchers persisted in the face of criticism from Rakic, that papers on adult neurogenesis began to appear.

The first evidence of neurogenesis in humans was published in 1998 by Peter Eriksson and colleagues. They reported neurogenesis in the hippocampus of the brains of older adults.[18] The hippocampus, a seahorse-shaped structure found deep in the temporal lobes, is essential for memory formation.[19] Consistent with Eriksson's finding, neural progenitor cells—an intermediate stage in the development of stem cells to neurons—have been seen in the sub-ventricular zone of the brains even of very old humans, some of them with neurological diseases.[20] Even in severe neurodegenerative diseases, including Huntington's disease, newborn neurons have been seen to migrate from their place of origin to the caudate nucleus.[21] If the brains of terminally ill older adults and those in the throes of devastating neurodegenerative disorders can give birth to new neurons, surely healthy older adults can do the same. There is recent evidence that they do.[22]

Neurogenesis and other mechanisms of brain plasticity appear to underlie the beneficial effects of many environmental factors aimed at ameliorating cognitive aging. Some apparent manifestations of brain aging can be reversed through such manipulations. For example, in a rat model of Alzheimer's disease, the chronic administration of nicotine prevented both memory loss and impairment in long-term potentiation (a well-studied neural signal associated with memory formation in the hippocampus).[23] Age-related loss of synapses (the regions of communication between individual neurons) can also be reversed by nicotine,[24] by estrogen,[25] and by exercise.[26] Housing mice in an "enriched environment" with toys, tunnels, running wheels, and other mice also increases the complexity of large pre-synaptic terminals in the hippocampus throughout life.[27]

In summary, the evidence emphasizes that the mind and the brain can remain plastic and continue to function well in old age. A better understanding of how these and other factors influence brain plasticity could help resolve the question of why only some individuals show age-related cognitive decline, and by doing so could suggest ways of limiting cognitive aging.

2.3 Individual Trajectories of Cognitive Aging

There is broad agreement that, on average, cognitive functioning declines with age. As we noted in the previous chapter, Denise Park and colleagues reported nearly linear declines from young adulthood to old age in a number of cognitive functions that they tested, including speed of processing (how fast mental operations occur), working memory (keeping things in mind for a short period of time), and long-term memory (recognizing or recalling previous events). The exceptions were vocabulary and verbal knowledge, which did not vary with age.[28] In their widely cited cross-sectional study, Park et al. compared groups of people of different ages, starting in early adulthood, and observed that cognitive decline followed a linear trajectory from the thirties to the seventies.

Cross-sectional studies can be criticized because they compare different cohorts of people (that is, people born at different times) and consequently it is not clear whether observed age-related changes are due to chronological age or to the experience of growing up in a specific era. For example, people who were teenagers during the 1940s had different social, cultural, and educational experiences than people who were teenagers during the 1960s. These cohort differences can affect cognitive functioning. Because cohort and chronological age are perfectly correlated, their effects cannot be separated. Consequently, many researchers have proposed that longitudinal studies can provide better evidence of age-related cognitive decline. In a longitudinal study, the people are followed as they age. Longitudinal studies also find a linear decline in speed of processing from young adulthood to old age. However, for other, arguably more important functions, including spatial orientation, reasoning, and verbal memory, average longitudinal decline is not evident until around age 60.[29] There is currently a vigorous debate as to when, on average, cognition begins to decline. Cross-sectional studies suggest the thirties; longitudinal analysis suggests the sixties.[30]

Irrespective of the specific age at which cognitive aging is detectable, proponents of both cross-sectional and longitudinal studies agree that, on average, older adults perform more poorly than young people on most cognitive tasks. At the same time, both groups of researchers would acknowledge that there are marked differences between individuals in the degree and the type of cognitive change that they exhibit as they age. Although the average age-related trend is clearly downward, there is considerable variability around the average. Figure 2.2 illustrates this variability. The triangles show the average trend of age-related cognitive decline, beginning in the twenties and extending until late in life. The arrows indicate the variability that surrounds the average. The solid, dashed, and dotted lines show the individual trajectories of three hypothetical individuals. The dashed line shows the trajectory of the average person: a progressive decline in cognitive functioning until age 90. The individual represented by the dotted line shows some fluctuation in cognitive functioning but otherwise is as cognitively vital in the nineties as in the

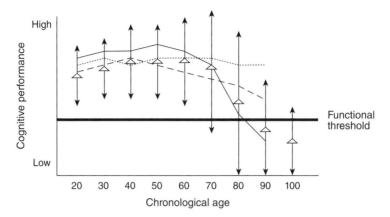

Figure 2.2

Hypothetical longitudinal trajectories of cognitive aging experienced by three fictional individuals plotted together with group performance at decade intervals. Triangles represent the mean cognitive performance in each age group. Vertical bars represent variation in cognitive performance about the mean of the group. The heavy horizontal line is the functional threshold below which individuals are greatly impaired in everyday cognitive functioning.

twenties. The individuals represented by the dashed and dotted lines, however, both enjoy a level of cognitive performance that is well above what is necessary for effective functioning in everyday life. The solid line represents the trajectory of a person who is not so fortunate, perhaps because of a degenerative disease such as Alzheimer's.

Figure 2.2 emphasizes our point that there are considerable individual differences buried in the average age-related function showing a decline in cognition. Individual trajectories of cognitive change can be such that many older individuals—human or animal—can function as well as young individuals.[31]

Many longitudinal studies, which according to some scholars provide the best index of within-individual change,[32] provide confirmatory evidence that the trajectories of cognitive aging do not inevitably point to decline. For example, in a large sample of healthy older people followed for 5 years by Robert Wilson and colleagues, some individuals declined, many remained stable, and some showed improvement on various cognitive tasks.[33] Similarly, Nicole Schupf and colleagues in Richard Mayeux's lab found that over at least 3 years about 25 percent of a large sample of older people showed no cognitive decline and another 15 percent showed only slow decline.[34]

Brain aging too shows wide individual variability. A large study of age-related changes in brain volume over 5 years found marked individual differences. Although most people showed decreased regional volume, a subgroup showed increased volume.[35] Furthermore, Jonas Persson and colleagues in Lars Nyberg's lab found that older people who hadn't declined cognitively over 10 years had a larger hippocampus (a brain structure critical for the ability to form new memories) and greater integrity of white matter (the collective axons of neurons in the brain) than those who had declined over the same interval.[36] Consistent with that finding, Paul Borghesani and colleagues found that people who improved in episodic memory in middle age—in contrast to those who declined or remained stable—showed greater hippocampal volume 14 years later in old age.[37]

What accounts for these individual differences in trajectories of cognitive aging? Some of the differences in apparent age-related cognitive and brain change may be due to chronic health problems. A recent study reported that in persons with cardiovascular problems, but not in persons in good health, white-matter lesions and shrinkage of the fusiform cortex were correlated with decline in working memory over 5 years.[38] Blood glucose levels—related to diabetes—have been found to be associated with both cerebral blood flow in the dentate gyrus of the hippocampus and memory performance.[39] Thus, not only are there substantial individual differences in studies of cognitive and brain aging; in addition, somatic health problems such as diabetes, hypertension, and heart disease may play a role in these individual differences. At the other end of the spectrum of individual variability (see figure 2.2), individuals who do not follow the typical trajectory of decline may have benefited from brain plasticity through their exposure to various experiential and lifestyle factors—something we discuss in greater detail in subsequent chapters.

2.4 Cognitive Aging Is Not Always Negative

Not only is cognitive aging not universal; when it does occur, it is not necessarily a negative development. Indeed, some apparently age-related cognitive deficits may confer benefits under specific circumstances. Shelley Carson and colleagues investigated the role of distractibility in creativity and found the two to be related. Young people categorized as creative were much less likely to screen irrelevant stimuli from conscious awareness, a finding that was confirmed by a meta-analysis of other related studies.[40] This is relevant to cognitive aging because older adults are known to be more distractible than young people. Lynn Hasher and Rose Zacks have long argued that a failure to inhibit irrelevant information is an aspect of cognitive aging.[41] Recently, however, Hasher's group (Healey, Campbell, and Hasher) reviewed the literature and concluded that, although older people cannot inhibit irrelevant information as well as young people, if that irrelevant information subsequently becomes relevant the old people perform better than young.[42] As will be discussed in greater detail in chapter 4, Hasher and colleagues have conducted empirical studies and obtained similar results.[43] Thus, the inability of the old to ignore distractors as well as the young do—a burden in tasks requiring closely focused attention and exclusion of distractor processing—becomes a benefit if those distractors are later made relevant in another context.

It is easy to think of real-world parallels to these basic research findings. Often it is not clear what information may or may not be relevant in the future, so there may be significant advantages to attending broadly to both task-relevant information and apparently task-irrelevant information. Such a broad allocation of attention could even be strategic and learned from experience, an aspect of what we call *cognitive plasticity*. Not only are older individuals likely to possess both greater experience than the young; the nature of age-related cognitive decline may make them disposed to distribute their attention broadly so as to increase processing of apparent irrelevancies.

From this perspective, it should be noted that much previous research on cognitive aging has considered abilities in isolation, rather than considering the more complex abilities involved in integrating across tasks. Consistent with this speculation is a recent finding from a flight-simulator study that the ability of older pilots to avoid other air traffic efficiently improved more with training than did the comparable ability of young pilots.[45]

Summary

Cognitive decline is not inevitable or universal. Although, on average, selected cognitive functions exhibit age-related decline, a substantial

proportion of aged individuals—animal and human—do not show significant decline in cognition or in brain function. Not only has the neural
substrate of cognitive aging proved elusive; the aged brain retains some
capability for plastic change, including neurogenesis. Some age-related
cognitive changes interpreted as "decline" may actually represent a change
in cognitive strategy. What happens to the mind and brain as people age
that might account for these observations? We discuss the relevant scientific evidence in the next chapter.

3 Brain Aging and Cognitive Aging

Does the brain deteriorate with age? That it does seems intuitively compelling. After all, the brain is an organ of the body. Most of our bodily structures show signs of debilitation as we age: bones become brittle, muscles wither, joints become less flexible, and the heart must work harder to supply the other organs with the blood-borne nutrients they need to keep us going. As we have seen in previous chapters, however, the decline in physical functioning with age is not invariably accompanied by a decline in mental functioning—not in healthy aging, anyway.

What happens to the structural integrity of the brain as we age? Many complex issues must be addressed in order to answer this seemingly simple question. A discussion of them necessarily involves many specialized matters related to the brain's anatomy, physiology, and chemistry. As a result, this is the longest and the technically most complex chapter in our book. (To better communicate the main ideas associated with these complex issues, we provide summaries in the form of questions and answers.)

That our minds grow dimmer with age because our brains shrivel and grows weaker, just as our muscles do, is a view that has dominated both scientific and popular thinking on the subject. Although most people will acknowledge that muscles can be strengthened by exercise at any age, often people don't think of the brain in the same way. The brain is thought to be subject to continuous deterioration in structure and function. As recently as 10 years ago, most neuroscientists held an unshakeable belief that it wasn't possible to grow new neurons in adulthood, much less in old age, and that age-related brain atrophy was unavoidable. Recent evidence has overturned this long-held view.

If older people perform more poorly, on average, than younger people, what is the neural substrate of that cognitive change? There is little agreement on an answer to this fundamental question. Although many alterations in the brain's structure, physiology, and chemistry occur with age, such brain changes have not been consistently linked with cognitive change. The best-documented evidence of brain change is the reduction in the brain's volume with age.

3.1   Changes in the Brain's Volume

At the midpoint of the twentieth century, our knowledge of the structure of the human brain was based largely on post-mortem studies of individuals. The neuroimaging revolution that followed, beginning with computed tomography (CT) and continuing with magnetic-resonance imaging (MRI), provided an unprecedented view into the living human brain. Relative to CT, MRI scans also provided for better visualization of specific brain regions, including the hippocampus and other small but important structures. More recently, MRI has allowed for quantification of both the brain's overall volume and the sizes of specific cortical and sub-cortical structures.

The brain shrinks throughout life

MRI studies have consistently shown that the brain loses volume with age in adulthood. However, changes in brain volume begin in childhood. Yet childhood volume loss in certain regions of the brain is considered developmentally appropriate and not regressive. The volume of cortical gray matter (the outer layers of the brain which contain cell bodies of neurons) begins to decrease linearly after the age of 11 or 12 years in frontal and parietal cortical areas, and after the age of 16 years in temporal cortex.[1] On the basis of autopsy studies, this change has been attributed to the pruning of synapses,[2] and has been found to correlate positively with the brain's overall growth.[3]

The loss of gray matter that begins in childhood continues linearly in adulthood. Using a cross-sectional sample of healthy adults and an automated method called voxel-based morphometry (VBM), Terry Jernigan and colleagues calculated the volumes of gray and white matter from MRI scans. They estimated that between the ages of 30 and 90 years there were volume losses of 14 percent in gray matter and 26 percent in white matter.[4] John Allen and colleagues in Hanna Damasio's lab, using hand tracing of high-resolution structural MRI images, found comparable results over an age range from 22 to 88 years—a 12 percent loss in cortex and a 23 percent loss in white matter.[5]

In addition to these overall estimates of brain shrinkage, regional changes have been assessed. The brain does not shrink evenly. Automated measures of brain volume using VBM or region-of-interest-based measures of high-resolution MRI images find linear cross-sectional decrease in gray matter over the adult life span, particularly in prefrontal and parietal regions.[6] Studies using the hand-tracing method find the most longitudinal shrinkage in prefrontal regions and none in primary visual cortex.[7]

Data from a representative cross-sectional MRI study of aging conducted by Naftali Raz and colleagues[8] are plotted here in figure 3.1. Raz et al. examined MRI images from 200 healthy participants, 119 of them women, with a mean age of about 47 years. Volumes of the cerebral hemispheres and of 13 regions of interest were measured by hand-tracing methods. As figure 3.1 shows, gray-matter volume decreased with age in both men and

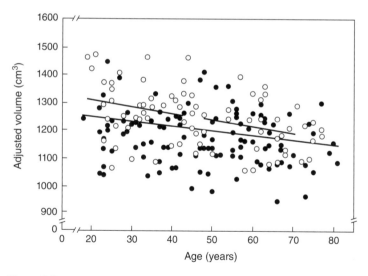

Figure 3.1

Scatter plot of volume of cerebral hemispheres, adjusted for height. Women are represented by filled circles and solid regression line; men are represented by open circles and broken regression line. Source: Naftali Raz et al., "Aging, sexual dimorphism, and hemispheric asymmetry of the cerebral cortex: Replicability of regional differences," *Neurobiology of Aging* 25 (2004): 377–396. Reprinted with permission of Elsevier (© 2004).

women. The overall correlation between brain volume and age was –0.24. Age-related volume loss was greatest in prefrontal cortical regions.

Most studies of the shrinking of gray matter with age have been done on living persons by structural imaging methods. Direct measurement of brain volume from post-mortem samples can be considered the "gold standard" against which the more indirect MRI measures can be compared, but such studies are understandably rare. However, in a recent study, Olivier Piguet and colleagues measured gray-matter and white-matter volume in post-mortem tissue from individuals ranging in age from 46 to 92 who had been carefully screened for neurological and neurodegenerative disease before their demise. Piguet et al. found no evidence of gray-matter shrinkage with age, although decreases in white-matter volume were seen.[9] The reason for this disparity between living and post-mortem assessments of gray-matter volume is not clear, but voxel-based MRI measures are subject to noise, particularly in older participants. Piguet et al., however, concluded that their findings indicate that healthy aging is associated more with white-matter changes than with gray-matter shrinkage. (We discuss white-matter changes with age in section 3.4.)

In summary, the bulk of the MRI evidence consistently points to gray-matter shrinkage with age, but such changes start in adolescence and are therefore part of a normal developmental process, at least initially. What this age-related gray-matter shrinkage signifies for cognitive performance is less clear.

Gray-matter shrinkage is not related to cognitive decline

There appears to be general agreement that gray matter shrinks with age, but what does that mean? At first glance there seems little to argue about. After all, as we have seen, cognition declines with age, on the average. And MRI methods reveal that the brain shrinks with age. Surely the two are causally linked? But it has proved surprisingly difficult to link brain shrinkage to cognitive decline. Many well-controlled studies finding significant reductions in brain volume with age nevertheless find that gray-matter shrinkage correlates either weakly or negatively[10] with cognitive performance in aged animals[11] and older humans.[12]

One way to examine the relation between brain volume and cognition is to conduct a study across the life span, beginning in childhood. If the "bigger is better" hypothesis is true (that is, if greater brain volume results in superior cognitive performance), one should be able to see the relationship by examining participants of widely different ages. Cyma Van Petten carried out such an examination and specifically examined the hypothesis with respect to the hippocampus and other temporal-lobe regions important for memory. Van Petten's study was a meta-analysis of investigations addressing that question in children, young adults, and older adults.[13] A meta-analysis is a "study of studies"—one looks at all the relevant papers on a topic and performs a formal quantitative analysis to determine how convincing the effects are overall. Contrary to the "bigger is better" hypothesis, hippocampal volume was found to be *inversely* related to memory performance in children and young adults in two other studies.[14] In older adults, the results were mixed. With regard to cortex, in a separate study Van Petten et al. assessed the relation between memory performance and neocortical volume in 65–85-year-olds and found correlations with prefrontal and temporal gyri, but in the *negative* direction—better memory was associated with lower volume in their sample.[15] Figure 3.2, which shows results for neocortical temporal volume, illustrates the negative relationship with memory performance. Although the individuals in the sample had above-average IQs, they had been selected for variability in performance on prefrontal tasks. David Salat and colleagues also reported better working memory with lower prefrontal gray-matter volumes in adults aged 72–94 with above-average scores on memory and IQ tests.[16] These results on older individuals are similar to results on young individuals that have shown an inverse relation between memory performance and hippocampal volume.[17] Thus, several different research groups using cross-sectional designs have found either no relation or an inverse relation between memory and regional brain volume.

The question of the relation between brain volume and cognition has also been addressed in longitudinal studies. Longitudinal designs have advantages for addressing this question, as cohort effects can be eliminated as a factor. If brain volume is in any way related to cognitive integrity, than age-related volume change should be related to cognitive change over the period of longitudinal testing. Karen Rodrigue and Naftali Raz related regional volume change to measures of episodic memory over 5 years in 95 healthy people aged between 26 and 82 years (mean age 58). Consistent with the cross-sectional evidence reviewed above, memory change over the 5 years was not correlated with longitudinal change in either the prefrontal

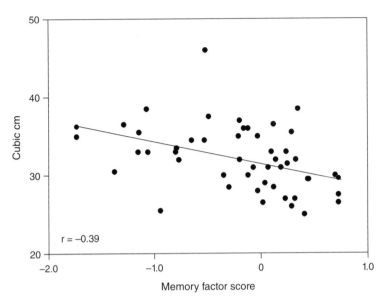

Figure 3.2
Volume of left temporal cortex (sum of superior, middle, and inferior temporal gyri plus fusiform gyrus) and memory-factor score (performance on five neuropsychological tests of memory). Source: Cyma Van Petten et al., "Memory and executive function in older adults: Relationships with temporal and prefrontal gray matter volumes and white matter hyperintensities," *Neuropsychology* 42: 1313–1335. Reprinted with permission of Elsevier (© 2004).

cortex or the hippocampus. The only region to show volume reduction related to memory change over the five-year testing period was the entorhinal cortex.[18] Entorhinal cortex lies close to the hippocampus and, along with that brain structure, is known to play an important role in memory formation.[19] These findings are notable for two reasons. First, consistent with the lack of relation observed by Van Petten[20] and Salat,[21] Rodrigue and Raz did not find a relation between memory performance and volume change in the prefrontal cortex (PFC) or in the hippocampus Second, the only region that did change in relation to cognition is known from extensive work to be the site of initial Alzheimer's-related pathology.

Regarding the second point, there is convincing evidence that structural changes in the entorhinal cortex are related to Alzheimer's disease and not to healthy aging. The neuroanatomists Heiko and Eva Braak conducted one of the earliest and best-known studies on pathological brain changes in Alzheimer's disease. Braak and Braak are well known for their careful study of a very large series of brains that came to autopsy at many different ages across the human life span. They found that the earliest pathology of Alzheimer's disease is seen in a specific region of the entorhinal cortex.[22] Consistent with that, Scott Small and colleagues conducted a neuroimaging study of 70 individuals aged 20–88 and found that some regions of the hippocampus do change with age in a manner that does not appear to be pathologically different from that seen in young people. However, the entorhinal cortex specifically was found to change pathologically with age (showing greatly reduced MRI-T1 signal) in 60 percent of the older group (aged 70–88). In those people, the reduced signal from the entorhinal cortex was correlated with decline in memory function on the Buschke Selective reminding task.[23] The findings that the entorhinal cortex is the site of early Alzheimer's pathology[24] and that it undergoes pathological change related to memory performance but only in a subset of apparently healthy older individuals[25] suggest that some of Rodrigue and Raz's participants may have had Alzheimer's pathology.[26]

Considered together, these findings suggest that regional brain shrinkage with healthy aging is not generally associated with cognitive decline. However, such an association may occur in a subset of apparently healthy groups of older people in the early stages of (undiagnosed) Alzheimer's-related pathology in entorhinal cortices. That age-related shrinkage of cortex and of the hippocampus does not result in cognitive change in healthy aging weakens the view that most older brains are irreversibly structurally impaired. Thus, although brains do appear to shrink regionally with age, the shrinkage is not necessarily pathological and may be related to cognition only in the presence of Alzheimer's pathology[27] and vascular disease.[28] If so, what is the significance of age-related brain change?

Dynamic structural brain change and successful cognitive aging

Although age-related changes in brain volume do not appear to be associated with cognition in healthy older adults, there have been several recent reports that longitudinal brain change in children is related to variation in cognitive functioning. Children with intelligence quotients (IQs) in the "superior" range show a non-linear pattern of longitudinal change in cortical thickness. Philip Shaw and colleagues in Jay Giedd's lab observed thickening of prefrontal cortex by age 7, followed by thinning in late adolescence. This was in contrast to a pattern of linear shrinkage observed in children with "average" IQs.[29] Elizabeth Sowell and colleagues also reported non-linear longitudinal changes related to cognitive functioning in children. Greater thinning in left dorsal frontal and parietal cortices has been observed in those children who showed greater increases in verbal IQ over 2 years.[30] Such data have been interpreted as reflecting dynamic remodeling of cortex in high-functioning children.[31] Does such cortical remodeling also occur in old age?

High-functioning older adults, like high-IQ children, show evidence of dynamic cortical thickening and thinning.[32] Anders Fjell and colleagues recently reported a non-linear pattern of cortical changes in adult aging. They divided their sample of older adults into "average cognitive functioning" and "high cognitive functioning," based on their scores of two subtests of fluid intelligence taken from the Wechsler Abbreviated Scale of Intelligence: the tests of block design and visual matrix reasoning. Cortical thickening across the cortical mantle was reported to occur in mid-life before later thinning, but only in individuals categorized as showing high cognitive functioning.[33] Average-cognitive-functioning individuals showed evidence only of cortical thinning. Figure 3.3 shows the results comparing average and high-functioning older adults. High-functioning adults had greater cortical thickness in the right hemisphere in the middle and inferior temporal gyrus, the precuneus, and the posterior cingulate gyrus. Fjell et al. also found a relation between cortical thickness in parietal and posterior cingulate, executive function, and amplitude of the P3a component of the event-related potential (ERP), a prominent brain wave.[34] Thomas Espeseth and colleagues have also observed patterns of age-related cortical thickening and thinning similar to those seen by Salat, although cortical thinning was accelerated in carriers of the APOE-e4 gene in some brain areas.[35] (The APOE gene is associated with increased risk of cognitive decline in healthy aging, and with increased risk of Alzheimer's disease.[36]) It should be noted that cortical thickening is not always beneficial— Espeseth et al. also found that cortical thickening related to the APOE-e4 gene was associated with lower target detection accuracy and lower amplitude in the ERP P3a component.[37]

Two major brain regions—the precuneus and the posterior cingulate gyrus—have been found to undergo early thickening and later thinning with age,[38] and in a manner correlated with amplitude of the ERP

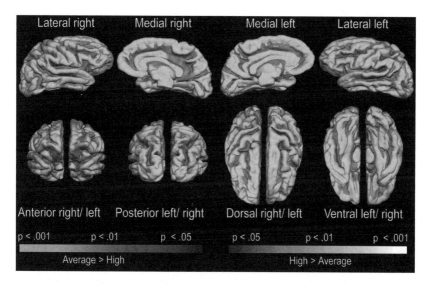

Figure 3.3

The cortical thickness of old participants with high and normal cognitive (fluid) functioning compared by general linear modeling across the entire cortical mantle. Red and yellow areas indicate where cortices were thicker in the high functioning group. In the right hemisphere, differences were seen in temporal middle gyrus and temporal inferior gyrus, gyrus cuneus, the gyrus and sulcus of the insula, gyrus rectus, the gyrus of the cingulate isthmus, and the posterior cingulate gyrus. Source: Anders M. Fjell et al., "Selective increase of cortical thickness in high-performing elderly—structural indices of optimal cognitive aging," *NeuroImage* 29: 984–994. Reprinted with permission of Elsevier (© 2006). This figure appears in color elsewhere in the book.

component P3a.[39] These are the same regions implicated by Randy Buckner and colleagues in the transition from healthy aging to Alzheimer's disease. On the basis of imaging of regional brain structure, imaging of regional cerebral metabolism, and the post-mortem literature, Buckner et al. concluded that dysfunction in the precuneus and in the posterior cingulate precede diagnosis of Alzheimer's disease.[40] Consistent with that, Kristine Walhovd and colleagues reported that cortical thickness of the left precuneus, the middle frontal gyrus, and the left parahippocampal sulcus predicted retention of learning on a verbal learning task over months.[41]

These results reporting non-linear longitudinal changes in cortical thickness in children and in adults echo the non-linear longitudinal changes seen in cognitive functioning from youth to old age.[42] Jay Giedd and colleagues interpreted the finding in children as revealing dynamic processes of cortical maturation.[43] The same explanation can be applied to cognitive aging. This suggests that in successful aging some process either promotes cortical thickness[44] or prevents cortical thinning.[45] Either appears to confer benefits for cognition.

Dynamic functional brain change and successful cognitive aging

Older people appear to use their brains differently than young people. PET and fMRI studies have shown that older people activate prefrontal cortices bilaterally (both left and right hemisphere) during a range of cognitive tasks, in contrast to the unilateral prefrontal activation (only one hemisphere) that is characteristic of young people.[46] What is the significance of this difference? Some investigators have suggested that bilateral activation is compensatory. Older people who were classified as "high memory functioning" activated both sides of their prefrontal cortex during a source memory task; those who were classified as "low memory functioning" did not.[47] Bilateral prefrontal activation during word classification was related to the success of encoding in the old group.[48] Thus, high-functioning older individuals use their brains more dynamically ("recruiting" both hemispheres) than average-functioning older individuals, who appear to activate only one hemisphere. Cheryl Grady and colleagues have argued that a more accurate characterization of young-old differences in regional brain activation patterns is that the old show a shift toward greater activity in regions mediating executive activity.[49]

We have argued that what is important for successful aging is not volume shrinkage or cortical thinning per se, but rather the ability of the aging brain to change dynamically—both structurally and functionally—in response to demands imposed on it by cognitive challenges and experiences. In this view, changed cognitive strategies in old age result in changed innervation.[50] This is consistent with the evidence (reviewed above) that the brains of high-functioning older adults show evidence of both cortical thickening and cortical thinning. Considered together, these findings suggest considerably more dynamism in the brain throughout development—including the later years of life—than was previously suspected. Such dynamism, which appears to reflect the brain's capability for plasticity, is likely to be important for successful aging.

Training-related plasticity in the adult brain

The view that dynamic brain change is important for successful brain aging implies that the adult brain must be capable of considerable plasticity, including adaptation in response to environmental stimulation and experience. A number of different aspects of brain plasticity have been investigated in recent years, but the strongest evidence concerns motor training.

Cortical plasticity is enhanced in response to motor training, as a substantial literature on animal studies shows.[51] There is evidence of cortical plasticity in humans too. Annette Sterr and colleagues in Edward Taub's lab reported that the hand area of human somatosensory cortex (the region of cortex that receives and processes sensory input from the hand) was enlarged in expert teachers of Braille. Whereas less skilled Braille readers "read" with one finger, experts read Braille with three fingers held together. Sterr et al. found that in the experts there was "smearing" of the finger representations in somatosensory cortex such that when one of the experts' three fingers was touched out of sight, the person was inaccurate at identifying which of the three had been touched. The novice students showed no such evidence of smearing. The interpretation was that in somatosensory cortex of the Braille experts, the representations of those three fingers had come to be merged into one after years of using the three fingers together during Braille reading.[52] Thus, the brain treated input from those three fingers as coming from one large finger.

Taub's group also found that players of stringed instruments show similar adaptation. Structural MRI scans revealed that in several players of stringed instruments the cortical representation of the fingers of the left hand was larger than in people who didn't play stringed instruments. Moreover, the extent of reorganization was greater in those who had started playing at an earlier age.[53] Reorganization of cortical maps may be a normal accompaniment to aging. That has been seen in healthy older people, who show an expansion of the hand representation in somatosensory cortex by 40 percent.[54]

The studies cited above certainly suggest that the cortical sensory and motor maps of the fingers are changed by experience. However, one drawback of these studies is that they were observational. For that reason, it cannot be ruled out that, for example, people who choose to become string players were born with larger finger-representation areas in the right hemisphere. A more convincing demonstration of brain plasticity requires a study that measures the brain before and after participants are randomly assigned to either a motor training condition or a control condition. Bogdan Draganski and colleagues conducted such a study. They took structural MRI scans of young people and then randomly assigned half of them to learn to juggle. For 3 months those people practiced a three-ball cascade (illustrated below in figure 9.4). They were scanned again after 3 months of training and then told to stop juggling. Three months after they were told to stop juggling, they were scanned a third time. After 3 months of practice, two areas were seen to have enlarged in the juggling group but not the control group: the middle temporal (MT) area and the intraparietal sulcus. The expansion had reversed somewhat—though not fully—at the third scan, 3 months after juggling practice had ceased.[55] Therefore, the Draganski study using random assignment confirmed the earlier observational studies showing that motor training can lead to change in cortical organization.

Although the focus of this book is on amelioration of healthy cognitive aging, it is relevant to note that sensory and motor training can also reduce deficits related to lesions through neuronal organization. That such training can reorganize the brains of older people indicates a retained capacity for plasticity. The research groups of Michael Merzenich and of Edward Taub[56,57] have conducted pioneering work in revealing adult neuroplasticity—the ability of the adult brain to remap itself in response to changed input, training, and damage. Merzenich and colleagues observed that after amputation of one digit of a monkey's paw there was rapid reorganization of cortical representation of the paw.[58] This was also seen after tactile stimulation of the digits of a paw.[59] Before this work, it had been long accepted that the adult brain was not able to reorganize itself after brain damage.

The important research field of adult neuroplasticity was developed by Edward Taub, who in the early 1980s began to study changes in cortical organization after the sensory input to monkey brains from forepaws had been interrupted surgically by cutting the dorsal roots of the spinal cord, thereby eliminating all sensory input from forepaw to brain. Motor output was not affected, so the monkeys could still move their numbed limbs. When trained, such monkeys learn to use the numbed limb effectively, though not fully normally. This important work was unfortunately halted for 10 years by the famous "Silver Spring monkeys" case. In 1981, Alex Pacheco, one of the founders of People for the Ethical Treatment of Animals, volunteered to work as a research assistant in Taub's lab. Apparently while Taub was away on vacation, Pacheco failed to clean the lab and care for the monkeys. Once the lab was in disarray, he called in the State Police, who confiscated the monkeys and charged Taub.[60] Although Taub was initially convicted of six misdemeanor counts of animal cruelty, those convictions were overturned on appeal. The fight over custody of the monkeys reached the US Supreme Court. Because PETA eventually lost that case, the monkeys could finally be examined. The surprising finding was that over time these adult monkey's brains had become "massively reorganized" as a result of the loss of sensory input.[61] For example, neurons that had originally responded to touch on the paw came to respond to touch on the face. The neurons in cortex still functioned, but while deprived of stimulation they had changed their jobs in order to obtain sensory input from another region of the skin. This evidence of adult neuronal plasticity stimulated new research, and V. S. Ramachandran and William Hirstein made similar observations in human amputees.[62]

Taub, his funding restored after the misdemeanor conviction was reversed, was finally able to take the important next step of extending the findings of the Silver Spring monkey studies to human stroke patients. Stroke (interruption of blood supply to a brain region caused by a clot or a hemorrhage) typically causes damage to the affected sensory and motor cortex. This results in weakness and numbness in the part of the body controlled by that cortical region (e.g., a leg or an arm). Taub hypothesized that because stroke patients typically come to avoid using such a paretic limb, "learned non-use" leads to further weakening and loss of function of the limb. The person tends to avoid using the paretic limb, and that leads to reduced functionality of the limb.[63] Reasoning from his work with the Silver Spring monkeys, Taub hypothesized that—just as training had induced the monkeys with numbed limbs to learn to use their limbs by reorganizing their brains—human stroke patients would be able to learn to use their paretic limbs by reorganizing their brains. To test this, Taub developed Constraint-Induced Movement Therapy (CIMT), which in 75 percent of stroke patients leads to substantially greater recovery of function in the parietic limb after a stroke than is seen with conventional therapy.[64] Typically a stroke affects a relatively small region of cortex in one hemisphere, so that only the contralateral limb is affected. In CIMT therapy, the *un*affected, normal limb is restrained for much of the waking day (for example, the normal arm is placed in a sling), and daily therapy is instituted to shape desired movement of the *more* affected limb using successive approximations to avoid frustration. CIMT has been shown to be effective in inducing cortical reorganization and motor recovery even in chronic stroke patients 4 years after their lesion.[65] Catherine Bütefisch and colleagues have shown that the critical factor in inducing such cortical reorganization is repeated training of the affected limb over weeks.[66]

What is the basis for this effect of motor training on recovery from brain damage? Randolph Nudo and colleagues conducted a series of studies in rats and monkeys showing that healthy cortex adjacent to lesioned cortex takes over some function from the damaged area, after training of the affected limb. An experimental ischemic infarct (stroke) in the paw area in one hemisphere of monkey motor cortex resulted in impairment in moving the contralateral paw. (In primates, the motor systems are generally crossed, such that the right hand or paw is controlled by the left hemisphere and vice versa.) In one study, Nudo and colleagues (Frost et al.) initiated motor training of the affected paw and after training made microelectrode recordings of the contralateral motor cortex near the lesion. These recordings revealed that cortex around the lesion began to participate in movement of the paw.[67]

In humans, a similar mechanism appears to underlie training-aided recovery of function after stroke. Liepert et al. in Taub's group used transcranial magnetic stimulation (TMS) to map the change in cortical control of a limb after motor training.[68] TMS is a non-invasive method for assessing cortical function by inducing a weak current in targeted brain areas through the intact scalp.[69] Inducing a weak current in the damaged hand area of motor cortex and surrounding cortex, results in stimulation of the neurons in that cortex. It is possible to observe which regions of cortex controlled the hand both before and after training. Liepert et al. mapped primary motor cortex in six patients with chronic stroke-induced paresis (weakness) of the upper limb. This mapping was done both before and after 2 weeks of CIMT. The cortical region that could induce a muscle response in the hand was much larger after CIMT than before it.[70] Similar results have been seen in acute stroke patients trained within 14 days of the stroke.[71] This line of research, which began with Taub's work with the Silver Spring monkeys, has led to greatly improved therapy for stroke patients.

The aforementioned work is a good example of direct application of basic research to clinical practice. It should be noted that at the time of the Supreme Court decision refusing to grant custody of the Silver Spring monkeys to PETA, the *New York Times* interviewed Dr. Neal D. Barnard, president of the Physicians Committee for Responsible Medicine, a group that opposes animal research. "This was a pointless experiment," Dr. Barnard stated, "and we learned nothing we didn't know. We don't see anything here that will help disabled people or lead toward a treatment."[72] To the contrary, this work has revolutionized therapy in brain-damaged patients. The American Stroke Association has called CIMT "revolutionary." In 2003, the Society for Neuroscience included CIMT among the top ten Translational Neuroscience Accomplishments of the twentieth century.

What are the mechanisms that allow neurons near a lesion to take over the functions previously carried out by the subsequently damaged cortex? In rats, skilled motor training leads to increased numbers of synapses per neuron, increased dendritic arborization (more extensive dendritic branching), strengthened synaptic responses,[73] and axonal sprouting.[74] Axonal sprouting is the actual growth of an axon of a neuron over some distance to make a new connection in a different region of the brain. Nudo's group has demonstrated such sprouting in monkeys. Five months after small experimental lesions were made in the paw area of monkey motor cortex, there was evidence of axonal sprouting and new connections from a distant target to a location near the lesion. Intracortical neurons were found to have grown around the lesion.[75] Moreover, such neuroplasticity appears to be heightened by experience. After a small infarct in the paw area of monkey motor cortex, monkeys undergoing post-lesion training of the affected paw showed expansion of paw representation into nearby cortical elbow and shoulder representation.[76] Thus, neurons near the lesion took over some of the function of the damaged cortex.

Even in the absence of brain damage, training in the use of tools can result in novel projections in the brain. This has been seen in adult monkeys trained to use a rake to pull in pieces of food that otherwise would be out of reach. This was a novel skill to these monkeys who do not use such tools in the wild. Before training, neurons in intraparietal cortex responded only to somatosensory input. After training, those neurons became bimodally active, responding to both touch and vision. Using tracer techniques, new axonal projections were found to have developed between temporo-parietal junction and intraparietal regions in the trained monkeys but not in the untrained ones.[77] Motor training has also been linked to dendritic spine remodeling. (Dendritic spines are protrusions from dendrites of neurons that help convey electrical signals to the neuron cell body. Spines are notably plastic in that they can appear and disappear and change their shape rapidly.) Training adult mice to reach for a seed using a forelimb was followed rapidly (within an hour) by formation of spines on dendrites of the output pyramidal neurons in contralateral motor cortex. The extent of spine formation was associated with the number of times the mouse reached successfully to grasp a seed (figure 3.4). The new spines persisted long after the training stopped.[78]

In summary, new synapses, dendritic spines, and axonal sprouting appear to underlie neuroplasticity in adult animals. These mechanisms can be stimulated by lesions but also by training alone. Learning of motor skills has been shown to expand the following:

the somatosensory representation of digits in Braille readers[79]

the volume of the posterior hippocampus in taxi drivers[80]

motor representation associated with improved motor function in the lesioned hemisphere of stroke patients[81]

a motion control area in parietal cortex of young people who learned to juggle[82]

axonal sprouting[83]

dentritic spine remodeling.[84]

As will be reviewed below, structural brain change in response to training has recently been extended from motor training to cognitive training.[85] Likewise, recent work from the cognitive training literature shows that regional blood flow pattern change—suggesting functional change—is also seen after cognitive training.[86]

Although there is substantial evidence that regions of cortex normally shrink with age, such shrinkage does not appear to have significance for cognitive performance in healthy individuals, but does appear to be related to cognitive performance when there is vascular disease.[87] Also, there is recent evidence that high-functioning children and older adults undergo

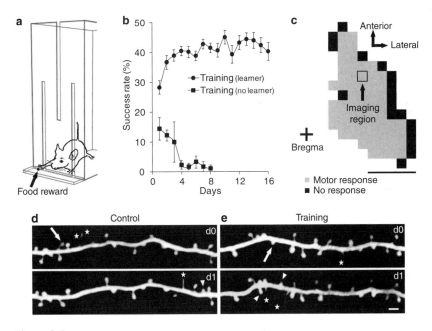

Figure 3.4
Training on seed reaching in mice is associated with rapid spine formation in contralateral motor cortex. (a) Training on seed reaching. (b) Success during training in trained and untrained mice. (c) Map of imaged region. (d and e) Imaging of same dendrite over two days, showing spine elimination (arrows) and formation (arrowheads), and filopodia (asterisks) in control (d) and trained (e) mice. Source: T. Xu et al., "Rapid formation and selective stabilization of synapses for enduring motor memories," *Nature* 462: 915–919. Reprinted with permission from Macmillan Publishers Ltd. (© 2009).

dynamic changes in cortical thickness over years,[88] and that the brain activation patterns of high-functioning adults differ from those of average-functioning adults.[89] Studies suggest that cortical shrinkage per se does not necessarily indicate pathology or regression, and that non-linearities in the thinning pattern and in the activation pattern may be associated with superior cognitive functioning.

The idea that age-related changes in cortical thickness and white matter integrity may be related to cognitive demand is supported by evidence that the microstructure of cortex changes with demand. Specifically, post-mortem studies in children indicate that synaptic pruning underlies the cortical thinning that occurs normally in childhood.[90] In adult animals, synaptic density appears to be very plastic, changing with administration of nicotine,[91] estrogen,[92] "environment enrichment,"[93] neurotrophins,[94] and exercise.[95] In adult humans, experience alone has been shown to change cortical structure and function,[96] even in older stroke patients.[97] It may be that the ability to remodel the brain in response to cognitive demands is important for successful aging.

Does the brain deteriorate with age?

Structural change in the brain does occur, mainly in the form of volume shrinkage. However, such structural change does not explain cognitive aging and therefore, it does not appear that the brain deteriorates structurally in healthy aging. Moreover, *functional* change does appear to reflect compensatory activation, especially of the prefrontal cortex. Cognitive and motor training can change the adult brain both structurally and functionally. Such evidence indicates that the aging brain has the capability to be remodeled in response to demand.

3.2   Neuron-Level Change and Cognitive Aging

If overall or regional brain shrinkage, at least in gray matter, cannot account for cognitive aging, perhaps a more microscopic examination of neurons (which make up the brain's gray matter) is in order. Does age-related change at the neuron level explain cognitive aging? An answer to this question has proved elusive. But many long-held views about neuronal health and its role in mental functioning have been overturned by recent scientific discoveries.

One such view was that neurons are progressively lost and damaged with age. One recent review referred to the "myth of brain ageing."[98] The misconception that aging results in neuron loss and dramatic changes in neuronal morphology came about because many early studies did not use appropriate stereological methods in their anatomical studies. As a result, aging was characterized as in panel a of figure 3.5, which compares the loss of dendrites in aged neurons against the loss of dendrites in young neurons in the dentate gyrus. However, as Burke and Barnes argued,[99] such figures do not accurately reflect the subtle and selective morphological alterations that occur in neurons as they age. Studies in young and old rats point to the more accurate depiction in shown in the remaining panels of figure 3.5. Panels b and d show dentate gyrus granule cells that were radio-actively labeled to reveal the dendritic structure in, respectively, a young and an old rat. There is no reduction in the number and extent of dendrites with aging, as the reconstructions shown in panel c also indicate.

It was long claimed and accepted that neurons died as age advanced, with estimates of losses ranging from 25 percent to 50 percent.[100] It is rather disturbing that for many years most claims of age-related neuron loss were based on only one study.[101] However, in the 1980s it was found that mea-suring neuron density rather than neuron number led to biased estimates. Subsequent development of more accurate stereological methods of count-ing neurons revealed little evidence for neuron loss with aging.[102] The sole exception to the conclusion that aging doesn't lead to meaningful neuron loss is one report of selective neuron loss in dorsolateral PFC, in Brodmann's Area (BA) 8A, that was correlated with performance of working memory in three aged monkeys. Neuron loss was not seen in adjacent prefrontal area, BA 46, also linked to working memory.[103]

Another long-held view is that the number of synaptic connections between neurons declines with age. One early study claimed to have observed linear decreases in synapse number with age.[104] However, recent studies using more accurate stereological methods to assess synaptic density have found no synapse loss until people are in their nineties.[105]

Thus, three long-held generalizations about brain aging can be discounted: that the number of neurons, the number of dendrites, and the number of synapses between neurons decrease inexorably with age. This is not to say that aging isn't associated with any change at the neuronal level. However, in general, as the following survey of the literature shows, age-related neuronal changes tend to be localized, subtle, and specific to certain layers and neuron types. Moreover, some changes are reversible.

Dendritic branching

Dendrites—the branched projections of a neuron that conduct electro-chemical stimulation received from other neurons to the neuron's cell body—are not static; they change in response to inputs from other neurons. Collectively, dendrites and their branches are referred to as the *dendritic arbor*. The tree analogy is appropriate because the branches of dendrites can grow with learning and in response to environmental stimulation, just

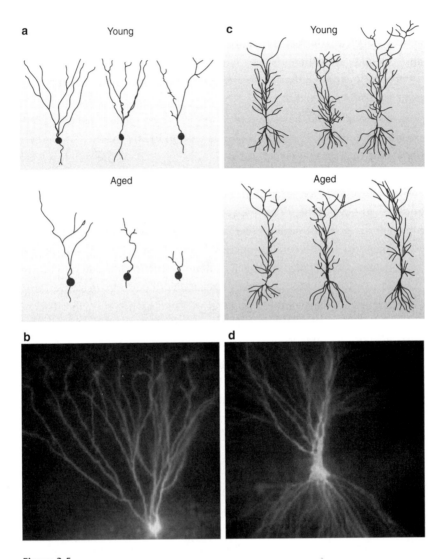

Figure 3.5

The evidence for brain aging is relatively weak. This composite figure (from S. N. Burke and C. A. Barnes, "Neural plasticity in the ageing brain," *Nature Reviews Neuroscience* 7, no. 1: 30–40) is reprinted with permission of Macmillan Publishers Ltd. (© 2006). (a) Apparent age-related loss of dendritic extent in the dentate gyrus and in CA1 was exaggerated by including demented individuals with healthy aged individuals in the same experimental group and by not using stereological controls. (Original source: M. Scheibel et al., "Progressive dendritic changes in the aging human limbic system," *Experimental Neurology* 53: 420–430.) (b) Two granule cells

continued

as the branches of a tree grow when nourished. Dendritic branching is therefore an important index of brain plasticity.

In humans, neurons in the hippocampal formation—important for memory—show either stable or increased dendritic branching with age.[106] As was discussed earlier, the long-held conviction that aging results in a reduction in the number and extent of neurons has been overturned. In parahippocampal gyrus, dendritic branching is even more extensive in very old people who were not demented at the time of death than in middle-aged or young people.[107] Dorothy Flood and colleagues found no age-related changes in dendritic branching in several subregions of the human hippocampus.[108] Likewise, in animals there is no evidence of reduced dendritic extent with age in the dentate gyrus[109] or in the CA1 field of the hippocampus.[110] In contrast, age-related decreases in dendritic branching in PFC have been reported in rats[111] and in humans.[112] However, these changes are confined to the basal dendrites of layer V pyramidal neurons.[113] Dendritic branching is a very plastic process. A new method of in vivo imaging shows non-pyramidal interneuron dendrites extending and retracting over months, even adding new branch tips.[114] As we will see in chapter 9, in aged mice only one month of exercise reversed age-related changes in synaptic structure at the neuromuscular junction.[115]

Figure 3.5

(continued)

from the dentate gyrus of an aged 24-month-old rat. Neurons in rat dentate gyrus show no significant age-related change in dendritic extent, but do show significant increases in electrotonic coupling. (Original source: T. C. Foster et al., "Increase in perforant path quantal size in aged F-344 rats," *Neurobiology of Aging* 12: 441–448.) (c) Reconstructions of neurons in hippocampal area CA1 from young rats (2 months) and old rats (24 months) showing no reduction in dendritic branching or length with age in area CA1. (Original source: G. K. Pyapali and D. A. Turner, "Increased dendritic extent in hippocampal CA1 neurons from aged F344 rats," *Neurobiology of Aging* 17: 601–611.) (d) A CA3 neuron filled with 5,6-carboxyfluorescein from a 24-month-old rat. There is no regression of the dendrites, but the aged neurons show a significant increase in the number of gap junctions compared with young neurons. (Original source: G. Rao et al., "Intracellular fluorescent staining with carboxyfluorescein: A rapid and reliable method for quantifying dye-coupling in mammalian central nervous system," *Journal of Neuroscience Methods* 16: 251–263. Reprinted with permission of Elsevier.) This figure appears in color elsewhere in the book.

Spine density

Spines protrude from dendrites to provide points of synaptic contact with other neurons. The dendrites of a single neuron can contain many such spines, numbering up to the thousands. Spines not only provide an anatomical substrate for synaptic transmission, they also serve to increase the number of possible contacts between neurons.

Spine densities in the CA1 subfield of the hippocampus and in the dentate gyrus appear (figure 2.1) to be generally stable in number in older humans[116] and rats.[117] We are aware of only one report of age-related reduction in spine density, and that was limited to one particular portion of monkey apical dendrites in neurons projecting from superior temporal cortex to prefrontal BA 46, with no other effects on dendritic morphology or extent and no other regions affected.[118] Thus, spine densities are typically stable in adult aging, with some localized exceptions.

Synaptic integrity

Hippocampal electrophysiology appears to be largely unaffected by aging. This includes resting membrane potentials[119] and action potential thresholds[120] and action potential amplitudes.[121] These measures refer to the neuron's ability to hold a charge at rest and its ability to fire normally. Age-related synaptic changes have been observed, but these do not lead to corresponding changes in neuronal transmission. Specifically, age-related increases in the refractory period when the neuron cannot fire do not translate to differences in firing rates in hippocampal CA1 pyramidal neurons in awake, behaving rats.[122] Nor do age-related reductions in EPSPs (small voltage changes in neurons that are not sufficient to trigger an action potential) in the dentate gyrus and in CA1 lead to alterations in the induction of long-term potentiation (LTP).[123] The electrophysiology of the PFC, including resting membrane potentials, action potential thresholds, and action potential duration, is also largely unaffected by aging.[124] There are some changes in characteristics of action potentials, but they are not related to cognitive performance.[125]

More evidence comes from studies of synaptic efficacy in cognitively impaired and intact older animals. LTP in the hippocampus is very important for new memory formation. LTP is a long-term increase in synaptic strength after high-frequency stimulation of a synapse. Studies comparing cognitively unimpaired and cognitively impaired aged rats find deficits in

LTP only in animals showing impaired spatial memory. The task commonly used to assess this—the Morris water maze[126] (figure 3.6)—has become a standard test of spatial memory, and is now used extensively in animal research.

Geoffrey Tombaugh and colleagues[127] used the Morris water maze to compare neuronal firing patterns in young and old rats grouped according to their performance on the maze. Tombaugh et al. observed deficits in LTP in neurons in the hippocampus, but only in the subgroup of aged rats that showed deficient spatial learning on the water maze. Rats without deficits in spatial learning had LTP properties similar to young controls. Thus, only rats that showed cognitive impairments also showed LTP deficits.

Edvard Moser and colleagues also observed differences in electrophysiology between age-impaired and age-unimpaired animals.[128] They recorded

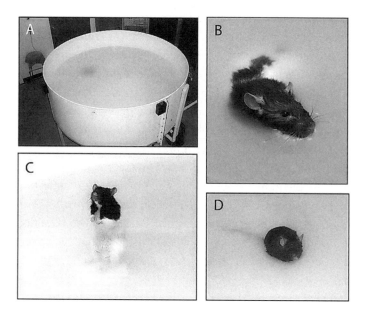

Figure 3.6
A Morris water maze. The task requires the rat or mouse to swim from a starting point and find a platform hidden just below the surface of the water of a circular pool (made opaque with a milky substance), with the starting point varied from trial to trial. The time to reach the platform (latency) and the path length are used as measures of performance. Learning is shown by a reduction in both latency and path length over successive sessions. Source: doi:10.4249/scholarpedia.6315. Copyright Richard Morris. Used with permission of Richard Morris and Scholarpedia.

from "place cells" (hippocampal neurons that encode specific locations in the rat's environment). This encoding is so accurate that the position of a rat in a maze can be determined by the firing patterns of the place cells alone.[129] Recordings from place cells in rat hippocampal subfield CA3—known to be important for rapid encoding of spatial information—revealed higher firing rates overall and failure to change firing rates in response to novel place fields in aged, memory-impaired rats. This was not seen in the aged, unimpaired rats.[130] Consistent with this, learning in aged impaired rats—but not in aged unimpaired rats—shows a marked reduction in numbers of specific types of synapses.[131] Hippocampal gene expression profiles differ little between young rats and high-cognitive-functioning old rats. Such profiles do differ significantly between high-functioning and low-functioning older rats, the impaired rats showing increased expression of genes controlling inflammation and demyelination.[132] Considered together, this evidence reveals that aging per se is not consistently associated with measures of synaptic integrity. In the subsets of animals that are not cognitively impaired, there is no reduction of synaptic effectiveness.

There is evidence of age-related decline in spatial navigation ability in humans,[133] monkeys,[134] dogs, and rats.[135] Nevertheless, there is also human evidence of hippocampal plasticity—related to spatial navigation experience, even later in life,[136] and related to verbal memory retention over 11 weeks.[137] Eleanor Maguire and colleagues found that time spent driving a taxi in London (ranging from 1.5 to 42 years) was correlated with the volume of the posterior hippocampus. This was interpreted as reflecting increased spatial knowledge of the city. The clear confound with age in this study (people who have been driving a cab longer tend to be older) cannot account for these results. A subsequent study looking at bus drivers who drove a set route over a range of years did not show the same evidence of hippocampal plasticity.[138]

Does age-related change at the neuron level explain cognitive aging?

Generally, no. There are some subtle and selective changes in neuron properties with age, but they do not consistently lead to behavioral change. Human brains, as they age, continue to show plasticity when cognitive demand requires it, and aged animals showing intact spatial memory ability have synaptic mechanisms similar to those of young animals. This is consistent with the evidence that aged animals and humans who do not show neuronal change also do not show cognitive change. Considered together, this literature suggests that it is not aging per se that causes the observed synaptic abnormalities, but rather some process that accompanies aging but is not inevitably associated with it.

3.3 Glucose Regulation and Cognitive Aging

One putative mechanism underlying the decline in cognitive functions seen in some—but not all—aged individuals (animals or humans) involves the effect of glucose regulation on the dentate gyrus of the hippocampus. Not only is the dentate gyrus important in the formation of new spatial memories[139]; it is one of the few sites of adult neurogenesis,[140] and it is the initial site of Alzheimer's pathology.[141]

The dentate gyrus also appears to be very vulnerable when glucose regulation is disordered, as in individuals who are diabetic and/or have the "metabolic syndrome." Even in non-diabetic older people, poor glucose regulation has been linked to cognitive impairment[142] and reduced hippocampal volume.[143] In individuals with diabetes, an inverse relation has been observed between blood glucose levels and blood flow in the dentate gyrus selectively and not in other hippocampal areas. Moreover, blood glucose levels are inversely correlated with the ability to recall verbal material.[144]

Antonio Convit and colleagues speculated on the mechanism underlying this relationship.[145] Animal studies show that during maze learning localized levels of glucose are reduced in the hippocampus in a manner that is related to the difficulty of the maze.[146] Moreover, the effect was stronger and lasted longer in older animals.[147]

Not only is the dentate gyrus selectively vulnerable to glucose levels, it benefits selectively from exercise. Three months of supervised aerobic training in young adult humans selectively increased cerebral blood volume in the dentate gyrus and improved cognition. Eleven people (mean age 33) participated in aerobic training, under supervision, four times a week for 12 weeks. They engaged in 40 minutes of cycling, treadmill running, elliptical trainer walking, or StairMaster walking. Hippocampal cerebral blood flow maps were obtained before and after exercise. Only the dentate gyrus showed increased blood flow related to exercise; other parts of the hippocampus showed no effects. Moreover, the increases in blood flow were related to improvements in cardiopulmonary function and in verbal memory.[148] (For more on the effects of physical exercise on the brain, see chapter 6.)

Does poor glucose regulation contribute to cognitive aging?

It may. Aged individuals are who are cognitively intact have synaptic mechanisms that are similar to those of young individuals. However, synaptic mechanisms are vulnerable to factors such glucose regulation that are associated with aging rather than being directly due to aging per se. Both animal studies and human studies implicate the health of the hippocampus, and especially the dentate gyrus, as important for maintaining cognitive integrity in old age. The dentate gyrus, which is particularly vulnerable to impaired glucose regulation, selectively benefits from aerobic exercise.

3.4 White-Matter Change and Cognitive Aging

Thus far we have discussed structural brain changes primarily involving neurons, which make up the gray matter of the brain. The axons of neurons, which constitute the white matter of the brain, may also be susceptible to age-related change. Does change in volume of white matter explain cognitive aging? During the first four decades of adulthood, when areas of cortex are shrinking, the volume of white matter increases. Though there is a general consensus that gray matter shrinks with age, there is considerable disagreement as to whether white-matter volume also declines in old healthy age. There are some reports of no decline[149] and some reports of decline only after age 70.[150] However, on the question of decline before old age there appears to be agreement that white-matter volume increases through adolescence[151] and up to about age 50,[152] after which there is shrinkage.[153] These findings, based on MRI measures in living human participants, were recently confirmed in an important post-mortem study which found that white matter, but notably not gray matter, showed shrinkage starting at age 46 up to age 92 in healthy individuals.[154]

Based on neuroimaging findings, George Bartzokis proposed that the demands on the oligodendrocytes (specialized glial cells) to manufacture cholesterol and to make and maintain myelin sheaths renders them vulnerable to age-related insults.[155] His model predicts a curvilinear relationship between white-matter volume and age, similar to what has been observed in neuroimaging studies.[156]

Of particular interest in the context of the present chapter is evidence that age-related change in white matter—in contrast to gray-matter volume and cortical thickness—has been fairly consistently related to cognitive change. However, there have also been fewer studies of white matter and cognitive aging. In aged monkeys, white matter under the PFC shows age-related alterations in structure, including splitting of the myelin sheaths, balloons, and blisters. Moreover, these myelin changes are related to memory performance in monkeys,[157] and there is some evidence from studies of cats that reduced integrity of myelin with aging leads to loss of conduction velocity.[158] Nevertheless, myelin increases even in old age in primates, as evidenced by myelin sheaths that are thicker, on average, in older monkeys than in younger ones.[159]

White matter and the integrity of the axons that constitute it can be measured in living persons with diffusion tensor imaging (DTI). This is a relatively new neuroimaging technique that makes it possible to measure the movement of water molecules in the axons that make up white matter. An index called *fractional anisotropy* (FA) measures the extent to which the flow of water molecules is restricted to one direction by the composition of various tissues, such as cell membranes.[160] FA ranges from 0 (water diffusion equal in all directions, or isotrophic) to 1 (diffusion occurring only in one direction, along the axon, or anisotrophic). Usually it is assumed that FA reflects the integrity of the myelin wrapped around myelinated axons. In multiple sclerosis patients, both FA and mean diffusivity correlate with the amount of myelin in the white matter.[161] Myelinated axons are anisotrophic (i.e., resist flow out of the walls of axons), but non-myelinated axons that are arranged in parallel and densely spaced can also present a barrier to water flow and thereby show anisotrophy. Anisotrophy in non-myelinated axons has been attributed to neuronal membranes and to neurofilaments.[162] Thus, FA reflects some aspects of the integrity of axons, but not necessarily only of myelinated axons.[163]

Age-related declines in FA[164] have been interpreted as reflecting reduced myelin integrity. If that is so, FA should be correlated with white-matter volume. Studying a large range of ages, Salat et al. observed positive correlations between white-matter volume and FA, but only for participants over age 40. FA did decline with age, but only in deep frontal areas and in the posterior limb of the internal capsule (made of axons from neurons in primary motor cortex), not in temporal or parietal areas.[165] FA in the anterior limb of the internal capsule has been reported to predict reaction time, but only in older adults.[166] Fjell et al. found, in a study of adults ranging in age from 40 to 60, that FA and white-matter volume were correlated only in specific cortical regions, including the cuneus, entorhinal, orbito-frontal, superior frontal, and inferior parietal cortices. Notably, most of the correlations were negative, suggesting that FA and white matter volume are sensitive to different aspects of white-matter integrity.[167] Consistent with that finding, FA in temporal-lobe structures (parahippocampal and superior temporal cortex) of MCI patients was negatively correlated with cortical thickness measures, suggesting that reductions in white-matter integrity do not necessarily correspond to reduced gray-matter cortical thickness.[168]

What are the consequences of age-related decline in axonal integrity for cognitive integrity? Reduced FA has been associated with impaired working memory,[169] impaired executive functioning,[170] slowed processing,[171] and global cognitive impairment.[172]

Investigating whether axonal integrity is a substrate for cognitive aging, David Ziegler and colleagues in Suzanne Corkin's lab asked whether cognitive performance in older adults was more closely associated with cortical thickness or with FA.[173] They observed that age-related cortical thinning was seen mainly in primary motor and sensory areas. Cortex in right posterior cingulate gyrus was thicker in older brains than in younger ones, which is consistent with findings of Fjell et al. that posterior cingulate undergoes thickening in high-functioning older people.[174] White-matter integrity, measured in FA, was reduced underneath cortical association areas. Correlations between both cortical thickness and FA measures and three aspects of cognition—cognitive control (Trails A and B, Digits Backward, Stroop interference), episodic memory (word lists, Wechsler Memory Scale Logical Memory), and semantic memory (Boston Naming, vocabulary)—were examined in the old group. No correlations between cognitive measures and cortical thickness were found but there were correlations between cognitive measures and FA. Cognitive control was correlated with frontal-lobe FA, whereas episodic memory was correlated with temporal-lobe and parietal-lobe FA. This suggests a role for axonal integrity in cognitive aging, but it was a cross-sectional study and it will be important for future work to relate longitudinal cognitive change to longitudinal change in FA and in other measures of axonal integrity. A role for white-matter integrity in cognitive aging appears to be particularly important in light of evidence, which we discuss in chapter 9, that FA can be increased by motor training[175] and by cognitive training,[176] even in old age.[177]

Another way to ask this question is to relate longitudinal cognitive change to FA. Jonas Persson and colleagues in Lars Nyberg's lab found that older people who had declined cognitively over 10 years showed both reduced hippocampal volume and lower FA in the corpus collosum relative to those who had not declined. Moreover, the reduced FA correlated negatively with right PFC activation, suggesting that compensation might have occurred by means of increasing PFC activation.[178]

Does white-matter volume change explain cognitive aging?

There is evidence that it may. However, white-matter change in relation to cognitive aging has been studied much less extensively than gray-matter change. Also, as will be discussed in chapter 9, white-matter integrity, measured in FA, can be increased by training, even in old age, which indicates that plasticity also influences this aspect of brain structure.

Summary

Cognitive aging occurs in rats, in monkeys, and in humans, but in each species not all individuals succumb to cognitive decline. Moreover, the neural substrate of cognitive aging remains uncertain. At a gross level, the best-documented change over the course of healthy aging is regional shrinkage in some regions of the brain, but shrinkage has not been consistently linked to cognitive loss. Age-related changes in white matter may contribute to cognitive change, but such changes appear to be relevant to cognitive change only late in life, in light of increased myelination until age 50 or so. However, the findings that white-matter integrity declines linearly from age 20 in cross-sectional studies and that FA can be increased by training in old age make white-matter change a complex topic in need of additional work.

At the synaptic level, effects of aging on biophysical properties of neurons are selective and subtle, seen only in certain regions and for specific cell types. Moreover, some of the age-related synaptic changes do not lead to corresponding changes in neuronal transmission, suggesting that caution should be exercised when interpolating from change in neuron properties to cognition. Also, synaptic changes are not seen in aged but unimpaired animals.

That 40–50 percent of older individuals remain cognitively unimpaired, with no synaptic dysfunction, raises the question of underlying age-related disease in individuals who do show decline. There is evidence that the hippocampus may play an important role, with alterations in hippocampal neuronal signaling occurring only in aged animals with deficits in spatial working memory. In humans, glucose regulation and the function of the dentate gyrus have been linked to memory performance, as has hypertension. Thus, chronic disease appears to make an important contribution to the pattern of individual differences in cognitive aging. If that is so, aggressive treatment in mid-life of diabetes and of the metabolic syndrome (a combination of excess weight, high blood pressure, and impaired glucose tolerance that increases the risks of diabetes and cardiovascular disease) might be important in limiting age-related cognitive decline.

4 Ameliorating Cognitive Aging: A Neurocognitive Framework

How can we explain the ability of many individuals to continue to function well late in life? It is well established that, on average, older adults perform more poorly than younger individuals on selected cognitive functions. Moreover, the aged brain undergoes cortical shrinkage[1] and dopamine receptor dysfunction, among more subtle neurochemical changes.[2] However, as was discussed in previous chapters, some 40–50 percent of older individuals—rodents, monkeys, or humans—do not show cognitive or brain decline.[3] And the brains of older individuals who remain cognitively unimpaired retain many characteristics of young brains.[4] In this chapter we advance a hypothesis aimed at explaining such successful cognitive aging.

4.1 Brain and Cognitive Plasticity

We recently advanced the hypothesis that the fundamental mechanism underlying successful cognitive aging is the interaction between brain plasticity and cognitive plasticity.[5] That hypothesis proposes that the brain has a capacity to respond to demand and insult by mechanisms of rejuvenation and repair, and that the mind has a capacity to adapt to the dual challenges of decline in nervous-system function (e.g., sensory loss, motor slowing) and to new experiences by changing the emphasis on the use of different cognitive functions. As was discussed in the preceding chapter, we acknowledge that changes in brain structure and function—regional shrinkage, neurotransmission change, and cortical hypometabolism—occur in aging, although it has proved hard to relate these brain changes to age-related cognitive change. We previously suggested that losses in regional brain integrity in old age result in functional reorganization through changes in cognitive processing strategies.[6] Here we extend that hypothesis to argue that the aging brain can be reorganized by cognitive demand (cognitive plasticity) but that reorganization depends on intact neural plasticity mechanisms (neuronal plasticity) and also on factors that enhance neuronal plasticity mechanisms (exposure to novelty, diet, exercise). "Neuronal plasticity" refers to durable changes at the neuronal level that have been shown by empirical work to be stimulated by experience, both cognitive and physical. Examples of neuronal plasticity are neurogenesis (birth and maturation of new neurons), synaptogenesis (expansion and growth of synapses), dendritic arborization (increased branching of dendrites), and network re-organization (alterations in connections between brain regions). "Cognitive plasticity" refers to age-related changes in the patterns of cognitive behavior—for example, increased susceptibility to task-irrelevant events, and greater use of executive control processes. Manifestations of cognitive plasticity depend on mechanisms of neural plasticity. We further argue that in the absence of disease, factors that enhance this interactive process (exposure to novelty, diet, exercise, etc.) can promote cognitive performance and preserve brain structure in healthy old age.

There is emerging evidence from many diverse research areas for each component of this hypothesis. In this chapter we systematically review the evidence in support of each component. We argue, from that evidence, that such remodeling can contribute to cognitive integrity in old age.

Mechanisms of neuronal plasticity in adulthood and old age

The idea that the brain is plastic in adulthood is relatively new. Even more recent is the view that such mechanisms of neuronal plasticity persist late in life. There is increasing evidence (some of it discussed in chapter 3 above) that the brain is a plastic organ capable of remodeling itself in adulthood, even in old age, in response to cognitive demands imposed on it. Manifestations of this remodeling process include neuronal structure, reorganization of cortical circuits, neurogenesis, and cortical structure.

Neuronal plasticity and synaptic structure

How does the brain rewire itself in response to new experiences during learning and training? Remarkable new techniques now allow imaging of neocortical structures in real time, thereby showing in great detail how cortex dynamically rewires itself in response to experience. Although the morphology of dendrites and axons is very stable over time, a subgroup of synaptic structures change rapidly in response to environmental demands.[7] Neurons that are within a few hundred micrometers of each other can form a synapse with each other simply by extending a dendritic spine or a terminal bouton.[8] After a change in experience, boutons and spines rapidly appear and disappear, forming and eliminating synapses within days to weeks.[9] New in vivo techniques allow dendritic structures to be imaged in real time in living animals. Small glass windows are implanted over cortex in mice that have been genetically engineered so that a subset of pyramidal neurons express a green fluorescent protein (eGFP). This fluorescence can be measured with two-photon laser-scanning microscopy.[10] By this method, spines can be imaged in neocortex. In a series of studies, Anthony Holtmaat and Karel Svoboda have observed dynamic changes in spine growth after whisker trimming. Whiskers are an important source of spatial and tactile information that have been studied extensively in rats and mice. Each whisker is represented by a "barrel" of thousands of densely packed neurons in the somatosensory cortex that code for the direction of whisker movement. After whisker trimming in mice, dendritic spines that had been stable over months (termed "persistent") retract and disappear, and new spines grow.[11]

Such dynamic changes appear to be the basis for experience-dependent plasticity in the cortex, allowing specific cortically regions to quickly and dynamically re-wire itself as the cognitive demands on the animal change. For example, plastic changes in neocortical spine formation have been seen after forelimb training in adult mice, during which spines form rapidly on dendrites of pyramidal neurons in contralateral motor cortex.[12]

It is important to our hypothesis that such synaptogenesis is promoted by novelty. In a study conducted by Patrizia Ambrogini and colleagues, rats that learned a maze—but also rats that merely swam in the maze without training—showed larger neuronal responses to stimulation and greater dendritic tree complexity than rats with no maze experience. Ambrogini et al. concluded that the novelty of having to swim and explore on a daily basis affected synaptogenesis in adult-generated neurons.[13]

Although most of the above-mentioned research was on rodents, dynamic changes in response to learning have also been seen in adult monkeys. Gregor Rainer and E. K. Miller investigated the effect of

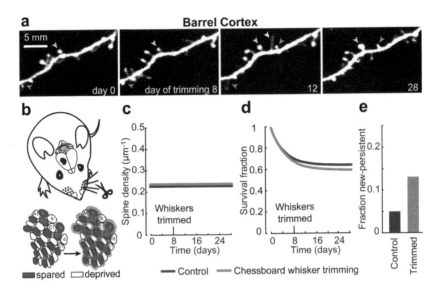

Figure 4.1
Spine plasticity in adult neocortex associated with experience. (a) Time-lapse image of dendritic spines in barrel cortex before and after "chessboard" whisker trimming at day 9 (indicated in plots c and d). Many spines were "persistent" (yellow arrowhead), but some appeared and disappeared (white arrowhead). After whisker trimming, new persistent spines (orange arrowhead) were more likely to grow and previously persistent spines (green arrowhead) were more likely to disappear. (b) The experimental paradigm. "Chessboard" whisker trimming was associated with changes in the whisker representational map in barrel cortex. (c) Spine density in barrel cortex remained unchanged after whisker trimming. (d) The fraction of surviving spines was slightly decreased after whisker trimming, owing to an increased loss of persistent spines. (e) New persistent spines increased by a factor of about 2.5 after whisker trimming. Source: A. Holtmaat and K. Svoboda, "Experience-dependent structural synaptic plasticity in the mammalian brain," *Nature Reviews Neuroscience* 10: 647–658. Reprinted with permission of Macmillan Publishers Ltd. (© 2009). This figure appears in color elsewhere in the book.

working-memory training on receptive field properties of prefrontal cortex (PFC) in monkeys. Monkeys were trained over days on a delayed match-to-sample task—a standard test of working memory. They were briefly shown target images (brightly colored photographs of faces, hands, fruits, and other things) before a delay was imposed. After the delay, a monkey was shown a test image that was either the same or different from the sample object. After the monkeys learned this task to 90 percent accuracy, the sample images were degraded by imposing a noise pattern on the images, so that a monkey had to compare a degraded sample image against a clear test image. During training, multi-unit recordings were carried out from 324 neurons in PFC. The undegraded images elicited differential firing patterns—neurons fired more strongly for "preferred" images than for "non-preferred" ones. As images became more degraded, this differential firing pattern dropped off. However, with experience, both neuronal responses and behavior showed increased discrimination of degraded images.[14] This study shows the effect of training on monkey neuronal firing patterns.

There is also human evidence that memory training alters neuronal activity. This was shown recently in an fMRI study by Pernille Olesen and colleagues in Torkel Klingberg's lab. Eight young participants practiced a working-memory task for 14 hours over 5 weeks and were scanned before training and five times during training. The fMRI BOLD signal in the middle frontal gyrus and in superior and inferior parietal cortices increased over the five scanning sessions in concert with improvements in the accuracy of working memory.[15] The same researchers also obtained direct evidence of synaptic plasticity in the dopamine system related to training of working memory in humans. They trained young adults on the same working-memory task, but scanned them in a PET scanner both before and after training. Dopamine receptor radioligands (radioactively unstable compounds chemically similar to dopamine receptors) were used to measure dopamine binding potential in several brain regions. Dopamine, a major neurotransmitter thought to mediate working memory, binds to several different receptors, designated D1–D5. A negative correlation was observed between D1 receptor binding and working memory in prefrontal and parietal cortices, with larger decreases in D1 binding associated with greater improvements in working memory.[16] These results are consistent with a previous finding of a negative correlation between capacity of working memory and D1 binding in schizophrenics.[17] This evidence that training of working memory affects synapse binding in a manner that persists over time is consistent with the other evidence reviewed in this section indicating that training affects neuronal plasticity in adult brains.

Reorganization of cortical circuits

Another potential mediator of neuronal plasticity is reorganization of cortical circuits. As described in chapter 3, the scientific world was surprised by the evidence of cortical reorganization seen in the Silver Spring monkeys after years of sensory loss.[18] The Silver Spring monkeys had their dorsal roots to the spinal cord cut, so that sensory input from the forelimbs could not be transmitted to the brain. Eventually neurons in somatosensory cortex that had originally responded to touch on the forelimbs came to process touch on the face, thereby gaining sensory input.[19] Michael Merzenich and colleagues used microelectrode mapping of cortex to observe this cortical re-organization over time. They found that after the digits of a monkey's hand were amputated, over a period of months the cortical representations of the intact digits were expanded into the deprived cortex that had previously represented the amputated digits.[20]

Evidence of cortical and white-matter reorganization has also been seen in humans. A man who spent 19 years in a "minimally conscious state" (a form of coma) after a head injury abruptly began to speak, despite continuing paralysis of all his limbs. A type of MRI scan aimed at white-matter integrity (diffusion tensor imagery, DTI, described in chapter 3) showed generally decreased white-matter integrity in this patient. However, there was an exception in the form of discrete regions of heightened white-matter integrity in the medial parietal and occipital regions.[21] As a single case study, this finding could be criticized for the lack of generality. However, other evidence from amputees and stroke patients also shows that training can reliably induce functional reorganization. V. S. Ramachandran conducted a pioneering series of studies with amputees who experienced a "phantom limb." Such individuals have a vivid sensation that their amputated limb is still present. Moreover, stimulation of the skin, commonly of the face (hand and face areas are near each other in somatosensory cortex), results in sensation in the phantom limb, often in a precisely organized manner. For example, each finger of the phantom hand would have a "map" on the face such that touch to that location on the face would result in sensation of touch to a specific finger of the phantom hand. Moreover, the sensation is specific to the modality of stimulation. For example, cold applied to the "finger region" of the face is sensed as cold applied to the finger.[22]

As was discussed in chapter 3, this evidence that neurons could "change jobs"—for example, from processing input on the hand to processing input on the face—led Edward Taub to build on his initial findings with the Silver Spring monkeys[23] and develop constraint-induced movement therapy (CIMT) for human stroke patients partially paralyzed on one side.[24] In a series of studies, Taub found that merely constraining the unaffected hand and arm of a hemiparetic stroke patient during the waking hours requires that the patient attempt to use the paretic hand and arm all day. In one representative study, comparisons before and after 14 days of CIMT treatment revealed (a) improved function of the paretic limb and (b) a larger cortical representation of the paretic limb in the damaged hemisphere.[25] Taub's group has found that even starting CIMT 15–21 months after a stroke yielded results similar to those seen when CIMT was initiated 3–9 months after a stroke.[26] Thus, even in aged, brain-damaged adults, considerable neuronal reorganization can be induced simply by deliberate limb use.

Studies conducted by Randolph Nudo and Kathleen Friel have revealed something about the mechanisms underlying such re-organization.[27] In monkeys, 5 months after an experimental lesion to the hand area of motor cortex, Numa Dancause and colleagues in Nudo's lab saw new connections in somatosensory cortex as well as new axonal projections between premotor cortex and somatosensory cortex.[28] New axonal projections have also been observed in response to training alone, in the absence of brain damage. Novel cortico-cortical projections with functionally active synapses have been found by Sayaka Hirara and colleagues in Atsushi Iriki's lab to arise when monkeys are trained in use of a new tool, but not in untrained monkeys.[29] This very important work shows that the brain can rewire itself by axonal sprouting and synaptogenesis not only after damage but also after training alone, even in aged adults.

Neurogenesis

Neurogenesis is another important mechanism of neural plasticity. Mammals produce thousands of new granule cells (small neurons) in the hippocampus each day.[30] This is seen even in older animals[31] and in aged and even terminally ill humans.[32] As was described in detail in chapter 2, the dogma among neuroscientists that after birth the brain doesn't generate neurons and only sheds them was so strong that initial evidence for neurogenesis in adult animals was swiftly discounted by leading authorities. Despite initial skepticism about the existence of neurogenesis[33] in adult animals,[34] there is now a consensus that neurogenesis occurs in adult mammals. The regions of the brain that have been found to be "neurogenic" in adults are the dentate gyrus of the hippocampus, the subventricular zone (the source of neocortical neurons in development), and the olfactory lobe. There is still debate as to whether neurogenesis takes place in adult mammalian neocortex.[35]

An important feature of neurogenesis is that newborn neurons are able to create functional connections. After all, growing new neurons would be of little use if they existed in only isolation and could not connect with existing neurons. Experimental approaches, including microscopic imaging and electrophysiology, have clearly demonstrated not only that these newly formed neurons exhibit all the characteristics of functional neurons but also that they are integrated into existing networks in which they participate actively, expressing postsynaptic potentials and spontaneous action potentials[36] and receiving excitatory and inhibitory inputs from the established neuronal network.[37] Although most studies have found more neurogenesis in younger animals than in older ones, nevertheless neurogenesis persists in aged animals and humans.[38] Although the initial studies of human neurogenesis were done with cancer patients, evidence of neurogenesis has recently been found in healthy people across the life span.[39] It is clear that old brains can grow new neurons. Such a view would have been heretical only 10 years ago, but now is increasingly supported by many emerging sources of evidence.

Cortical structure

Experience can also change the very structure of the brain. It has long been common in everyday parlance to say that experience shapes a developing brain, but only recently has it been recognized that this is literally true. Cortical mass can be changed by experience in a way that affects cognitive performance. "Environmental enrichment" (typically, housing large groups of animals in cages with toys, tunnels, hidden food, and running wheels) is associated with changed cortical structure and better cognitive performance. Macroscopic cortical changes that have been observed include increased cortical thickness; microscopic changes include increased dendritic extent and synaptic size.[40] Such changes can occur even in old animals.[41]

In contrast to animal studies, children cannot be randomly assigned to an "enriched" or an "impoverished" environment. However, the brains of people who differ in their level of cognitive functioning can be compared in order to examine how environmental factors may drive plasticity in humans. The brains of high-IQ children, but not those of average-IQ children, undergo episodes of cortical thickening and thinning during childhood.[42] Consistent with that finding, Fjell and colleagues found that the brains of high-functioning older people, but not those of average-functioning older people, appear to undergo cortical thickening in mid-life.[43] Likewise, Kristine Walhovd and colleagues found that older people who exhibit good retention of new memories over months showed stable cortical thickness whereas those with poorer retention showed cortical thinning.[44] Of course, these studies cannot reveal the source of the brain differences. Higher-IQ children may have been exposed to a more stimulating environment, as may have higher-functioning older adults. Another way to look at this is to consider how controlled experience affects brain change.

Learning experiences also change the human brain. The morphology of the human cortex has been found to change in response to many different types of perceptual-motor motor training, notably Braille reading[45] and the playing of stringed instruments.[46] Morphological changes have been demonstrated to occur as a result of training in juggling; this has been done by comparing MRI scan-measured cortical volume changes before and after the training.[47] Nor are such effects limited to motor training. Structural brain change has been seen both during and after 3 months of extensive knowledge acquisition undertaken by medical students preparing for board exams.[48] Comparisons of MRI scans before, during, and after training showed enlargement of posterior and lateral parietal cortex and of the posterior hippocampus bilaterally. This shows structural cortical change associated with extensive studying of physiology and medicine in young adults.

Do neuronal plasticity mechanisms persist late in life?

Yes, clearly. Adult and older animal and human brains are capable of experience-related plasticity, neurogenesis, and functional recruitment of neurons adjacent to a lesion. Human brains also change structurally as a consequence of motor and cognitive training, and the change is related to motor and cognitive performance. Thus, even in the aged adult, the human brain retains the capability to be dynamically re-wired—a capability that may be exploited to help limit age-related cognitive decline.

Cognitive plasticity in old age

Does the capability for cognitive plasticity persist late in life? Older people appear to undergo some change in cognitive strategy. This change, which we have characterized as "cognitive plasticity," could be considered compensatory or strategic, or both.

First, older people are known to be more susceptible to distracting events. This observation is the basis for the well-known "Inhibition" theory of cognitive aging,[49] which states that older people have difficulty ignoring task-irrelevant distractors on a range of tasks and that this difficulty reduces their ability to process task-relevant events. More recently, however, Lynn Hasher and colleagues found evidence that reduced inhibitory processes can confer a *benefit* on older adults. They observed a relation between the ability to ignore apparently irrelevant distractors and the use of those distractors in a subsequent task.[50] While reading passages after being instructed to ignore imbedded italicized words, older people were slowed more than the young by the presence of those irrelevant words. However, in a subsequent task in which the irrelevant words were the solution to a problem (three weakly related words could be linked if a fourth word was supplied), older people performed better than young people.[51]

In a subsequent study, the same research group gave young and old participants an "*n*-back" task in which they were given a series of word-picture pairs and asked to ignore the words and respond if the same picture had been presented on the previous trial (i.e., one trial back). Ten minutes later, the participants were given a "paired-associate task" in which words and pictures were paired. In the paired-associate task, people were trained by presenting the pairs and tested by presenting the picture and asking the participant to supply the word. Some of the word-picture pairs had been used in the *n*-back task (preserved pairs), some had been paired differently since the *n*-back task (disrupted pairs), and some were new. Older adults showed better memory for preserved pairs and worse memory for disrupted pairs. Younger adults showed no difference. That finding was interpreted as revealing "hyper-binding" in the older people, whereby they automatically form a memory of events that occur together. Although all participants were told to report on the pictures and ignore the words in the "one-back" task, the old participants had formed a memory of the word-picture pairs; the young ones had not.[52]

It has long been considered a deficit that older people are less able to ignore distractors then the young. However, when viewed in the context of other tasks and demands, such increased distractibility may confer a benefit. Are old people less able to ignore apparently irrelevant information, or have they "learned" over the course of a long life that sometimes apparently irrelevant information may be relevant at another time and/or in another context? Hasher and colleagues speculate that older adults who show this "hyper-binding" have greater knowledge of which events vary together in daily life—knowledge that is thought to underlie everyday reasoning and problem solving.[53] We propose that the ability to bind together apparently extraneous covarying events suggests some degree of cognitive plasticity in old age.

Second, neuroimaging evidence indicates that older adults use their brains differently than young adults. On a range of tasks in which young people show unilateral activation of either the left or the right cerebral hemisphere, old people tend to show bilateral activation, especially of PFC.[54] For example, in an early study, Roberto Cabeza and colleagues measured regional blood flow with PET during both memory encoding (e.g., remembering the pair of words "parents" and "piano") and memory retrieval (e.g., recalling the second word of the pair).[55] In young people, prefrontal cortex showed greater blood flow in the left hemisphere than in the right during encoding (because the pair of words was encoded in the left hemisphere) but greater blood flow in the right hemisphere than in the left during retrieval (when the second word of the pair was retrieved). This is a well-established pattern in young people—left-lateralized activity during encoding and right-lateralized activity during retrieval.[56] However, Cabeza et al. observed a different pattern in older people: they showed little prefrontal activity during encoding of the word pair, but bilateral activity during retrieval of the pair.[57]

In addition, older people show weaker activity of posterior brain regions than younger people but greater activation of anterior regions (notably

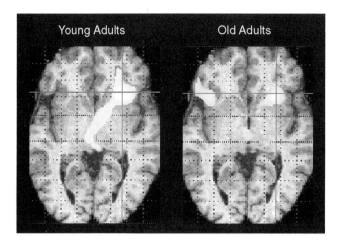

Figure 4.2
Prefrontal activity during episodic memory retrieval was lateralized to the right hemisphere in younger adults but was bilateral in older adults. Source: R. Cabeza, "Hemispheric asymmetry reduction in older adults: The HAROLD model," *Psychology and Aging* 17, no. 1: 85–100. This figure appears in color elsewhere in the book.

PFC) across a range of tasks.[58] Patricia Reuter-Lorenz and colleagues interpreted such data as indicating a compensatory process in older people, who must use both hemispheres to carry out a task that one hemisphere can do in young people.[59]

There have been some reports that the bilateral brain activations seen in older people are linked to the effectiveness of processing. In an incidental memory paradigm, Alexa Morcom and colleagues in Richard Frackowiak's lab found that young people activated PFC only on the left during word classifications but old people activated PFC bilaterally. Older people showed bilateral activation during a word-classification task with the degree of PFC activation related to the success of encoding.[60] Moreover, reliance on PFC varies with education level. In young adults, frontal activity was weaker in those with more years of education; in old adults, frontal activity was greater in those with more years of education.[61] This suggests some benefit to the old from the additional right PFC recruitment. Older people who were good at comprehending complex sentences showed less left temporal-parietal activation but greater left dorsal inferior frontal and right temporo-parietal activation during the task than young people. These regions have been associated with rehearsal of verbal information in working memory.[62]

In contrast, older people who were poorer at comprehension showed greater activation of dorsolateral PFC,[63] a region associated with general problem solving.[64] Considered together, these results indicate a shift with age to greater use of both right and left prefrontal regions during task performance. However, such a shift in activation patterns may simply reflect a shift in strategy toward more "controlled" or effortful processing.

It is something of a paradox that the PFC—consistently found to shrink with age[65]—also shows greater task-related activation in old age.[66] Cheryl Grady has interpreted findings of unilateral prefrontal activation in the young but bilateral prefrontal activation in the old as reflecting a greater need by older people for controlled executive processing,[67] known to be mediated prefrontally. Consistent with that interpretation, Bart Rypma and colleagues found that when young people performed a simple reaction time task, those who were slower to respond also experienced a greater influence (seen in neuroimaging studies) from prefrontal areas during the task. Rypma interprets this finding as indicating that people who are slower to respond require increased executive control from PFC to do the task.[68] Moreover, the well-established association between self-control and higher IQ[69] is mediated by activity in anterior PFC.[70] These studies are consistent in linking superior cognitive performance with more controlled processing and greater use of prefrontal regions.

Therefore, although it seems established that older people use prefrontal and parietal cortices more than young people do, it is less clear what that difference means. It could be compensation, as Cabeza has speculated. However, in light of evidence that increased reliance on the PFC is associated in the young with better working-memory performance[71] and higher IQ,[72] increased prefrontal use may also reflect a change in cognitive strategy. In 2007, we speculated that a gradual shift from a bottom-up, stimulus-driven strategy in the young to a more top-down, controlled strategy directed from prefrontal areas in the old is the normal developmental course.[73] The most effective cognitive approach in a young individual possessing relatively little knowledge and experience may be one that is influenced largely by environmental input, which leads to rapid, stimulus-driven processing. In contrast, the most effective cognitive approach in an older individual who has acquired considerable knowledge and experience may be one that is more deliberative and involves controlled integration with existing cognitive constructs. If so, the failure of old people to inhibit processing of apparently irrelevant events[74] may be a strategic change (perhaps conscious, but not necessarily) that induces them to attend more broadly. There is some evidence for this view. Older adults are capable of showing the unilateral pattern of brain activation that is characteristic of young people if the older adults are instructed to use "deep encoding"[75] or are "high-functioning" in cognitive performance.[76] These findings provides further evidence that the age-related change in patterns of regional brain use is strategic, consistent with our characterization of cognitive plasticity.

Do mechanisms of cognitive plasticity persist late in life?

Yes. There is emerging evidence that old people do not simply show poorer cognitive performance than young people. First, a subset of older people perform as well as young people. Second, older people appear to rely more on PFC and on executive control processes during performance of cognitive tasks. Our *neuronal and cognitive plasticity* hypothesis argues that sensory losses and even processing losses in healthy old age lead to greater reliance on executive processes. In turn, plasticity mechanisms that persist late in life both underwrite and reify cognitive plasticity. Use of top-down, executive functions may slow processing overall, but may result in greater accuracy. For example, remembering that you pulled the plug on the curling iron or turned off the kettle can occur automatically as you do each, or you can make a point to remember doing so. The latter strategy may slow you down; however, it may result in greater confidence in your memory, so that later, when you are driving to work, you do not have to worry about your house burning down.

For a schematic representation of our framework for understanding how environmental and experiential factors can ameliorate cognitive aging, see figure 4.3.

Interactions between neural and cognitive plasticity

Do neural plasticity and cognitive plasticity interact? We have hypothesized that successful cognitive aging requires interactions between neuronal and cognitive plasticity—interactions stimulated by environmental demands and supported by factors that enhance brain integrity. We have argued that, in the absence of disease, factors that enhance this interactive process can promote both cognitive integrity (preserved cognitive ability)

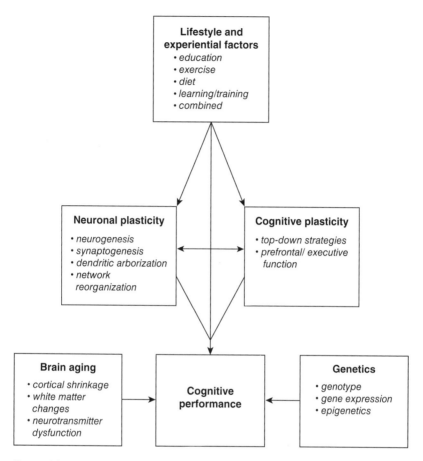

Figure 4.3
A schematic diagram of our framework for understanding how environmental and experiential factors can interact to ameliorate cognitive aging.

and brain integrity (preserved brain structure) in healthy old age. What evidence is there that these two types of plasticity interact?

The interaction between neuronal and cognitive plasticity appears to be bidirectional. First, neurogenesis, synaptogenesis, and other mechanisms of neural plasticity underpin cognitive plasticity, which is reflected in learning, in greater executive control, and related processes. At the same time, and in turn, neural plasticity is enhanced by cognitive plasticity. Both neurogenesis and synaptogenesis are correlated with learning and training. For example, older rats that retain good spatial memory as they age generate more new neurons than rats with impaired spatial memory.[77] Also, new neurons must be present for new long-term memories to be formed.[78] In turn, new learning heightens neurogenesis[79] and enhances the survival of newborn neurons.[80]

We speculate that this interactive process can sustain cognitive and brain integrity in healthy aging. Gerd Kempermann, one of the pioneers of the study of adult neurogenesis in rats, speculated that ongoing adult neurogenesis in the hippocampus provides brain networks the ongoing capability to adapt to novelty and complexity.[81] We extend that view by proposing that the relation between the two forms of plasticity is interactive, with exposure to novelty needed to heighten and maintain plasticity mechanisms. Consistent with that claim is evidence that environmental enrichment has no additional brain or cognitive effects after 6 months,[82] which indicates that the novelty of the new environment is important in inducing the brain and cognitive changes. This is consistent with the evidence that novel experiences or learning must occur for newborn neurons to survive and become integrated with existing neurons.[83] To the extent that cognitive plasticity requires changed cognitive strategies and learning experiences, it may also be subject to individual motivation and circumstances. If that is so, then to age successfully an individual may have to be exposed to novel cognitive demands (e.g., learning new material) so as to benefit fully from mechanisms of neural plasticity.

Do neural plasticity and cognitive plasticity interact?

Yes. At least three sources of evidence can be cited. One source is evidence from studies of human brain volume. Better memory performance in older adults is associated with reductions in the volumes of prefrontal and temporal gyri. Another source is human neuroimaging studies, reviewed in section 4.1. Several research groups have observed that older people show greater activation of anterior regions (notably PFC) and weaker activation of posterior regions across a range of tasks, perhaps reflecting compensatory processes in older people. This evidence suggests that both structural remodeling and greater use of prefrontal cortex are linked with successful cognitive aging. A third source is the finding that neurogenesis is stimulated by environmental enrichment and exposure to novelty. Likewise, learning new skills is associated with heightened survival of the newborn neurons. These three sources of evidence point to interactions between neural and cognitive plasticity in aging.

4.2 Applying the Framework

What are the factors that influence interactions between cognitive plasticity and neuronal plasticity? If our hypothesis that successful aging depends on interactions between neuronal and cognitive plasticity is correct, then factors that appear to support cognitive integrity in aging should be related to these two mechanisms or to their interactions.

Among the factors that have been shown to influence neural plasticity are novel experiences in childhood, physical exercise, diet and nutrition, estrogen around menopause, cognition-enhancing drugs, and learning and cognitive training. We will consider each of these in the following chapters. As will be seen, there is evidence that most of these factors affect neural mechanisms associated with cognition. In the last chapter, we will discuss how such evidence can be exploited so that individuals are able to maintain their cognitive vitality in late adulthood.

Cognitive challenge in youth (chapter 5)

Exposure to particular environmental factors and demands can be serendipitous, as when a new learning opportunity comes along. At other times, benefiting from experience may require a person to make deliberate attempts—for example, to increase physical activity, or to change dietary habits. In either case, experience can drive brain plasticity, both structurally and functionally. The evidence suggests that experience-driven plasticity is not just a feature of youth; it can occur in old age.

Novel experiences appear to be an important trigger in stimulating both learning and brain plasticity.[84] Rats placed in a novel, "enriched" environment showed more than doubling of the number of synapses in the hippocampal CA1b field.[85] Environmental enrichment (that is, removing animals from standard laboratory housing and placing them in large enclosures with cage mates, toys, tunnels, and hidden food) is associated with improved cognitive performance, increased brain weight, and changed cortical structure.[86] Despite this, 6 months after initiation of the enriched experience there are no further brain or cognitive effects,[87] which suggests that it is the novelty of the experience that is important in triggering the effects.

Education in humans is roughly analogous to environmental enrichment in animals. People who receive greater education in childhood undergo less cognitive decline in old age[88] and are at lower risk of

Alzheimer's disease.[89] Higher educational attainment has been associated with greater integrity of white matter associated with medial temporal-lobe areas and of fiber tracts in those regions.[90] Male twins who participated in novel leisure-time activities (e.g., studying for courses) in mid-life were at reduced risk of dementia late in life relative to their twins who participated in passive leisure-time activities such as watching television.[91] It is important for the present thesis that formal education appears to affect both cognitive strategies and patterns of brain activity in ways that vary with age. In fMRI scans during encoding and recognition memory tasks, more education was associated in young adults with less frontal activation, whereas more education was associated in older adults with greater bilateral frontal activity.[92] Considered together, this evidence suggests that novel experiences, including education, protect against brain and cognitive decline, perhaps by stimulating plasticity mechanisms.

Physical exercise (chapter 6)

Both animal studies and human studies have linked physical activity, and aerobic exercise specifically, to improvements in cognition.[93] Several mechanisms have been identified. One is neurogenesis, identified as a mechanism on the basis of evidence that voluntary running increases proliferation of new neurons[94] and survival of new neurons[95] in rats. The effect of running on neurogenesis is seen even in aged rats.[96] Another mechanism is synaptic plasticity, identified as a mechanism on the basis of evidence that voluntary running heightens long-term potentiation during learning[97] in rats. Increased regional cerebral blood flow from exercise also exerts a benefit. In animals, exercise-related increase in cerebral blood flow has been correlated with post-mortem measurements of neurogenesis in the dentate gyrus. In the same study, exercise in humans was likewise found to selectively increase blood flow in the dentate gyrus in a manner that correlated with both cognition and cardiopulmonary function.[98] Considered together, this evidence suggests that aerobic exercise stimulates a range of plasticity mechanisms, perhaps in part through effects on cerebral blood flow.

Diet and nutrition (chapter 7)

Diet can affect cognitive performance through the amount of specific nutrients consumed but also through the pattern of consumption. Poor glucose regulation is linked to hippocampal atrophy and poor memory performance, even in non-diabetics,[99] and obesity is a risk factor for type 2 diabetes. Dietary restriction (DR) exerts positive effects on insulin and blood glucose levels in rodents and in primates (including humans). Resveratrol, a polyphenol found abundantly in red grapes and red wine, appears to exert its effects via some of the same mechanisms as DR and has been found to protect the hippocampus from brain damage.[100] Thus, dietary restriction and resveratrol both appear to heighten hippocampal neuroprotection, and both can influence neural plasticity. On the other hand, a high fat diet exerts negative effects on cognition.[101] A high-fat diet has been shown to increase insulin resistance,[102] and insulin resistance has been linked to hippocampal dysfunction.[103] The effects on brain and cognition of energy intake levels (DR, obesity) and content (type of fat, polyunsaturated or saturated fat) are well documented in animals.

Estrogen and cognition-enhancing drugs (chapter 8)

The use of drugs to help people alleviate the symptoms of cognitive disorders such as Attention Deficit Hyperactivity Disorder, Mild Cognitive Impairment, and Alzheimer's disease is well accepted. Are there cognitive-enhancing drugs that could be used as a treatment for cognitive aging? Before 2002, menopausal and post-menopausal women "replaced" the estrogen their bodies no longer produced to alleviate a range of symptoms, some cognitive. Epidemiological studies using neuropsychological assessment[104] and information-processing tasks[105] have consistently observed positive effects of estrogen on cognition in women. In animal studies, estrogen has been found to rapidly induce neurogenesis[106] and to alter synaptic transmission.[107] However, randomized trials of hormone replacement in older women taking either estrogen alone or estrogen and progesterone have not consistently shown benefits to health or cognition. Moreover, when hormone treatment is begun late (e.g., after age 65) rather than around menopause, the risks of cancer and dementia increase. Thus, the use of estrogen to boost cognition in older women involves the balancing of costs and benefits.

The use of psychoactive drugs to treat cognitive aging in both men and women can be as controversial as the replacement of estrogen in post-menopausal women. Nevertheless, it is valuable to examine what is known about the effects of cognition-enhancing drugs. Several drugs, most notably cholinesterase inhibitors, have been found to benefit cognition in older people. Drugs in this class have typically been used to treat people with suspected or diagnosed Alzheimer's disease, in whom they typically have only small effects on cognitive functioning. These drugs operate in the synapse, where they act to slow the breakdown of the neurotransmitter acetylcholine after release. Such drugs have not been widely studied in healthy aging, but a few studies have found a benefit.[108]

Another drug that a few studies have shown to exert a selective effect on cognition is modafinil, which is approved for treatment of daytime sleepiness. This drug's precise mechanism of action is not known, but there is animal evidence that it elevates both dopamine and norepinephrine levels in the synapse.[109] Modafinil has been studied mainly with regard to sleep deprivation, but a few studies have found it to benefit aspects of cognition in healthy young adults (who showed improvements on some information-processing tasks but worse performance on other tasks[110]), in middle-aged people,[111] and in emergency-room physicians.[112] However, these studies do not consistently find benefits and costs within the same tasks, and the drug has not been studied in the context of healthy aging. The effects of these other cognition-enhancing drugs are described further in chapter 8. That chapter also examines the practical and ethical issues associated with the use of cognition-enhancing drugs to ameliorate cognitive aging.

Learning, cognitive training, and cognitive stimulation (chapter 9)

As we have noted, one of the strongest drivers of neuronal plasticity is learning new information and/or skills. The effect begins soon after birth and continues throughout adulthood. Many newly generated neurons do not survive,[113] but new learning appears to rescue new neurons from programmed cell death.[114] Moreover, LTP induction (the cellular mechanism underlying memory) improves the survival rate of newly formed neurons in the dentate gyrus.[115] Several studies have reported a positive correlation between neurogenesis and learning.[116] Learning can benefit from neurogenesis.[117] In humans, there is ample evidence that new learning heightens regional brain activation, particularly of prefrontal cortex.[118] As was reviewed above, whereas young people show unilateral prefrontal activation during encoding of new information, older people show bilateral activation.[119] The age-related increase in dependence on executive processes mediated in prefrontal cortex may be a manifestation of cognitive plasticity mechanisms—the capacity to benefit from experience by changes in the way prefrontal and parietal cortices are used.

Summary

In this chapter we proposed the hypothesis that brain plasticity and cognitive plasticity are involved in ameliorating cognitive aging. These can act independently as well as interactively. The neurocognitive framework we have outlined can be used to understand the beneficial effects on cognition of five general classes of experiential factors: novel experiences in childhood, physical exercise, diet and nutrition, estrogen and cognition-enhancing drugs, and learning and cognitive training. We consider the scientific evidence for each of these factors in turn in the succeeding chapters.

5 Cognitive Challenge in Youth

And these I see—these sparkling eyes,
These stores of mystic meaning—these young lives,
Building, equipping, like a fleet of ships—immortal ships!
Soon to sail out over the measureless seas,
On the Soul's voyage.
—Walt Whitman, "An Old Man's Thought of School"

In the previous chapter we proposed the hypothesis that experiential and lifestyle factors that promote neuronal and cognitive plasticity can allow for enhanced cognitive performance in old age. In the subsequent six chapters, we consider the evidence pertaining to each of these factors. However, we begin our examination of the evidence in an unusual place: childhood and adolescence. It is rare to find a description of the cognitive experiences of children in a book on aging in older adults. Yet such a discussion is appropriate because such experiences—including both formal education and other opportunities for intellectual challenge—may play an important role in determining the extent to which certain individuals are able to maintain cognitive vitality as they age.

Formal education is widely considered to confer economic, social, and personal benefits. There are many fine schools in the United States, and many children are lucky enough to be the recipients of a good education. Yet increasingly we are failing to educate young people who attend urban schools. The "on-time" graduation rates of the public school systems in the District of Columbia and in Philadelphia, for example, hover around 50 percent.[1] According to 2010 data, 30 percent of the nation's high school students are not graduating. Poorly performing schools deprive future adults of earning power and lower their quality of life. They also deprive future adults of a healthy old age. The cost of a poor education in youth extends to old age. As we discuss in this chapter, a good education early in life appears to afford some protection against cognitive decline, including Alzheimer's disease, late in life. Substandard education is contributing to a public health problem.

The effects of cognitive stimulation in youth can be seen in animals. Since laboratory animals are raised in captivity, they are deprived of the normal cognitive stimulation of their wild counterparts and experience an impoverished environment. In laboratory animals, cognitive challenge can be manipulated experimentally by "environmental complexity" or "enrichment" (described in chapter 5). It has been known since the work of Donald Hebb that housing laboratory animals in a "complex" environment improves their cognition.[2] For Hebb that meant giving lab rats the run of his house. But starting with the work of Mark Rosenzweig,[3] environmental "enrichment" or "complexity" has been manipulated experimentally by means of toys, tunnels, and social interaction with cage mates.

5.1 Cognitive Challenge in Animals

How do animals benefit from living in a cognitively challenging environment? Wild rats have to live by their wits. They must find their own food, avoid being poisoned, avoid getting eaten by hawks and other predators, attract a mate, create a nest in a safe place, and so on. Rats living in a laboratory have a very cognitively impoverished environment. However, their environment can be manipulated somewhat to offer greater stimulation. The "environmental enrichment" manipulation in animals described briefly in chapter 5 has been consistently associated with gains in both brain measures and cognitive measures.[4] Of course, even an enriched laboratory environment is less complex than the environment experienced by a wild rat. Nevertheless, environmental enrichment in laboratory rats is associated with enhanced memory, reduced age-related memory decline,[5] and increased exploratory activity.[6] Reliable brain changes—increased cortical thickness, greater dendritic extent, and synaptic size—also occur in response to enrichment.[7] The observation that enrichment benefits cognition even in old rats[8] suggests that some capacity for experience-related brain plasticity is retained even in old age.

Cognitive enrichment per se has also been found to protect mice destined to develop Alzheimer's pathology. Wild mice don't get Alzheimer's disease, but some well-studied strains of transgenic (genetically altered) mice develop amyloid deposition (a characteristic of Alzheimer's disease) in the hippocampus and in cortex and show cognitive decline. When such transgenic mice are housed in an enriched environment, their brains show a dramatic reduction in amyloid deposition. As has been noted, environmental enrichment is a complex manipulation, exposing the animal to more social interactions, more physical activity, and more cognitive stimulation. Jennifer Cracchiolo and colleagues in Gary Arendash's lab sought to isolated the various effects of environmental enrichment on Alzheimer's transgenic mice. The mice were raised from 1.5 months to 9 months of age in one of the following conditions (see figure 5.1): impoverished (one mouse per cage), social (two to four mice per cage); physical (running wheels in cage), social + physical (two to four mice with wheels), "complete" (social + physical with toys, tunnels, novel experiences). Their rigorous study found that only mice in the "complete" environment showed reduced amyloid deposits and improved cognition.[9]

One mechanism underlying effects of environmental enrichment is adult neurogenesis. Adult mammals are known to produce thousands of new hippocampal granule cells each day, even in old age.[10] Neurogenesis has also been observed in aged humans.[11] In 1998, Peter Eriksson and colleagues[12] published the first report of neurogenesis in humans. Terminal cancer patients were given the thymidine analog bromodeoxyuridine (BrdU), a drug used to label rapidly dividing cells for the purpose of determining the growth of tumors. BrdU was concurrently being used to label newborn neurons in adult tree shrews[13] and rodents.[14] The patients, aged 57–72 at the time of death, had agreed to donate their brains so that the hippocampus and the subventricular zone could be examined for evidence of neurogenesis. From 500 to 1,000 newly generated cells had been formed each day in the brains of those terminally ill older adults. Further, these new cells were found to undergo cell division and to differentiate into neurons.[15] Consistent results have been obtained in subsequent human work.[16]

Although the number of new cells developing into neurons in the aged cancer patients studied by Eriksson et al. was lower than had previously been observed in young adult monkeys,[17] that observation is consistent with evidence that the rate of neurogenesis declines with age[18] (although

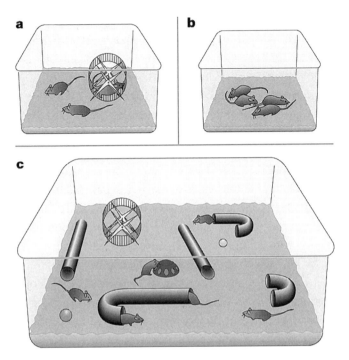

Figure 5.1
A range of environmental manipulations for mice: (a) exercise only, (b) impover-ished, (c) enriched. Source: H. van Praag, G. Kempermann, and F. H. Gage, "Neural consequences of environmental enrichment," *Nature Reviews Neuroscience* 1, no. 3: 191–198; reprinted with permission of Macmillan Publishers Ltd.; © 2000.)

in mice that decline can be reduced with voluntary wheel-running[19]). Nevertheless, neurogenesis can be increased, even in aged animals, by an "enriched environment." Aged mice that had been raised in standard housing (three mice to a cage, with bedding, food, and water only) were moved in old age to "enriched" housing with tunnels, toys, exercise wheels, and cage mates for social interaction. After 40 days, the mice in the enriched housing showed a threefold increase in numbers of new cells that had differentiated into neurons.[20] Thus, not only do aged brains remain capable of neurogenesis; the process can be heightened by exercise and environmental complexity.

Adult newborn neurons can create functional connections. Within 3 weeks of differentiation, newborn neurons have become integrated in the existing circuitry of the dentate gyrus and extend their axons to hippocampal area CA3.[21] Several experimental approaches, including microscopic imaging and electrophysiology have clearly demonstrated that newborn neurons not only exhibit characteristics of functional neurons but also are integrated into existing networks in which they actively participate, expressing postsynaptic potentials and spontaneous action potentials.[22] Further, it was recently shown that newborn neurons in adult mouse hippocampus form functional synapses. Nicolas Toni and colleagues in Fred Gage's lab demonstrated that newborn granule cells in the mouse dentate gyrus form functional glutamatergic synapses with hilar interneurons and CA3 pyramidal cells, and that they receive excitatory and inhibitory inputs from the established neuronal network.[23]

It has been harder to show a causal relationship between learning and neurogenesis. A growing literature associates neurogenesis and migration of new neurons with benefits on skill learning and cognitive performance.[24] Also suggestive of such a relation is the finding that novel-object recognition in rats is impaired after administration of an agent that decreases cell division.[25] (Rodents will explore a novel object more than a previously explored object if they remember the familiar one.) Dar Meshi and colleagues limited hippocampal neurogenesis by directing focal radiation to that structure. After this treatment, spatial learning still benefited from enrichment,[26] which suggests that the enrichment manipulation influenced brain function in other ways.

It has been hypothesized that adult neurogenesis in the dentate gyrus is important for integrating novel information after a task changed.[27] To test this, Alexander Garthe and colleagues administered temozolomide—a drug known to suppress adult neurogenesis up to 80 percent—to mice. The goal was to suppress adult-generated granule cells in the dentate gyrus before training and testing in a Morris water maze. When long-term potentiation from the dentate gyrus and hippocampal region CA1 was measured 4 weeks after treatment, the treated mice (a) showed no evidence of a current from the newborn granule cells in the dentate gyrus (consistent with suppression of those cells) and (b) had trouble with water maze performance, but only under "reversal" conditions when the platform was moved 3 days after training.[28] These findings are consistent with claims that adult-generated granule cells play a role in integrating novel cues when the learning context has changed.[29] Whatever the relation between environmental enrichment and neurogenesis, it is well documented that a complex environment results in smarter animals with more plastic syn-apses and better cognitive functioning in old age.

How do animals benefit from living in a cognitively stimulating environment?

By becoming smarter, by showing less cognitive decline in old age, by showing structural brain changes producing more neuronal connections, and by generating more neurons in adulthood.

5.2   Cognitive Challenge in Young Humans

In humans, education in youth is roughly analogous to "environmental enrichment" in animals. (Cognitive training in adulthood, which also confers cognitive benefits, will be discussed in chapter 9.) Consistent with animal studies, education appears to change the brain. It has been reported that in healthy older people more years of education are associated with greater integrity of white matter in medial temporal lobe and associated subcortical white-matter fiber tracts.[30] In brains that came to autopsy, there was a large and consistent increase in length of cortical dendrites (branched projections of a neuron which conduct neuronal signals) with more years of education.[31]

There is long-standing epidemiological evidence that education protects against Alzheimer's disease. Robert Katzman and colleagues conducted an important study of the effect of formal education on development of dementia late in life.[32] Some 5,000 randomly selected older people living in Shanghai were interviewed and were given a Chinese-language version of the Mini-Mental State Exam (the 30-item test that assesses memory and orientation to place and time). The presence of a cognitive impairment was related to age, gender, and educational attainment. In the sample, formal education ranged from essentially none (people who had never been taught to express themselves with a brush or pen) to advanced degrees (PhD and MD). Approximately 4 percent of the participants were found to be severely demented, and 14 percent mildly demented. A strong inverse relation was observed between educational attainment and cogni-tive status. Prevalence of dementia was greatest in those with fewer years of education. A number of other studies have confirmed the relation of education to risk of Alzheimer's disease.[33]

Fewer years of education have also been associated with faster age-related decline in memory,[34] in mental status,[35] and in verbal ability.[36] In view of such evidence, Yaakov Stern has advanced the concept of "cognitive reserve,"[37] according to which individual differences in efficiency or flexibility of cognitive processing—perhaps built up through early educational experiences—can provide a reserve against the effects of later age-related or Alzheimer's-related brain pathology. Stern's concept has been invoked to explain why better-educated Alzheimer's patients appear to be in a more advanced state of neuropathology when diagnosed with the disease: cognitive reserve has allowed them to function longer in the presence of pathological change.[38] The notion is that individuals possess both "brain reserve" (cortical volume) and "cognitive reserve" (ability to recruit brain networks optimal for a task).

Alexandra Foubert-Samier and colleagues assessed "brain reserve" in a large cohort of 9,294 people, aged 65 and older, randomly selected from the electoral records of three cities in the Bordeaux region of France. This cohort were assessed annually on various factors, including cognition and vascular health. In a recent report from this ongoing study, Foubert-Samier related cortical volume to education in 331 healthy older people who had remained non-demented for 4 years. They were categorized by occupation, social activities, and one of five levels of education, ranging from primary school without a diploma to university graduate. Each participant had undergone a structural MRI scan at the beginning of the study and another after 4 years. Although occupation and social activities were found to be associated with cognitive performance, only education was found to be correlated with gray-matter and white-matter volume. The strongest effects were seen in left temporoparietal lobes and in bilateral orbitofrontal lobes.[39]

Mellanie Springer and colleagues in Cheryl Grady's lab attempted to test the concept of cognitive reserve experimentally. If cognitive reserve is an important factor, they reasoned, more education may enable older people to use alternate circuits for task performance when the previously used circuits become dysfunctional.[40] Using functional MRI, Springer et al. found that better-educated young adults showed less frontal activation during encoding and recognition memory tasks while those will fewer years of education showed greater frontal activation associated with worse recognition performance. In contrast, better-educated older adults showed greater bilateral frontal activity than their less-well-educated counterparts. The opposite relation was found for medial temporal activity. These results are consistent with imaging evidence that poorer-performing

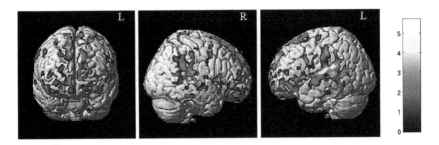

Figure 5.2
Areas in which gray-matter volume differed significantly between healthy aged
groups categorized as having attained high vs. low education. Source: A. Foubert-
Samier et al., *Education, Occupation, Leisure Activities, and Brain Reserve: A Population-
Based Study*. Reprinted with permission of Elsevier (© 2010). This figure appears in
color elsewhere in the book.

young individuals produce more bilateral prefrontal activity than better-
performing young adults.[41] Also consistent with that, young individuals
with higher IQs showed stronger lateral prefrontal cortical activation
during an *n*-back (working-memory) task than young individuals with
average IQs.[42] Young people also showed increased activation in middle
and inferior frontal gyrus in concert with improvements in performance
on a working-memory training task. These changes in the fMRI signal were
due not to activation of additional areas but rather to increases in the
extent of cortex activated.[43] A number of studies have found that higher-
performing older people show more bilateral activation of prefrontal
cortex.[44] Cheryl Grady has argued that, because most older participants in
imaging studies are relatively high functioning (as evidenced by their
having volunteered to participate in the research), the numerous studies
reporting increased prefrontal activity in older participants on a range of
tasks[45] indicate that executive strategies—which depend on prefrontal
cortex—are adopted during demanding tasks, particularly in high-functioning older adults. Consistent with that, as was mentioned above, several
authors have observed that better-educated Alzheimer's patients are diagnosed in a more advanced state of pathology relative to those with less
education.[46] Also in concert with the compensation view is evidence that
several months after a left-hemisphere stroke producing aphasia (impairment in speaking or understanding speech), better recovery of language
function was seen in those patients who exhibited bilateral activation

of language areas in the brain than in those who exhibited unilateral activation.[47]

Reviewing these and related findings, Greenwood argued that adopting a more top-down, controlled cognitive strategy—in contrast to a stimulus-driven, bottom-up strategy—may be the optimal developmental course for efficient cognitive functioning in adulthood.[48] Higher educational attainment in youth may encourage shifting to a top-down, executive strategy in old age, a shift reflected in more neural activity in prefrontal regions. This could, in turn, provide protection from age-related decline, and perhaps delay the diagnosis of Alzheimer's disease.

How do humans benefit from exposure to a cognitively stimulating environment in youth?

By showing better cognitive functioning late in life (perhaps as a result of greater use of executive processing) and by lowering their risk of Alzheimer's disease. Neither environmental enrichment in animals nor education in humans appears to have any direct effect on the pathology of Alzheimer's disease, but both environmental enrichment in animals and education in humans influence the ability of the mind and the brain to function well despite that pathology.

Summary

There is considerable evidence from animals that cognitive challenge in youth improves brain function and cognitive function in adulthood, even in old age. Because children cannot be randomly assigned to enriched or impoverished environments, as animals can, the human evidence is more epidemiological. Nevertheless, higher educational attainment in youth is associated with better cognitive functioning and reduced risk of Alzheimer's disease late in life. Higher educational attainment is also associated with greater reliance on prefrontal cortex, interpreted as successful compensation. Yaakov Stern has advanced the concept of "cognitive reserve" to explain why early cognitive enrichment is protective. If the brain changes associated with cognitive enrichment in animals (for example, increases in dentritic extent, number of dendritic spines, and density of synapses) also are associated with higher education in children, such brain changes may persist throughout adulthood and may be a bulwark against later loss and decline.

6 Physical Exercise

Exercise ferments the humors, casts them into their proper channels, throws off redundancies, and helps nature in those secret distributions, without which the body cannot subsist in its vigor, nor the soul act with cheerfulness.

—Joseph Addison, *On Exercise*

To get back my youth I would do anything in the world, except take exercise, get up early, or be respectable.

—Oscar Wilde, *The Picture of Dorian Gray*

Americans are increasingly sedentary. Many of us rarely engage in any kind of physical activity other than walking a few feet to and from our cars. This fact, coupled with the propensity to eat large amounts of high-fat foods, has led to what is being called an epidemic of obesity. In adults, obesity is defined as a body mass index (BMI) of 30 kilograms per square meter or greater; being overweight is defined as having a BMI of 25 kg/m^2 or more. By these standards, more than half of the US population is currently estimated to be overweight or obese.[1] Perhaps more troubling are recent figures indicating that children are increasingly likely to be overweight and unfit. Being overweight in childhood is a strong risk factor for being overweight in adulthood and for developing diabetes in adulthood.[2] Unless they can be encouraged to increase their physical activity, these future adults will add even more to the numbers of the overweight and obese in the United States.

Physical inactivity can have significant health and economic costs.[3] The obesity epidemic has been accompanied by an increase in the incidence and in the age of onset of Type 2 diabetes. There is strong evidence that physical activity can reduce the incidence or the severity of this type of diabetes and of other illnesses, including heart disease and various cancers. Yet participation in exercise and other forms of physical activity across the US population remains modest and sporadic. The federal Department of Health and Human Services[4] recommends that adults spend at least 2 hours and 30 minutes involved in moderately intense physical activity per week. The guidelines additionally call for aerobic activity to be performed in episodes of at least 10 minutes, preferably spread throughout the week. More than 74 percent of the US population does not meet these guidelines.

Physical exercise is associated with reduced morbidity and mortality from a number of causes, including cancer and cardiovascular disease. Whereas exercise's beneficial effects on physical health are well known, its beneficial effects on cognitive functioning are less well appreciated. Yet there is a long tradition of animal research showing the beneficial effects of physical exercise on cognition and brain function. In particular, rats that are given access to running wheels and other activity devices show improvement in various learning and memory tasks; moreover, these effects are linked to neural changes, including neural growth.[5] More recently, studies conducted in humans, including older adults, have also pointed to beneficial effects of aerobic activity on cognitive and brain function.[6] In this chapter, we briefly review the literature on animal studies, take a more detailed look at the human studies, and describe the implications of engaging in physical exercise for cognitive aging. The scientific evidence reviewed here indicates that exercise—aerobic exercise in particular—can delay or minimize age-related brain and cognitive decline.

6.1 Physical Exercise and Brain Function in Animals

How does physical exercise affect the brain and cognition in animals? Even in old age, physical exercise is associated with improved learning and cognitive performance in healthy animals.[7] Rodents that exercise either voluntarily or involuntarily typically show superior learning on a range of learning and memory tasks.[8] A good exemplar of this kind of research is a study in which Henriette van Praag and colleagues compared control groups of young and old mice against young and old mice given access to a running wheel.[9] The young and the old running mice did not differ significantly in the amount of time they spent voluntarily running. All four groups were then tested on the Morris water maze. Figure 6.1 shows the performance results. Latency (time to reach the hidden platform in the maze) and path length decreased over days, indicating that all groups of mice learned the task. Old control mice were considerably slower to find the platform and took a more circuitous route than young control mice, but this age effect was eliminated in the old runners. As figure 6.1 shows, the old runners performed as well as the young control mice on both measures of spatial memory. These results provide clear evidence of the beneficial effects of exercise for cognitive performance in older animals. Earlier in this chapter we discussed the problem of obesity in the US population and the contributory role of physical inactivity and large intake of fatty foods. It is therefore of interest that other animal studies using the Morris water maze have found that exercise reduces the degree of impairment in spatial learning in rats fed a high-fat diet.[10]

A related line of animal research indicates that cardiovascular exercise is neuroprotective against brain damage, specifically experimental stroke.[11] In these studies, effects of exercise are assessed before and after a stroke is induced experimentally in the animal. Most rats and mice will run voluntarily when given access to a wheel but are not fond of swimming, which they have to do in the Morris water maze. But animals that exercised on a running wheel before an experimental stroke—even if the exercise was

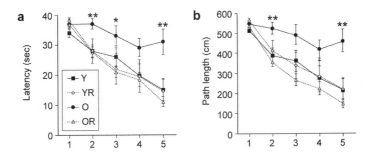

Figure 6.1
Results of a probe test in the Morris water maze at the end of the fifth day of training. The platform finding of the young running (YR) and old running (OR) rats was better than that of the non-running rats (Y and O). (a) Latency. (b) Path length. Source: H. van Praagand et al., "Exercise enhances learning and hippocampal neurogenesis in aged mice," *Journal of Neuroscience* 25: 8680–8685. Reprinted with permission.

forced and not voluntary[12]—showed reduced infarct volume (area of tissue death due to lack of oxygen) and better function after a transient stroke.[13] There are also benefits when the exercise occurs after rather than before the stroke.[14] These several lines of research all point to the significant benefits that exercise confers for learning, memory, and recovery from brain damage in animals.

Monkeys also show cognitive benefits of exercise. Middle-aged and older monkeys were randomly assigned to either run on a treadmill ($n = 16$) or sit on an immobile treadmill ($n = 8$) for 1 hour a day, 5 days a week, for 5 weeks. Beginning in the fifth week of training, monkeys were tested on a working-memory task. Regardless of age, exercising monkeys learned the task twice as fast as non-exercising monkeys. In addition, the exercising monkeys showed increased blood flow to the motor cortex, but that effect had reversed to baseline 3 months after the monkeys stopped exercising.[15]

How does physical exercise affect the brain and cognition in animals?

By improving cognitive performance, by reducing cognitive decline in old age and after a stroke, and by increasing adult neurogenesis in a manner correlated with improved memory performance.

6.2 Physical Exercise and Human Brain Function

What are the cognitive benefits of exercise in humans? It is easy to conduct rigorous studies of exercise with animals, since they can be randomly assigned to either exercise or control conditions. Random assignment is harder to achieve in humans, who must be observed carefully to verify that the assigned intervention is followed. For this reason, most studies of the effects of physical exercise on humans are observational—that is, cognitive functioning is compared between groups of people who choose to exercise or not, or between groups of people who exercise often and groups of people who exercise only occasionally. Such studies are not ideal. There can be many uncontrolled reasons why people choose whether or not to exercise—financial factors, mental health, physical health, and so on—so it can be hard to determine which factors are contributing to any effect on cognition that is found. Despite these limitations, human observational studies, like the better-controlled animal studies that use random assignment, find exercise to be associated with cognitive benefits. This has been confirmed by several recent meta-analyses in which a large number of observational studies using many different cognitive tasks have been considered together.[16] Stanley Colcombe and Arthur Kramer[17] hypothesized that effects of aerobic exercise on cognition are specific to executive functioning (a somewhat loosely defined set of functions characterized by ability to form and carry out plans, inhibit certain functions, resolve conflicts, and so on). Colcombe and Kramer used a meta-analysis to assess four "theoretical positions" of the basis for cognitive aging: (1) speed of processing, (2) visuospatial vs. verbal processing, (3) controlled vs. automatic processing, and (4) executive processes of inhibition, planning, and working memory. Their analysis involved categorizing the tasks used in the empirical papers on exercise into one of the corresponding hypotheses. Figure 6.2 shows the results of their analysis. Interestingly, all four task categories showed beneficial effects of exercise. However, the largest effects of exercise were seen for tasks of inhibition, planning, and working memory. Colcombe and Kramer therefore concluded that exercise has the greatest benefits for tasks that invoke executive processes.

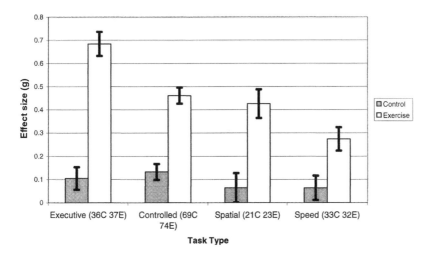

Figure 6.2
Results of meta-analysis of studies examining the effects of physical exercise on different cognitive tasks. Effect sizes are shown as a function of four different cognitive task categories. Source: S. J. Colcombe and A. F. Kramer, "Fitness effects on the cognitive function of older adults: A meta-analytic study," *Psychological Science* 14, no. 2: 125–130, © 2003 S. J. Colcombe and A. F. Kramer. Reprinted with permission of Sage Publications.

Kirk Erickson and colleagues in Art Kramer's lab hypothesized that the well-documented shrinkage of the brain with age would be reduced by exercise. They used structural scans to measure hippocampal volume in 165 healthy older people who varied in their level of physical fitness. The hippocampus is a structure deep in the temporal lobes long known to be critical for forming new memories. They found that people who showed higher aerobic fitness had larger hippocampi bilaterally. Moreover, the fitter group also showed better spatial memory performance than the less fit group.[18] Erickson and colleagues also examined the relation between brain volume and exercise in a large group of older people in an observational study. The Cardiovascular Health Study Cognition Study[19] is an extensive study begun in 1988 to investigate factors affecting cardiovascular health specifically in older people. Participants in that study were assessed on physical measures at study onset and scanned structurally 9 years later by Erickson and colleagues,[20] who analyzed gray-matter volume as a function of self-reported walking over the preceding 9 years. Participants were grouped according to how many miles they reported walking each week. MRI scans were analyzed using voxel-based morphometry to measure brain volume in a point-by-point manner in a large number of brain regions. In people who routinely walked 6–9 miles a week, there was significantly greater gray-matter volume in a number of brain regions, including the precuneus and the hippocampus (both associated with Alzheimer's pathology in a number of studies).[21] Moreover, some people in the study had declined cognitively over the 9 years and had become demented or had developed Mild Cognitive Impairment (MCI), a condition of cognitive decline that is not severe enough to warrant the label of dementia but which increases the risk of becoming demented later. When Erickson and colleagues examined the volume of different brain regions, they found that volume of inferior frontal gyrus, hippocampus, and supplementary motor regions predicted cognitive change to dementia or MCI over the 9 years.

The above-reviewed human studies of effects of exercise on cognition were all observational. Such studies have advantages but also have the important limitation that participants choose their "treatment." People who choose to attain a higher level of physical fitness may differ in important ways from people who choose to accept a lower level of fitness. A less confounded approach to investigating effects of exercise on cognition is to randomly assign people to aerobic or non-aerobic exercise conditions and assess cognition before and after the exercise manipulation. Kramer and colleagues have conducted such a study, randomly assigning healthy older people to either aerobic (walking briskly for 45 minutes under supervision) or non-aerobic (stretching and toning) exercise sessions 3 days a week for 6 months.[22] After 6 months, people assigned to aerobic exercise showed better performance on a task of executive attention than people assigned to non-aerobic exercise. The task was a "flanker" task in which participants were asked to judge the orientation of a central arrow in an array of arrows. On some trials, the central arrow was surrounded by arrows (flankers) pointing in the same direction, termed "congruent" (e.g., >>>>>). On other trials, the central arrow was surrounded by arrows pointing in the opposite direction, termed "incongruent" (e.g., >><>>). The typical result is that people are slowed under the incongruent condition because they cannot ignore the irrelevant flanker arrows. Performance is considered to be better when the flanker arrows have less of an effect. In this study, people who exercised aerobically showed a reduced flanker effect on the behavioral task.[23]

Other groups have found similar effects of exercise in randomized designs. Ann Smiley-Oyen and colleagues randomly assigned older people to 10 months of either aerobic training or strength training and found

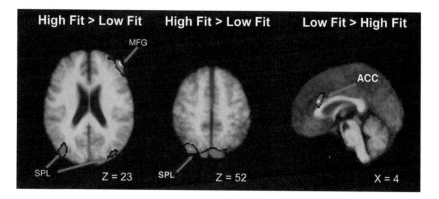

Figure 6.3
Cardiovascular fitness is associated with different patterns of cortical activation. Source: S. J. Colcombe et al., "Cardiovascular fitness, cortical plasticity, and aging," *Proceedings of the National Academy of Sciences* 10: 3316–3321 (© 2004 National Academy of Sciences). Reprinted with permission. This figure appears in color elsewhere in the book.

benefits only of the aerobic training on tasks dependent on executive functioning, although their test battery was weighted toward such tasks.[24] This finding is consistent with findings that young adults showed increased working-memory capacity only 30 minutes after completing randomly assigned aerobic training, but not strength training.[25]

Randomly assigned exercise also appears to affect brain function. Kramer and colleagues found not only reduced flanker effects after aerobic training, but also greater activation of both the superior parietal cortex and the middle frontal gyrus of the brain, as assessed by functional MRI. Both these regions are associated with working memory (see figure 6.3). Those people who were non-aerobically trained showed greater activation in anterior cingulate cortex—a region associated with resolution of response conflicts, as in the flanker task.[26] These exercise-related differences in task performance and in patterns of brain activation suggest that the increased aerobic fitness allowed people to better ignore irrelevant distractors, an ability interpreted as indicating better executive function. However, they assessed performance on only one task, so other abilities could have been affected.

Kramer's research group also looked at brain volume changes as a function of exercise in the same participants studied by Colcombe and colleagues.[27] Participants underwent structural MRI scanning both before and after the 6 months of aerobic training or stretching and toning, depending on random assignment. In the group randomly assigned to the 6 months of aerobic exercise, regions in the frontal lobes showed an increase in regional brain volume relative to the non-aerobic group—specifically in dorsal anterior cingulate, supplementary motor cortex, and middle frontal gyrus, bilaterally. (Colcombe et al. did not report cognitive results.[28])

Kramer et al. have also looked at benefits of aerobic exercise over a longer period of time in older people randomly assigned to regular walking (moderate intensity) or to stretching and toning (using dumbbells or resistance bands) for a year. Relative to pre-exercise measures of hippocampal volume, those assigned to walking showed increased volume of anterior hippocampus, better spatial memory, and greater serum levels of BDNF (a neurotrophin with a role in memory formation) relative to those assigned to stretching and toning. The improvements in aerobic fitness levels (VO_2 max) were associated with greater increases in global hippocampal volume. Volumes of caudate and thalamus were not affected, which suggests some selectivity in the effect of aerobic exercise.[29]

Nicola Lautenschlager and colleagues looked at effects of randomly assigned exercise on cognition in healthy older people who were, on the basis of self-assessment of memory problems, at increased risk of dementia. People who had been assigned to an aerobic exercise intervention (increase in number of steps taken per week over the 24-week intervention period, measured by a pedometer) showed a significant increase in a standardized test of global cognitive function (ADAS-Cog). Another sample that had been randomized to "usual care" showed a slight decline in the ADAS-Cog over the same period.[30] Although Alzheimer's disease is not the primary focus of this book, it is relevant to note that cognitive decline was slower in Alzheimer's patients who exercised[31] and that the incidence of Alzheimer's disease was lower in people who chose to exercise.[32]

How does physical exercise affect the brain and cognition in humans?

By improving cognitive performance (with the strongest evidence for executive-function tasks), by increasing regional brain volume, and by changing the patterns of brain activation. Both animal evidence and human evidence supports the view that aerobic exercise has a strong positive effect on cognitive performance and brain function in both healthy and demented older adults.

6.3   Neural Mechanisms of Exercise Benefits

What neural mechanisms underlie exercise's beneficial effects on brain and mind? Several mechanisms have been identified: neurogenesis and the survival of new neurons, synaptic plasticity, and cerebral blood flow.

Effect of exercise on neurogenesis

As was described in chapter 5, new neurons are formed in the brain throughout adulthood. This process of neurogenesis is best studied in the hippocampus, a structure deep inside the temporal lobe that has been known since the 1950s to play a major role in the formation of new conscious memories.[33] Moreover, the hippocampus is one of the main sources of new neurons in the adult brain. Studies have consistently found that exercise increases proliferation and survival of new neurons in the dentate gyrus of adult rodents.[34] Rats and mice will run voluntarily as much as 8 kilometers a night if given access to a running wheel. Doubling or tripling of the number of newborn cells in the dentate gyrus has been has been reported in wheel-running rodents.[35] Voluntary wheel running not only increases the number of newborn neurons in adult animals; it also increases the survival of new cells in the process of becoming neurons—that is, later-stage progenitor cells and newly formed (early post-mitotic) neurons.[36] This effect may extend beyond the hippocampus, as running rats also showed significantly higher numbers of cholinergic neurons in the diagonal band of Broca.[37]

Exercise-induced growth of neurons in the hippocampus can occur at any age, including young adulthood[38] and old age.[39] Remarkably, the positive effect of exercise can occur even in the womb. Hong Kim and colleagues found that rat pups born to mothers that had exercised aerobically (by running on a treadmill) during pregnancy had a greater number of surviving hippocampal neurons than did pups born to mothers that had not exercised.[40] Finally, in mice, continued voluntary exercise has been found to reduce the typical age-dependent decline in adult hippocampal neurogenesis that occurs in sedentary animals.[41]

Does the exercise have to be voluntary for the benefit to be obtained? Few studies have controlled for voluntary versus forced exercise. As has been noted, rodents will readily run voluntarily but do not like to swim. Chun Xia Luo and colleagues found that voluntary wheel running, but not forced swimming, enhanced the survival of progenitor cells in the dentate gyrus after experimental strokes affecting cortex and striatum in mice.[42] The animals were then examined for neurogenesis in the ischemic dentate gyrus and for improvement in spatial learning. Mice that ran voluntarily showed better survival of newborn neurons in the dentate gyrus, greater upregulation of CREB (a protein with well-established links to formation of long-term memory and neuronal plasticity[43]), and improved spatial memory. The number of newborn neurons in the dentate gyrus that survived was correlated with ability to find the platform in the Morris water maze. Voluntary running, but not forced swimming, was found to be associated with reversal of the spatial-memory impairment that had occurred as a consequence of the experimental stroke. Running after an episode of focal ischemia promoted new neuron survival in the dentate gyrus, but did not affect neuron proliferation, infarct volume, or neuron degeneration in the dentate gyrus. In contrast, forced swimming was associated with decreased cell proliferation, perhaps due to stress.

Synaptic plasticity

Some of exercise's beneficial effects on learning may be attributable to its effects on mechanisms of synaptic plasticity. Exercise has been observed to alter the length and the complexity of dendrites and the density of the spines on the dendrites.[44] Both of these exercise-induced dendritic changes can improve the efficiency of communication between neurons.

Exercise also appears able to effect partial reversal of age-related changes in synaptic structures. Such changes can be observed at the neuro-muscular junction—the synapse between motor neurons and muscle fibers. In contrast to the brain, the neuro-muscular junction can easily be repeatedly imaged so that changes due to manipulations can be easily assessed. In one study, young and old mice raised under standard laboratory conditions (small cages, no running wheels) underwent a procedure that allowed repeated imaging of the hind-leg muscles. This was done by exposing the muscles and labeling cholinergic receptors with fluorescent alpha-bunga-rotoxin, a potent neurotoxin (naturally produced by Taiwanese banded krait snakes) that binds irreversibly to acetylcholine receptors in the neuromuscular junction. The fluorescence reveals cholinergic receptors. After a small subset of receptors were labeled with fluorescent alpha-bungarotoxin, the neuromuscular junction was imaged; the mice were then returned to their cages to heal. Half of the mice were then given access to a running wheel for one month. After the month of running, the neuromuscular junctions were again imaged. Before the exercise, the sedentary aged mice showed a number of abnormalities, including axons that didn't occupy the postsynaptic structures fully, fragmented receptors, decreased receptor density, and misshapen axon terminals. After the month of exercise, the aged mice still showed more abnormalities than the young mice but fewer abnormalities than the sedentary old mice. After the exercise, receptors that had appeared fragmented in the aged mice, had re-assembled together to form continuous branches. Moreover, after the exercise there were fewer denervated postsynaptic sites.[45] These changes were seen only in exercised muscles and not in muscles minimally affected by exercise. Nor were effects of exercise seen in young mice. Although these findings were not obtained from synapses in the brain, they do provide a model of the way in which effects of exercise might be manifested between neurons in the brain. Because cortical neurons can be imaged in the living, intact rodent brain over time through a "cranial window,"[46] a similar study looking at the effects of exercise on neuronal structure in the brain is feasible, although has not as yet been reported.

Synaptic changes related to exercise are also seen in long-term potentiation. As was discussed in previous chapters, LTP is a durable increase in the strength of a synapse after repeated stimulation of the synapse. LTP appears to be the basis for memory formation, and it is induced by learning alone.[47] In the dentate gyrus, beneficial effects of exercise on LTP have been observed.[48] Representative results on exercise-related modulation of LTP are plotted in figure 6.4. J. Farmer and colleagues in Fred Gage's lab used two methods to induce LTP in the dentate gyrus: a weak-patterned stimulus (wTPS) consisting of bursts of patterned conditioning stimuli at 100 hertz and a strong-patterned (sTPS) protocol at 400 hertz. The higher frequency of the sTPS method resulted in significantly greater LTP in the dentate gyrus. As figure 6.4 shows, sTPS resulted in significant induction of LTP, but to a greater degree in runners (open circles) than in controls. The weak-patterned stimulus induced LTP in runners but not in controls.[49]

Neurotrophins

Evidence is emerging that neurotrophins have some role in mediating effects of exercise on cognition. Neurotrophins are naturally occurring

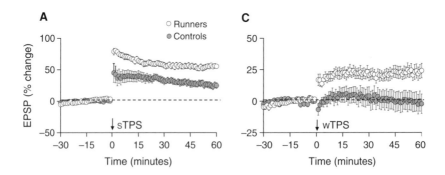

Figure 6.4
Increased synaptic plasticity in animals exercising voluntarily. (A) When administered the strong stimulation condition (sTPS), Runners (open circles) showed more LTP induction than Controls (dark circles). (Panels B and D are not reproduced here.) (C) When administered a weaker stimulation condition (wTPS), Runners showed stronger LTP induction than non-running Controls. Source: J. Farmer et al., "Effects of voluntary exercise on synaptic plasticity and gene expression in the dentate gyrus of adult male Sprague-Dawley rats in vivo," Neuroscience 124, no. 1: 71–79. Reprinted with permission of Elsevier (© 2004).

secreted proteins that act as growth factors. Found in brain and blood, they were originally discovered to be important in neuronal development. The major neurotrophins include nerve growth factor (NGF), brain-derived neurotrophic factor (BDNF), neurotrophin-3 (NT-3), and neurotrophin-4 (NT-4), all of which are structurally related to one another. They interfere with initiation of programmed cell death (apotosis) by neurons; they also promote formation of progenitor cells during neurogenesis. The effects of exercise on neurogenesis and on the formation of new capillary blood vessels (angiogenesis) are associated with upregulation of the neurotrophic agents BDNF[50] and NGF[51] as well as other growth factors, viz., insulin-like growth factor (IGF-1)[52] and vascular endothelial growth factor (VEGF).[53]

More recently, neurotrophins have also been found to act as synaptic modulators, and BDNF appears to be the workhorse among neurotrophins. Specifically, BDNF modulates LTP.[54] LTP is considered an important cellular mechanism underlying memory formation. Evidence that the effect of exercise on memory can be mediated by BDNF comes from a study in which voluntary exercise (wheel running) was found to improve spatial memory in mice with genetically reduced neurogenesis. Moreover, the duration of running was associated with increased BDNF levels.[55]

BDNF also appears to contribute to exercise's beneficial effects on cognition. A substantial literature shows that serum concentrations of BDNF increase at least twofold after acute exercise in young adult humans. In one impressive study, Peter Rasmussen and colleagues sampled BDNF in venous blood collected from the internal jugular veins of volunteers during 4 hours of rowing (presumably the participants were paid well). They estimated that the brain contributed 70–80 percent of BDNF in the peripheral circulation.[56] This was confirmed by a recent review reporting that 69 percent of 24 high-quality studies on healthy people found increased serum or plasma BDNF concentration after aerobic exercise. None of the studies showed a long-duration BDNF response to exercise.[57]

But what is the role of BDNF in the effect of exercise on cognitive aging? There is human evidence showing a relation among serum BDNF, memory, and age. In humans, a comparison of BDNF in cerebrospinal fluid from spinal taps of healthy older people revealed that BDNF concentration in the cerebrospinal fluid decreased with age and that a lower concentration of BDNF was associated with poorer memory recall.[58]

The effect of BDNF appears to be mediated through the hippocampus, which is critical for in memory formation. Evidence of a relationship among age, memory, and BDNF levels has also been found in humans. Kirk Erickson and colleagues looked at structural MRI scans in 165 healthy older people who had been screened for aerobic fitness by measuring VO_2 peak with exercise testing on a treadmill, and found that higher fitness levels were associated with larger left and right hippocampi and with better performance on a task of spatial memory (remembering one, two, or three locations for 3 seconds).[59] These findings are consistent with a report from the same group that people who reported walking 6–9 miles a week showed a reduced incidence of cognitive impairment 9 years later.[60] Further, increasing age was associated with smaller hippocampal volumes, reduced levels of serum BDNF, and poorer performance on a spatial working-memory task. The relationship among serum BDNF, hippocampal size, and working memory was found even when age-related variation was controlled for. In a preliminary mediation statistical analysis, Erickson et al. found that hippocampal volume mediated the decline in spatial memory with age whereas BDNF mediated the decline in hippocampal volume with age.[61]

However, neither the age-related decrease in BDNF nor the association between low BDNF levels and poorer memory has been found to be associated with biomarkers in the cerebrospinal fluid, such as Aβ and tau, previously linked with increased risk of Alzheimer's disease.[62] This suggests that the BDNF neurotrophin modulates normal aging but doesn't modulate risk of Alzheimer's disease.

In summary, there is convincing evidence from animal studies that the neurotrophin BDNF plays a role in the beneficial effect of aerobic exercise on cognition. Several large-scale human studies have shown that aerobic fitness affects the hippocampus in a way that is related to serum BDNF levels and to memory. It has yet to be shown in a human study that BDNF is the mechanism that underlies the relation between exercise and memory in aging.

Cerebral blood flow

Because the brain needs a constant supply of well-oxygenated blood, the supply of blood to the brain probably figures in the effects of exercise on cognition. It is known from animal and human studies that age reduces cerebral blood flow and capillary density, and that exercise increases capillary density.[63]

A human neuroimaging study recently revealed the importance of exercise to brain and cognition. Ana Pereira and colleagues in Scott Small's lab found that aerobic exercise increased the flow of blood to the dentate gyrus. A group of 11 low-fitness adults (so categorized on the basis of the American Heart Association's guidelines of VO_2 max > 43 for men and >37 for women, VO_2 max measuring aerobic capacity) of mean age 33 underwent 12 weeks of supervised exercise for one hour four times a week, of which 40 minutes was aerobic (cycling, running, or climbing stairs). Pereira et al. found that cerebral blood flow was increased selectively in the dentate gyrus—that is, the blood flow was increased in the dentate gyrus more than in any other hippocampal region. That increase was related to both cardiopulmonary function and cognitive function. A similar exercise manipulation in mice, reported in the same paper, also resulted in increased blood flow in the dentate gyrus, which was correlated with neurogenesis in the same structure.[64] Thus, Pereira et al. found in both animals and humans that an important site of adult neurogenesis and a region important in memory formation—the dentate gyrus—experienced increased blood flow with exercise. That finding is consistent with a recent report that blood glucose levels in older people were related to both blood flow in the dentate gyrus and memory performance.[65]

William Wu and colleagues in Scott Small's lab hypothesized that the hippocampus—long known to be highly vulnerable to disease and insult— might be affected differentially by two common diseases of old age: diabetes and stroke (ischemia). They sought to link diabetes and ischemia to specific subregions of the hippocampus by using steady-state basal cerebral blood flow with fMRI. They generated high-resolution cerebral blood flow maps in 240 older individuals who also were assessed for dementia, for type 2 diabetes, and for the presence of infarcts. They found that dysfunction in the dentate gyrus was selectively linked both to diabetes and to blood glucose levels. In other studies, poor glucose regulation has been linked to cognitive impairment.[66] That finding is consistent with another study by the same group described above which found, in both mice and humans, that exercise-related increase in blood flow in the dentate gyrus was correlated with both fitness and cognitive function. Further, in the mice the increased blood flow in the dentate gyrus was associated with increased neurogenesis.[67] Of all the hippocampal subregions, the exercise had the strongest effect on blood flow in the dentate gyrus, the only hippocampal region known to undergo adult neurogenesis. Considered together, this evidence indicates that aerobic exercise has the potential to counteract the apparent negative influence of high blood glucose levels on the integrity of the dentate gyrus, on which memory formation depends.

What neural mechanisms underlie the effects of exercise on cognition?

A number of neural mechanisms appear to be involved, including neurogenesis, synaptic plasticity, the neurotrophin BDNF, and cerebral blood flow. It probably is meaningful that the dentate gyrus is strongly linked to mechanisms underlying the effects of exercise on cognition. The dentate gyrus is one of the few sites of adult neurogenesis. It is negatively affected by reduced blood flow and poor glucose regulation, but positively affected by increased blood flow induced by aerobic exercise. It may be the highly plastic nature of this structure—important for the formation of new memories—that makes it vulnerable to variations in blood flow.

Summary

Of the various experiential and lifestyle factors in cognitive aging reviewed in this book, physical exercise is probably the one whose effects are best understood. There is strong evidence that aerobic exercise can reduce and in some cases eliminate cognitive deficits associated with healthy aging, and there is growing evidence of the neural mechanisms that underlie such benefits. These mechanisms appear to be centered on the dentate gyrus, important for formation of new memories.

7 Diet and Nutrition

Men dig their Graves with their own Teeth and die more by those fated Instruments than by the Weapons of their Enemies.
—Thomas Moffett (seventeenth-century English physician)

The adage "you are what you eat" is well accepted in regard to physical functioning and health. That diet and nutrition can affect mental functioning is also thought to be true, but is perhaps not so widely accepted. Certainly the effects of diet and nutrition are subjects of debate and controversy. It doesn't help that the popular media are full of misinformation and evidence-free assertions about diet and cognitive functioning. There are many hundreds of books claiming that certain diets or nutritional compounds can enhance physical and mental health, and many more such claims can be found on the Internet. Unfortunately, many of the claims about nutrition made in popular books and in the mass media don't meet even the minimal requirements of scientific evidence.

This is not to say that diet and nutrition don't affect mental functioning. In fact, there is considerable evidence of dietary influences, positive and negative, on both cognition and cognitive aging. The effects for which we have the best evidence involve total caloric intake and fat consumption. An excess of calories in the diet is associated with harmful effects on cognition, as are certain types of foods, notably those with saturated fat. Conversely, restricting calories in one's diet improves health and cognition. At the same time, resveratrol and polyunsaturated fatty acids appear to promote cardiovascular health and perhaps also cognitive integrity. Effects on cognitive functioning of specific types of diets, such as the so-called Mediterranean diet, or of particular nutritional compounds, are less well established. One reason for that is the difficulty of conducting randomized-controlled trials on the effects of nutrition on cognition in which all food must be provided. Though there are many relevant observational and epidemiological studies (these designs are discussed in section 3.1 of chapter 3), such methods provide weaker evidence for the specific influence of diet on maintaining cognitive vitality in older adults than do randomized trials.

Faced with these issues and with the many unsubstantiated claims made in popular books and on the Internet, the ordinary layperson will have great difficulty in sorting the wheat from the chaff. What is likely to be true and what is not?

This chapter was one of the more difficult ones for us to write. We have tried to strike a balance between being overly conservative (holding all studies to the high methodological standard of randomized clinical trials) and too permissive (making recommendations for diet or nutrition based on poorly controlled observational studies). Our goal is to describe the nutritional factors for which the best scientific evidence exists, while also discussing less well-supported factors that might benefit from future research that might establish a basis for their efficacy.

What dietary practices appear to influence overall health and cognitive functioning? In answering this question, we consider the effects of restricting calories, the effects of excess calories, the effects of resveratrol (which occurs naturally in red wines and in some foods), the effects of dietary fats, and the effects of antioxidants.

7.1 Dietary Restriction

Dietary restriction—that is, restricting calories in a well-balanced diet—has been consistently associated with positive effects on health and cognition. This may be due to the physical stress caused by such restriction. It may be surprising to some that moderate exposure to physical stress can *improve* health and longevity. Such an effect seems counter-intuitive because we generally think of stressors as harmful. But a weak stressor to an organism induces a protective response against further stress. This hypothesis, which has been termed *hormesis*,[1] is based on observations that mildly stressed animals outlive unstressed members of the same species.[2] Survivors of the nuclear bombing of Nagasaki provide a human example of hormesis: those who experienced low to intermediate doses of radiation had longer life spans than those who experienced no radiation.[3]

One stressor common to all species is famine, and a sustained reduction of calories is a powerful way to increase life span. Mark Mattson, Chief of the National Institute on Aging's Laboratory of Neurosciences and a prolific researcher on diet and aging, has argued that the benefits for health and cognition of a low-calorie diet might arise from such a hormetic response.[4] Dietary restriction (DR) typically refers to the reduction of calories by up to 50 percent in the context of an otherwise nutritious balanced diet. In animal studies, a DR condition is usually compared against "ad libitum" feeding (in which the animal eats the amount it chooses) or against feeding a standard amount of food. Across a range of animal species, DR has been observed to extend life span by 20–40 percent. Its strongest effects occur when it is initiated in youth, but effects can be seen even if it is initiated in mid-life.[5] To demonstrate the life-span effect of DR experimentally in humans will be very difficult, but an ongoing study of rhesus monkeys by Navin Maswood and colleagues has shown promising effects after 5 years of random assignment to DR or ad libitum feeding.[6]

Increased life span is not the only positive effect of dietary restriction. It has been known for some time that DR reduces the incidence of age-related disease, maintains youthful physiological measures, increases resistance to stress, and protects neurons against toxic insults.[7] For example, in the study mentioned above, Maswood et al. randomly assigned monkeys to either a DR diet or an ad libitum diet for 6 months. After 6 months on the DR diet, the monkeys were exposed to a neurotoxin that is known to produce symptoms of Parkinson's disease. Parkinson's disease arises when a marked loss of dopamine neurons in the substantia nigra, a region in the basal ganglia of the midbrain, leads to disorders of movement (including tremors, rigidity, and posture problems). Use of a neurotoxin to induce the disease by causing selective degeneration of dopamine neurons is valid because there is increasing evidence of a link between exposure to neurotoxins in insecticides and Parkinson's disease.[8] Roth's group compared movement activity in the monkeys fed an ad libitum diet against movement activity in the monkeys on a DR diet. DR monkeys with Parkinson's symptoms showed significantly higher levels of movement than the monkeys on the standard diet. The monkeys in the DR condition also had higher levels of dopamine in the striatal region (known to be depleted in Parkinson's disease), and higher levels of neurotrophic factor derived from glial cells.[9] As was discussed in chapter 6, neurotrophins have wide-ranging roles in promoting the health of neurons. This study is very relevant to humans, as it shows that DR can protect the primate brain against environmentally related insults and damage.

Even if initiated when monkeys are old, dietary restriction is associated with improvement in some markers of health, including levels of insulin and triglycerides, although not with improvement in total serum cholesterol or blood pressure.[10] Recently, Ricki Colman and colleagues found that random assignment of adult rhesus monkeys to DR at age 7–14 years (out of a 27-year average life span) lowered the incidence of aging-related deaths, but not the incidence of deaths for any cause. DR monkeys showed other benefits as they aged, including preservation of lean muscle mass, prevention of diabetes, cancer, cardiovascular disease, and reduced atrophy of subcortical brain regions (caudate, putamen, insula).[11]

Dietary restriction's effects on health are impressive in their own right, but when considered together with its beneficial effects on cognition they are remarkable. In rats, lifelong DR prevented age-related deficits in cognition[12] and in long-term potentiation.[13] Also in rats, age-related impairment in spatial discrimination measured in the Morris water maze was lessened in animals assigned to DR. However, the DR rats also exhibited greater physical activity, a common confound in DR studies.[14] We are aware of only one study that has reported negative effects of DR on cognitive performance in animals. DR rats performed more poorly on the Morris water maze at 17–18 and 24–27 months, although an injection of glucose raised performance of the DR animals to that of the ad libitum animals.[15]

What about effects of dietary restriction in humans? For obvious reasons, studies of DR in humans are difficult to conduct, but nevertheless some data have been obtained. There are individuals who voluntarily restrict calories for numbers of years, largely for health reasons,.[16] Observational studies of such people find benefits for physiology that are similar to the benefits seen in rodents randomly assigned to DR.[17] However, evidence from such epidemiological designs is limited in that the participants themselves select which diet to follow—people who follow healthier diets may have been healthier to begin with. Probably the best observational data are from Biosphere 2, an enclosed artificial ecological system in Arizona. The participants in that experiment in self-sufficiency did not choose to restrict calories but inadvertently experienced about 30 percent calorie reduction as a result of crop problems. Nevertheless, they showed the

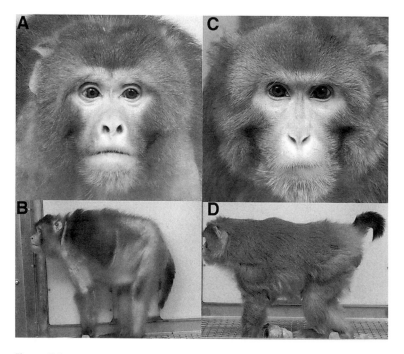

Figure 7.1
Rhesus monkeys aged 27.6 years (the average life span). Panels A and B show a
monkey maintained on a standard diet; panels C and D show a monkey maintained
on a DR diet. Source: R. J. Colman et al., "Caloric restriction delays disease onset
and mortality in rhesus monkeys," *Science* 325, no. 5937: 201–204. Reprinted with
permission of American Association for the Advancement of Science. This figure
appears in color elsewhere in the book.

benefits associated with DR in animals: declines in metabolic rate, body
temperature, blood pressure, blood glucose, and insulin.[18]

As is discussed throughout this book, randomized assignment to a treatment group or a control group provides the best method for assessing the
effect of a factor thought to ameliorate brain and/or cognitive aging. Only
a few empirical human studies have been conducted using random assignment to a DR diet or a control diet. One such study found that 6 months
of randomly assigned DR improved fasting insulin levels and body
temperature—both considered to be markers of longevity—in healthy sedentary young adults. DNA damage was also reduced in the DR group.[19]
This study also assessed properties of cells in culture collected from humans
who had undergone 6 months of DR. These cells showed greater resistance

to stress (increased heat resistance) and also upregulation of the sirtuin 1 (SIRT1) gene, which has been linked most strongly to protection against age-related diseases.

What about dietary restriction and human cognition? Only a few studies have assessed cognition in the context of DR in humans. Self-reported dieting has been associated with cognitive deficits in some[20] but not all[21] studies. Of course, self-designed diets are not necessarily well balanced, and deficiencies could contribute to the observed cognitive deficits. Researchers at the Pennington Biomedical Research Center in Baton Rouge looked at the cognitive effects of 6 months of a well-balanced DR diet provided in the form of a low-calorie liquid. This was a well-designed randomized clinical trial, limited mainly by the sample sizes (12 participants in each group). The participants—overweight adults aged 25–50—were randomly assigned for 6 months to either a diet consistent with weight maintenance, DR, DR plus exercise, or a low-calorie diet until 15 percent weight loss and then maintenance. No effects on cognition, as assessed by tests of verbal and visual memory and attention, were seen after 6 months of DR.[22] However, a well-designed study with a larger sample found an improvement in memory after 3 months of randomized assignment to DR.[23] In that study, A. V. Witte and colleagues assigned normal-to-overweight older adults to 3 months of one of the following conditions: 20 to "caloric restriction" (instructed to achieve 30 percent reduction of calories), 20 to "UFA" (instructed to achieve 20 percent greater consumption of unsaturated fatty acids, UFA), and 10 to "Control" (instructed not to change eating habits). Dietary consumption was assessed by self-report on daily nutritional records. In the calorie-restricted group, verbal memory improved by 20 percent in a manner correlated with decreased plasma level of insulin. There were no memory changes in the other dietary groups. This observed relation between memory performance and insulin levels is consistent with a recent finding in humans: blood glucose levels were found to be inversely correlated with cerebral blood flow to the dentate gyrus (important in memory formation) and with recall on the Buschke selective reminder test.[24] This evidence suggests that the effects of DR on cognition may arise from its modulation of functioning of the hippocampus.

If dietary restriction were shown to be beneficial for cognitive integrity in aged humans, as has already been shown in aged animals, would people be willing to restrict calories in the manner described above? In their study,

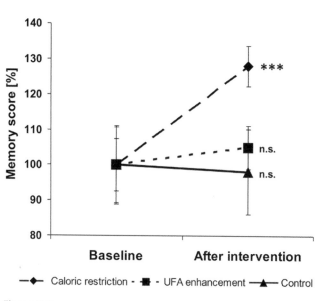

Figure 7.2

Effect of dietary restriction on verbal memory. Verbal memory increased by 20 percent in the group randomly assigned to 3 months of calorie restriction. Source: A. V. Witte et al., "Caloric restriction improves memory in elderly humans," *Proceedings of the National Academy of Sciences* 106: 1255–1260. Reprinted with permission of National Academy of Sciences (© 2009).

Witte et al. required their older participants to maintain a 30 percent reduction of calories over 3 months by means of chronically reduced calorie intake.[25] Although it was not reported in the paper, personal communication with the senior author indicated that the diet was tolerated well by the participants.

Mark Mattson of the National Institute on Aging explored another approach to implementing DR in humans. He conducted a pilot study of the effects of a one-meal-a-day diet in healthy middle-aged people (40–50 years old). Only 15 people completed the study. All food was provided in a randomized "crossover" design such that participants experienced both treatments, one after the other. In one treatment condition, participants consumed all their calories in a single meal per day. In the other condition, all calories were consumed in three meals per day. There were "modest" but significant changes in physiological measures on the one-meal-per-day diet: lowered systolic pressure, raised diastolic pressure, reduced fat mass, and reduced weight (65.9 vs. 67.3 kilograms). This small study did not find major health benefits from increasing fasting time each day without reducing calories overall.[26]

Another approach to dietary restriction is alternate-day fasting—fasting one day and eating normally the next. In animal models of chronic disease, this form of DR leads to improvements in insulin levels, glucose concentrations, and incidence of diabetes.[27] Alternate-day fasting has been found to be as effective for inducing health benefits as daily DR in animals[28] and in humans.[29] This is important because in rodents alternate-day fasting doesn't lead to overall calorie reduction, and the animals don't lose weight.[30] A note of caution was sounded recently, with a report that alternate-day fasting for 6 months in rats was associated with some reductions in cardiac function.[31] A trial of alternate-day fasting with 16 non-obese human participants (between 23 and 53 years old) found that the participants lost weight and also increased their fat oxidation.[32] Moreover, this study found that self-reported hunger remained elevated throughout the 22-day trial of alternate-day fasting, reducing the feasibility of this method for achieving benefits of DR in humans. It is therefore an open question whether older people would tolerate the fasting needed to obtain the benefits of DR.

What dietary practices appear to influence overall health and cognitive functioning?

Although we do not yet have much evidence on the effects of dietary restriction on human cognition, there is general agreement that what is good for the cardiovascular system is good for the brain and mind. Animal studies with dietary restriction have shown robust effects on life span, resistance to neuropathology, and age-related increases in health and cognition in a range of species, including infrahuman primates. The amount of human research is limited, but initial results suggest beneficial effects of dietary restriction on health, on markers of longevity, and, in one study, on cognition. But whether many older adults would embrace dietary restriction remains unclear, at least in the absence of any long-term educational program that might encourage participation in future randomized controlled studies. For the present, therefore, we conclude that, although there is good evidence for the efficacy of dietary restriction, its practicality is open to question.

7.2   Excess Energy Intake

The opposite of calorie restriction is excessive consumption of food. Unfortunately, and particularly in the United States, the number of people who practice dietary restriction is far exceeded by the number of overeaters. Furthermore, those who eat too much are growing in number. There is increasing evidence that excess energy intake, whether from fat or from sugar, impairs cognitive function and neural plasticity. A number of large-scale epidemiological studies have found that long-term obesity—assumed to be a result of chronic excess energy intake—is associated with cognitive impairment in mid-life,[33] notably in executive function.[34] (However, one large-scale study found better cognitive function in older people who were overweight.[35]) Moreover, obesity and diabetes are consistently found to be major risk factors for Alzheimer's disease.[36] Animal studies have also consistently found obesity to be linked to lower cognitive performance. For example, obese mice were impaired in forming new memories in three different cognitive tests: a T-maze that required the mice to actively avoid obstacles, the Morris water maze, and a task in which a mouse had to press a lever to receive a food reward.[37]

Is there a specific component of food that is responsible for the deleterious effects of excess energy intake on cognitive functioning? Fat has been implicated as the main culprit. A number of observational studies of humans have linked fat intake with increased risk of Alzheimer's disease late in life. One of the longest running of such longitudinal studies found an association between fat intake (from butter, margarine, and milk—all rich in saturated fats) in mid-life and risk of both dementia and Alzheimer's disease 21 years later. Even a moderate intake of saturated fats was linked with greater risk.[38] A number of other longitudinal studies have reported similar findings over shorter time intervals (4–8.5 years).[39] A recent randomized controlled trial appears to confirm these observational findings. Jennifer Bayer-Carter and colleagues in the laboratory of Suzanne Craft randomly assigned 49 older adults to one of two diets. The participants were either healthy older adults ($n = 20$) or older adults diagnosed with amnestic Mild Cognitive Impairment (an intermediate stage between normal aging and Alzheimer's disease). Participants were assigned for 4 weeks either to a diet high in fat (45 percent) and high in simple carbohydrates (which mimics diets that promote diabetes and insulin resistance) or a diet low in fat (25 percent) and low in simple carbohydrates. All meals were provided, and the diets were not designed to induce weight loss. Outcome measures were CSF biomarkers associated with Alzheimer's disease. Low levels of Aβ42 are typically seen in Alzheimer's disease, widely attributed to reduced clearance of the substance from the brain.[40] In the MCI group, the low-fat diet increased CSF levels of Aβ42, suggesting that Aβ42 was more rapidly cleared from the brain. There was no effect of the high-fat diet in the MCI group on this measure. In contrast, in the healthy older group the low-fat diet decreased CSF levels while the high-fat diet increased CSF levels of Aβ42. A critical finding was that the low-fat diet improved delayed episodic memory for both the healthy control and MCI groups.[41] Although the dietary manipulation used in this study was complex, as both fat and sugar were varied, this is the strongest human evidence to date of a link between an unhealthful diet and risk of Alzheimer's disease.

Animal research shows that a diet high in fat reduces both neuronal plasticity and the capacity of the rodent brain for learning. This has been seen in two different laboratory tasks of learning and memory: the delayed alternation task[42] and the water radial arm maze.[43] Moreover, the "dose" of the fat seems to matter. Paul Pistell and colleagues[44] in Annadora Bruce-Keller's lab found that mice fed a diet consisting of 60 percent saturated fat gained weight and showed impaired cognition, increased brain inflammation, and reduced levels of the neurotrophin BDNF. Mice on a 41 percent fat diet showed weaker effects. One source of the negative effect of a high-fat diet on cognition may be the development of insulin resistance and its effect on the hippocampus.[45] This is consistent with recent findings by Small and colleagues that high levels of blood glucose are associated with hippocampal pathogenesis in humans.[46]

An important characteristic of the negative effects of a high-fat diet is that they develop rapidly. Rats fed a diet high in saturated fat for only 9 days were impaired in treadmill speed and in radial maze performance.[47] Somewhat surprisingly, fat in the diet of a rat's mother also had a detrimental effect. Offspring of rats fed a high-fat diet showed reduced neurogenesis.[48] Christy White and colleagues found that the rats most strongly influenced by a high-fat diet were offspring of mothers who were themselves fed a high-fat diet. Offspring fed a normal diet were less affected, regardless of the mother's diet. The offspring of mothers fed a high-fat diet who were themselves fed a high-fat diet showed poorer retention in a water-maze memory task, though they did not show poorer acquisition.[49]

7.3 Mechanisms

The mechanisms that underlie the costs of a high-fat diet and the benefits of dietary restriction are not fully understood. However, much research is under way, and a number of possible mechanisms have been implicated: glucose regulation, hippocampal plasticity, oxidative stress, neuroprotective mechanisms, neurogenesis.

Glucose regulation

The strength of the effect of dietary restriction has been associated with glucose regulation. Poor regulation of blood glucose levels has been linked to poorer memory performance and hippocampal atrophy, even in older people who do not have diabetes.[50] In animal and human studies, DR has been found to improve glucose regulation.[51] These benefits have been observed even when weight loss did not occur, as is typical for rodents fasting on alternate days.[52]

Plasticity

Another mechanism that appears to underlie the benefits of dietary restriction is increased plasticity in hippocampal circuits, specifically those involving NMDA receptors.[53] NMDA receptors are an important component in the control of synaptic plasticity in the brain.[54] Induction of long-term potentiation at granule-cell synapses in the dentate gyrus depends on activation of NMDA receptors.[55] DR eliminated the previously observed age-related loss in numbers of glutamate NMDA and AMPA receptor subunits in the hippocampus.[56]

Oxidative stress

A more fundamental mechanism may be oxidative stress, defined as the balance between the production of peroxides and free radicals (called "reactive oxygen species") from oxidation during aerobic metabolism and the ability of the body to repair damage to neurons resulting from oxidation. Such damage can occur to all components of neurons, including proteins, lipids, and DNA. Normally, most of the reactive oxygen species produced are generated at a low level through aerobic metabolism, with the damage to cells being continuously repaired. However, severe levels of reactive oxygen species can cause cells to die by necrosis.[57] Several studies have found that a high-fat diet increases oxidative stress, even in the offspring of rats that were fed a high-fat diet.[58] The effect of oxidation on such offspring can be measured in common products of protein oxidation— protein carbonylation and protein nitration in brain tissue. Markers of inflammation can be assessed in the reactivity of glial cells (support cells for neurons), which use reactive oxygen and nitrogen species during inflammatory reactions. Measuring reactivity of glial cells, White et al. found evidence of increased oxidative stress and inflammation in the brains of the female rats fed a high-fat diet and in their offspring.[59]

Dietary restriction appears to reduce the effects of a high-fat diet on oxidative stress and inflammation. Mice on an alternate-day fast exhibited less inflammation in several brain regions than mice on an ad libitum diet.[60] Similarly, negative effects of a high-fat diet were found on both BDNF levels and cognition, but those changes were reversed by the anti-oxidant vitamin E in the diet. These findings suggest that the negative effects of a high-fat diet on both BDNF and cognition may be related to oxidative stress.[61]

Moreover, there is human evidence that oxidative stress and inflammation are important contributors to some of the negative consequences of obesity. James Johnson and colleagues[62] found that nine overweight human asthmatics who were put on an alternate-day fasting diet for 8 weeks (eating as usual [ad libitum] on one day, but consuming only 20 percent of their normal calories on alternate days) showed benefits for serum cholesterol and triglycerides. This finding points to reductions in markers of oxidative stress and inflammation as important for proper weight maintenance.

Neuroprotection mechanisms

The benefit of alternate-day fasting appears to be exerted on neuroprotective mechanisms. After suffering an experimental stroke, rats on an alternate-day fasting diet showed better performance on a radial arm maze than rats on an ad libitum diet.[63] That diet did not prevent loss of hippocampal neurons (in regions CA1 and CA3), which suggests that the low-energy diet heightened the functioning of the remaining neurons although it did not prevent neuron loss. The actual mechanisms are suggested by evidence that alternate-day fasting upregulates the expression of a range of neuroprotective proteins in brains of young mice, including, but not limited to, BDNF. Further, that diet suppressed the expression of pro-inflammatory cytokines TNFα, IL-1β, and IL-6.[64] Cytokines are signaling molecules secreted by glial cells and pro-inflammatory cytokines signal a need for inflammation. This effect of alternate-day fasting on cytokines was greater in young than in old mice. Thus, a high-energy diet increases markers of oxidative stress and inflammation while alternate-day fasting reduces these markers and stimulates neuroprotective functions without affecting neuron mortality.

The finding that alternate-day fasting was associated with increased neurotrophin expression[65] points to the role of neurotrophins in promoting neural plasticity, even after experimental stroke. As described above, a neurotrophin that has been found to be particularly important in the brain is BDNF, which is produced by neurons. Neuronal activity regulates the transcription of the BDNF gene and the secretion of BDNF protein from neurons. BDNF's receptor, TrkB, is found in the plasma membranes of dendrites and presynaptic terminals and appears to be important in synaptic plasticity and neuron survival.[66] BDNF also has a role in stimulating neurogenesis—the birth of new brain cells.

There appear to be specific synaptic changes associated with diet that are modulated by BDNF. Rats fed a high-fat, high-sugar diet for 8 months were impaired in learning and memory, in synaptic strength, and in density of hippocampal dendritic spines. The diet-related reductions in BDNF were seen in association with reduced LTP at hippocampal CA1 synapses and reduced hippocampal dendritic spine density.[67] Dendritic spines are protrusions along the dendrite which typically receive synaptic input from other neurons. Reduced spine density means fewer synapses.

Neurogenesis

Another mechanism for the effects of a high-energy diet involves neurogenesis. It is now known that the hippocampus of mammals continues to produce new brain cells throughout life—a process termed neurogenesis (and described in chapter 4). Several groups have reported that high-fat diets impair hippocampal neurogenesis in rodents. This was seen in male but not female rats, even when the animals did not become obese.[68] Seven weeks of a high-fat diet in mice significantly reduced the levels of BDNF (known to have a role in stimulating neurogenesis[69]) and reduced the numbers of newborn neurons in the dentate gyrus.[70]

Overlap in mechanisms

There appear to be overlapping mechanisms among some of these factors. Cognitive deficits in spatial working memory associated with a high-fat diet were avoided in animals allowed access to wheel running but were not avoided in sedentary animals.[71] Because BDNF levels increased in the running animals but decreased in the sedentary animals, the authors concluded that the high-fat diet and the exercise influenced the same mechanisms of synaptic plasticity but in opposite directions. As reviewed in the next section, another mechanism may involve the Sirt1 enzyme, which is thought to be involved in the positive effects of resveratrol on cerebrovascular health and cognition. Mice with overexpression of the SIRT1 gene (which produces the enzyme Sirt1) showed fewer consequences of a high-fat diet—lower lipid-induced inflammation and better glucose tolerance.[72] These findings suggest a common mechanism underlying effects of resveratrol, dietary fat, and dietary restriction on cognition.

What dietary practices can influence overall health and cognitive functioning?

In addition to dietary restriction, a diet low in calories and in fat. A high-calorie, high-fat diet is detrimental to both health and cognitive functioning—especially if it results in obesity. Several related mechanisms have been implicated: oxidative stress, impaired glucose regulation, plasticity, and neuronal protection, suppressed neurogenesis. There is evidence of interactions among some of these mechanisms.

7.4   Resveratrol

Resveratrol, a natural polyphenol notably abundant in grapes, grape skins, and wine, has been found to have broad beneficial effects on health. Resveratrol exerts wide-ranging effects on cancer, angiogenesis, drug metabolism, heart disease, platelet aggregation, antioxidant activity, stress, and aging.[73] In a previous section we discussed the beneficial effects on health and cognition of dietary restriction. It is therefore of considerable interest that there is evidence that resveratrol acts by mimicking the effects of DR.

If this claim is substantiated with further research, then reducing caloric intake by 30 percent—an unpalatable prospect for many—could be countered with the more readily acceptable recommendation to older adults to have one or two glasses of red wine with dinner.

A consistent theme of the evidence on resveratrol is that it exerts broad effects on health and cognition. We will briefly describe several recent animal studies that point to the breadth of its effects. Mice on a high-calorie diet that was supplemented with resveratrol survived longer and with better general health than mice without the supplementation.[74] Another recent study found that aged mice fed a standard diet supplemented with resveratrol showed decreased inflammation, increased elasticity of the aorta, better motor coordination, lessened cataract formation, and preserved bone density relative to aged mice not receiving resveratrol.[75] In a study of middle-aged mice, a resveratrol-supplemented diet was compared with a control diet. The mice under long-term dietary supplementation with resveratrol showed preserved spatial Y-maze performance and greater density of small blood vessels relative to mice on the control diet.[76]

As was mentioned above, there is increasing evidence that dietary restriction and resveratrol stimulate mechanisms in common. A recent study directly compared the effect of a control diet with a DR diet and a resveratrol-supplemented diet. They found that both the DR diet and the resveratrol diet inhibited the expression of genes associated with aging in brain, cardiac muscle, and skeletal muscle relative to the control diet. Moreover, both diets prevented age-related cardiac dysfunction and affected the uptake of glucose in muscle mediated by insulin.[77]

Resveratrol has been shown to exert therapeutic benefits not only in animal models of human disease (cancer, cardiovascular disease, brain injury from ischemic interruptions to the blood supply) but also in human disease itself. Most of this work has been done with human cells in culture. A few of the many studies that have been conducted can be summarized briefly as follows:

• Resveratrol was found to protect human endothelial cells from oxidative stress from cigarette smoke.[78]

• The protein Aβ42 is thought to play an important role in the development of the degenerative disorder of Alzheimer's disease. Resveratrol was also found in cell culture studies to inhibit both Aβ-42 fibril formation and cell toxicity in a dose-dependent manner. However, resveratrol did not prevent formation of Aβ42 oligomers.[79] These findings suggest that resveratrol may act by preventing Aβ42 fibrils from aggregating (clumping together), which is thought to have some role in their known toxicity to neurons.

• Pre-treatment for 3 hours with resveratrol or quercetin (a polyphenol found in green tea) reduced the death of neurons induced by the neurotoxin MPP.[80] MPP is known to target dopamine cells and to cause Parkinson's disease.

This is only a sampling of the many studies that are being conducted on the potential health benefits of resveratrol. A number of Phase I and Phase II clinical trials of resveratrol on a range of chronic diseases are in progress, including trials of its effects on cognitive aging.[81] Although there is much reason for optimism, future studies may not consistently find evidence of efficacy. Of course, many testimonials to the benefits of resveratrol (which is available in extract form over the counter) have appeared in the popular press. However, because supplement manufacturers are not required to meet standards for purity or quality, over-the-counter formulations may be neither pure nor of high quality.

Resveratrol appears to exert its effects by activating the "SIRT1 gene" (that is, the gene that controls production of Sirtuin 1, an enzyme previously claimed to have a role in longevity). Though recent work has called that into question, there remains strong evidence that the SIRT1 gene has a role in protecting against age-related disease and physiological decline.[82] The SIRT1 gene is one of a family of genes with roles in gene silencing (a process of gene regulation), in apoptosis (programmed cell death), in mitochondrial function, energy homeostasis, in DNA repair, in apoptosis, and possibly in longevity.[83] Leonard Guarente, one of the pioneers of SIRT1 investigation, has postulated that sirtuin proteins may help to ensure survival in the face of stress and adversity.[84]

The SIRT1 gene was found to inhibit neurodegeneration in mouse models of Alzheimer's disease. A "slow" virus (a lentivirus) was altered so that it produced Sirtuin 1 and was then injected into the left or the right hemisphere of transgenic mice that had been genetically engineered to develop human-like Alzheimer's pathology. This was an elegant design in that both sides of the brain developed the Alzheimer's pathology but only one side held the virus producing Sirt1. After 2 weeks, the virus-treated side of the brain was compared against the control side. Greater numbers of neurons, neurons that were "morphologically healthier," and neurons that were expressing SIRT1 were found on the virus-treated side.[85]

To test the claim that resveratrol and dietary restriction have mechanisms in common, Kevin Pearson and colleagues in David Sinclair's lab directly compared transcriptional changes associated with resveratrol against changes associated with dietary restriction in mice. The effects of resveratrol were compared against those of a standard diet, those of alternate-day fasting, and those of a high-calorie diet. Resveratrol induced gene expression patterns in multiple tissues (liver, skeletal muscle, adipose, and heart) that paralleled those induced by DR. The animals on resveratrol had less heart disease, fewer cataracts, and greater mobility as they aged. However, their life spans were not extended.[86]

All this evidence raises the pleasing prospect that the considerable benefits of dietary restriction can be obtained without daily deprivation by substituting sources of resveratrol instead. The growing experimental literature on resveratrol is consistent with observations of the "French Paradox" (low risk of cardiovascular disease despite a diet high in saturated fat with regular consumption of red wine).[87] Thus, it could be that consuming red wine somehow counteracts the negative effects of saturated fat in the diet.

However, caution should be exercised when using wine as a source of resveratrol. Recent estimates indicate that drinking about two glasses of red wine a day can provide a pharmacologically relevant dose of resveratrol,[88] but that having more than four alcoholic drinks a day of any sort eliminates the protective effect conferred by alcohol in general on the risk of myocardial infarction.[89] However, the widely accepted view that moderate alcohol consumption leads to reduced cardiovascular disease and mortality has been questioned. That view is based largely on prospective studies of alcohol use and mortality in which a "J-shaped" curve is observed—non-drinkers and the heaviest drinkers have the highest cardiovascular mortality. Several authors have argued that the results of prospective studies are attributable to a tendency of less healthy people to be more likely to reduce their alcohol consumption.[90] In a meta-analysis, Kaye Fillmore and colleagues grouped 54 prospective studies according to whether they had included in the "abstainer" category individuals who had stopped drinking because of ill health. Such a practice would spuriously reduce the apparent health of the abstainers. Fillmore et al. found that studies in which care was taken not to classify former drinkers as abstainers did not observe that light drinking conferred a significant benefit for mortality from any cause or from cardiovascular disease.[91] Nevertheless, several studies have shown that moderate alcohol consumption upregulates the antioxidant enzyme paraoxonase1 in liver and serum, whereas heavy alcohol consumption has the opposite effect in humans and animal models.[92] No large-scale randomized trials in humans have been conducted with alcohol. This is due largely to the ethical and practical challenges of double-blind administration of alcohol. (Participants would know from their symptoms that they had been given alcohol.)

Of course, none of these concerns about alcohol and health negates the findings from resveratrol, although to date most of the work with resveratrol supplements has been conducted with animals. At present, resveratrol is not available in reliable form for humans except in diet. Moreover, there is evidence that other substances in red wine (among them quercetin, also found in green tea) heighten the effects of resveratrol,[93] which suggests that resveratrol supplements may not have the same desired physiological effects as resveratrol in food and wine.

At present it is difficult to separate the benefits of resveratrol from the risks associated with alcohol.

Does including resveratrol in the diet have a positive influence on overall health and cognitive functioning?

Consumption of foods containing resveratrol may confer benefits on health and cognition that are similar to those of dietary restriction. There is some evidence that the same mechanisms are involved in both, making the choice between red wine and dietary restriction an easy one. Moreover, it may be important to consume resveratrol in food (peanuts, red grapes, wine) rather than as a supplement, as there are other active ingredients, notably in wine, that could heighten the benefits of resveratrol. An important caveat is that the benefits of wine can be countered by the costs of excessive alcohol consumption. The National Institute of Alcohol Abuse and Alcoholism suggests no more than two drinks a day for men, no more than one for women. In view of the known dangers of alcoholism, there are risks to encouraging people to consume wine.

7.5 Fatty Acids

Dietary fatty acids play a dual role in the brain, functioning both as energy substrates and as cell membrane components essential for neuronal function. Triglycerides provide a substrate for energy metabolism when glucose is low, while polyunsaturated fatty acids contribute to cell membrane structure and function. Fatty acids are components of fat: hydrocarbon chains that vary in number of carbon atoms and in number of double bonds. Saturated fatty acids have no double bonds; monounsaturated fatty acids (MUFAs) have one double bond; polyunsaturated fatty acids (PUFAs) have two or more double bonds. Omega-3 and omega-6 PUFA vary in the location of the first double bond—at carbon 3 or carbon 6 from the methyl end, respectively. Fatty acids are an integral part of phospholipid cell membranes, and the double bonds of the PUFA appear to confer flexibility to the cell membrane, although that is not their only benefit. PUFA appear to have detectable effects on physiology. Specifically, high consumption of omega-6 PUFA results in a state that promotes formation of blood clots, with greater concentration of red blood cells, constriction of blood vessels, and fast clotting time—all of which increase the risk of stroke. High consumption of omega-3 PUFA produces an anti-inflammatory state, with dilation of blood vessels, inhibition of clotting, and lowering of triglycerides.[94]

Docosahexaenoic acid (DHA) and eicosapentaenoic acid (EPA) are important omega-3 fatty acids that are obtained mainly through diet. DHA is manufactured in microalgae and moves up the food chain, becoming concentrated in the oil of cold-water fish. It is also found in a few types of nuts and seeds, notably walnuts and flax seeds. Animals make very little DHA, although some is produced through consumption of alpha-linolenic acid in milk.

One consistently observed benefit of omega-3 fatty acids is on the health of the heart. It has been confirmed in a meta-analysis of more than 60 randomized controlled trials that PUFA lower the ratio of total/ high-density lipoprotein cholesterol.[95] That ratio has been found to be the best predictor of heart disease.[96] In 2004, the US Food and Drug Administration gave a "qualified health claim" status to EPA and DHA omega-3 fatty acids, claiming that "supportive but not conclusive research shows that consumption of EPA and DHA omega-3 fatty acids may reduce the risk of coronary heart disease."

PUFA may also confer a benefit on the brain. Michael Crawford and colleagues have advanced a theory that the evolution of the large hominid brain depended on the consumption of fish,[97] and omega-3 fatty acids are obtained largely by eating fish. DHA, the most abundant omega-3 fatty acid in the brain, may affect neural function by enhancing synaptic membrane fluidity and function,[98] by regulating gene expression and inhibition of programmed cell death (apoptosis),[99] by mediating cell signaling,[100] by enhancing LTP.[101] Since there is evidence for each of these mechanisms, researchers have speculated that a combination of effects may underlie the beneficial health properties of DHA.[102] (An excellent overview of the influence of diet on neural and epigenetic mechanisms is provided in an article by Fernando Gómez-Pinilla, one of the pioneers in this field.[103])

In agreement with the evidence mentioned above, animal studies have consistently found benefits of DHA supplementation for cognitive and brain integrity. A DHA-enriched diet significantly increased spatial learning performance.[104] Treatment of rats with fish oil before causing an experimental stroke did not prevent hippocampal damage. However the administration of fish oil did gradually reverse memory retention deficits in a radial arm maze after the stroke.[105] Dietary administration of EPA to rats before infusion of Aβ (associated with development of Alzheimer's disease in humans) into the brain resulted in a lower number of errors in reference-memory and working-memory tasks.[106] Aged mice whose diets were supplemented with DHA over weeks showed better maze performance, higher levels of BDNF in the hippocampus, and higher levels of dopamine in striatum relative to aged mice on unsupplemented diets.[107]

The beneficial effects of DHA extend to rats exposed to "disadvantaged" environments. It is well established that rats raised in an enriched environment (with toys, tunnels, running wheels, and other rats) showed better cognitive performance in adulthood. Rats raised in an impoverished environment (one animal per cage, with bedding and food only) with a diet supplemented with both DHA and phosphatide precursors showed increased total brain phospholipid (a major component of neuron membranes) and better performance on the Morris water maze under the difficult "hidden platform" condition relative to rats without DHA supplementation. Rats raised in an enriched environment did not show the dietary benefits of DHA,[108] which suggests that dietary DHA and environmental enrichment have some mechanisms in common. In summary, there is fairly consistent animal evidence that PUFAs, and omega-3 fatty acids in particular, are important in health and in cognitive functioning. A number of groups have observed improved cognitive and brain functioning associated with DHA supplementation in animals.[109]

The human evidence is more mixed, perhaps because the studies are not as well controlled. Observational studies find benefits of PUFA and/ or fish consumption on brain and cognition, although not consistently. People who consumed greater amounts of fish showed fewer subclinical infarcts in MRI scans than people consuming lesser amounts of fish (n = 3,660).[110] Marie-Noel Vercambre and colleagues found that in a large sample of women 65 and older (n = 2,551), higher intake of monounsaturated fat and PUFA over 5 years was found to be inversely related to cognitive decline, but only in the oldest participants.[111] Dementia, broadly defined, has also been associated with low plasma levels of omega-3 fatty acids.[112] In contrast to these positive results, a 6-year follow-up study of more than 1,000 participants in the Veterans Affairs Normative Aging Study found no association between omega-3 PUFA intake and cognitive function (assessed on tests of memory, language, response speed, and visuospatial attention.[113] Consistent with a real benefit of PUFA, Gregory Cole recently reviewed the literature, concluding that more than twelve epidemiological studies show that lower consumption of dietary DHA or fish was associated with greater risk of age-related cognitive decline.[114]

Other epidemiological studies have taken a somewhat more rigorous approach by measuring biomarkers of omega-3 fatty acids in blood, in addition to cognitive performance. One study measured alpha-linolenic acid (ALA), EPA, and DHA in blood and related those measures to cognitive performance in 280 volunteers between 35 and 54 years of age. Participants were not taking PUFA supplements. Higher DHA, but not EPA nor ALA, was associated with better performance on non-verbal reasoning and mental flexibility, working memory, and vocabulary.[115] A study of 210 men aged 70–89 found that those who self-identified as frequent consumers of fish had significantly less cognitive decline over 5 years than those who did not eat fish.[116] A deficiency of DHA is also associated with depression,[117] which is suggestive of the broad effects of this PUFA. There is some evidence bearing on the possible mechanism underlying effects of PUFA on cognition and brain aging. Aβ is strongly linked to Alzheimer's pathology, and DHA decreased Aβ peptide secretion in aged neural brain cells in culture.[118]

These findings are of interest but face the same general limitations of all observational studies that do not randomly assign participants to treatment and control groups. A better way to evaluate the effect of dietary DHA and EPA on cognitive function is with randomized trials. However, few such trials have been conducted. A Cochrane Database review of this topic concluded that, as of 2008, no qualifying randomized trials on the effects of omega-3 fatty acids and cognitive functioning in older people had been conducted.[119] Since that time, several randomized trials have been published. A trial of 485 older people across 19 clinical sites using an oral dose of 900 milligrams of DHA per day or a matching placebo was administered for 24 weeks.[120] Plasma DHA levels doubled over the trial, and that increase was correlated with improved paired associate performance and improved verbal recognition memory performance (on the CANTAB battery of cognitive tests). Working memory and executive functioning were not affected. In contrast, other groups have not found a similar benefit with different doses and assessments. Alan Dangour and colleagues enrolled 867 older people (mean age 75) in a trial across 20 private practices in England and Wales. They administered a lower dose (500 mg DHA plus 200 mg EPA) but for a longer period of 20 months. Using mainly the California Verbal Learning Test, which assesses verbal memory for items in lists, they observed no decline in the placebo group and no benefit from the PUFA treatment.[121] Ondine van de Rest and colleagues compared the effects of different formulations of DHA and EPA. They randomly assigned 302 people 65 and older to either 1,800 mg/d of EPA-DHA, 400 mg/d EPA-DHA, or placebo capsules for 26 weeks. Although plasma concentrations of EPA-DHA increased substantially in all treatment groups over the duration, no cognitive changes were observed.[122] Moreover, in a separate report, the same group found no evidence of effect of these PUFA supplements on quality of life in older people.[123]

Thus, to date, although the benefits of PUFA for health and cognitive function in animals are clear, there is only conflicting evidence to support the use of DHA supplementation to enhance cognitive functioning in humans. However, what is known of the role of PUFA in preventing coronary heart disease (CHD) may shed some light on the matter. Recently a meta-analysis was conducted of eight randomized controlled trials involving 1,042 CHD events (myocardial infarction and/or cardiac death) among 13,614 participants. In most of those trials, diets were fully provided to participants with PUFA substituted for saturated fat. Overall, there was a 10 percent reduced risk of CHD for each 5 percent energy of increased PUFA.[124] This finding is consistent with a recent comparison of popular weight-loss diets which revealed that diets high in saturated fat, such as the Atkins Diet, increase the risk of cardiovascular disease relative to diets low in saturated fat and high in PUFA.[125] This impressive analysis assessed studies in which diets were provided, PUFA was substituted for saturated fat, and actual cardiovascular events (myocardial infarction and/or cardiac death) were measured. Thus, it may not be sufficient to simply add DHA to a diet—as was done in the randomized controlled trials described above—if saturated fat isn't reduced. Additional research will be needed to fully understand the implications of these findings.

There is convincing evidence that increasing DHA in the diet results in increased incorporation of PUFA into cell membranes at the expense of saturated fatty acids, although the mechanism by which that results in health benefits is not known.[126] Because substituting PUFA for saturated fats appears to reduce incidence of coronary heart disease, such a dietary change might at the very least prevent cognitive decline associated with coronary heart disease. Therefore, it will be necessary to conduct trials of the effects of PUFA on cognition in which the PUFA are *substituted* for saturated fat in the diet, not merely added in supplements.

What are the effects of polyunsaturated fatty acids on overall health and cognitive functioning?
Substitution of polyunsaturated fatty acids for saturated fat in the diet has convincing benefits for health and cognition in animals and for reducing the risk of coronary heart disease in humans. The evidence on human cognition is inconclusive.

7.6   B Vitamins

High levels of plasma homocysteine, an amino acid with a role in essential metabolic pathways, are consistently associated with blood vessel endothelial dysfunction, with stroke, and with coronary heart disease.[127] Homocysteine is found in increased levels in blood when there is deficiency of vitamins B12 and folate, which are required for conversion of homocysteine into cysteine. Insofar as homocysteine levels can be reduced if there is sufficient dietary intake of B vitamins and folate, what is their effect on cognition?

Much of the relevant evidence comes from examining the role of these compounds in the development of dementia. The risk of Alzheimer's disease has been found to increase with plasma homocysteine level in a graded manner.[128] Cognitive decline in older adults in the absence of Alzheimer's disease is also associated with deficiencies of serum folate and B12.[129] Such observations have led to efforts to treat dementing disorders with supplements of B vitamins. Fortification of flour with folic acid has been mandated since 1998 in the United States and Canada in a largely successful attempt to prevent neural tube defects.[130] This fortification has also reduced the incidence of hyperhomocystenemia (elevated levels of homocysteine in the blood[131]). (The incidence of Alzheimer's disease hasn't declined noticeably over that time despite the folate fortification.)

High concentrations of unmetabolized folic acid can have negative consequences. Such concentrations are associated with reductions in the innate immune response to infection and malignancy due to reduced natural killer cell cytotoxicity. Consistent with that, folic acid supplementation has been linked to colon cancer.[132]

Because B12 deficiency is now the remaining significant cause of homocysteine-related disease (cardiovascular disease and Alzheimer's disease), clinical trials of B-vitamin supplementation have been undertaken for both conditions. A number of clinical trials have failed to show that superphysiological doses of folate and B vitamins reduce major cardiovascular events[133] or alter the rate of cognitive decline in healthy older women, except in those with low baseline dietary intake of B vitamins[134] or in Alzheimer's patients.[135]

More positive results have been observed in individuals classified as having Mild Cognitive Impairment, thought to be an intermediate stage between normal aging and Alzheimer's disease. MCI individuals are at high risk of cognitive decline; about half of them receive a diagnosis of Alzheimer's disease in 5 years. David Smith and colleagues administered 168 participants either high-dose B vitamins (0.8 mg/day folic acid, 0.5 mg/day B12, 20 mg/day B6) or placebo for 2 years. They reported only on the rate of brain atrophy as measured by MRI scans, though they stated that cognitive results would be published separately. B-vitamin treatment decreased brain shrinkage from more than 1 percent per year in the placebo group to an average of 0.75 percent per year in treated participants. Those with initial homocysteine levels below 9.5 μmol/L obtained no benefit from the B vitamins; those with higher levels of homocysteine above 13 μmol/L saw a significantly different 53 percent reduction in their rate of atrophy.[136] This evidence is consistent with the finding of Jae Hee Kang and colleagues,[137] in Francine Grodstein's laboratory who reported cognitive change in older women with a low baseline dietary intake of B vitamins.

A recent review concluded that there is no consistent evidence that folic acid, with or without vitamin B12, has a beneficial effect on cognitive function in healthy or cognitively impaired older people.[138] Therefore, although there appears to be no doubt that high levels of circulating homocysteine represent a risk factor for diseases (such as Alzheimer's disease) that produce late-life cognitive decline, the evidence that supplementation is the solution is mixed.

Do B-vitamin supplements have beneficial effects on overall health and cognition?

There is little evidence that B-vitamin supplementation exerts any benefits on brain function or cognition in healthy people, although there may be some benefit in the case of deficiency.

7.7 Other Nutrients and Foods

The nutritional factors we have discussed above are those for which the scientific evidence is particular strong, or at least those that have been studied extensively. Among the factors we have considered are calorie intake, saturated fat, polyunsaturated fatty acids, resveratrol, B vitamins, and folates. Of course, this does not exhaust the list of nutrients and foods that may affect cognitive function. Hundreds of other macro- and micronutrients, foods, spices, herbs, etc. have been said to promote health and cognitive vitality in older adults.

The effects on cognition of some of these nutrients have been tested, at least in observational studies, but randomized controlled studies remain rare. We cannot review the vast literature here, but we can briefly mention some nutrients that may be efficacious and on which further research may be profitable. Table 7.1[139] summarizes what is currently known about the effects of various nutrients on brain function and cognition. Several

Table 7.1

Reprinted, with permission, from F. Gómez-Pinilla, "Brain foods: The effects of nutrients on brain function," *Nature Reviews Neuroscience* 9, no. 7: 568–578 (© 2008 Macmillan Publishers Ltd.).

Nutrient	Effects on cognition and emotion	Food sources
Omega-3 fatty acids (for example, docosahexaenoic acid)	Amelioration of cognitive decline in the elderly; basis for treatment in patients with mood disorders; improvement of cognition in traumatic brain injury in rodents; amelioration of cognitive decay in mouse model of Alzheimer's disease	Fish (salmon), flax seeds, krill, chia, kiwi fruit, butternuts, walnuts
Curcumin	Amelioration of cognitive decay in mouse model of Alzheimer's disease; amelioration of cognitive decay in traumatic brain injury in rodents	Turmeric (curry spice)
Flavonoids	Cognitive enhancement in combination with exercise in rodents; improvement of cognitive function in the elderly	Cocoa, green tea, Ginkgo tree, citrus fruits, wine (higher in red wine), dark chocolate
Saturated fat	Promotion of cognitive decline in adult rodents; aggravation of cognitive impairment after brain trauma in rodents; exacerbation of cognitive decline in aging humans	Butter, ghee, suet, lard, coconut oil, cottonseed oil, palm kernel oil, dairy products (cream, cheese), meat
B vitamins	Supplementation with vitamin B6, vitamin B12 or folate has positive effects on memory performance in women of various ages; vitamin B12 improves cognitive impairment in rats fed a choline-deficient diet	Various natural sources. Vitamin B12 is not available from plant products
Vitamin D	Important for preserving cognition in the elderly	Fish liver, fatty fish, mushrooms, fortified products, milk, soy milk, cereal grains
Vitamin E	Amelioration of cognitive impairment after brain trauma in rodents; reduces cognitive decay in the elderly	Asparagus, avocado, nuts, peanuts, olives, red palm oil, seeds, spinach, vegetable oils, wheat germ

of the items on this list have already been discussed in this chapter. Below we will briefly discuss two that have not been mentioned so far: beta carotene and curcumin.

As we noted earlier in this chapter, oxidative stress has been identified as a mechanism underlying the negative effects of many nutritional factors on health and cognition. Consequently, there has been considerable interest in examining the possible beneficial effects of various antioxidants on cognitive and brain functions. Various dietary interventions rich in antioxidants may slow age-related cognitive decline and reduce the risk of Alzheimer's disease.[140] One such antioxidant is beta carotene, found in various fruits and vegetables. A recent randomized controlled investigation compared the effects of 50-mg beta carotene supplements with placebo in more than 4,000 participants in the Physicians Health Study II.[141] A small positive effect of beta carotene on cognitive function (as assessed using a telephone interview) was found in those participants who had maintained the supplement regimen for a long period (18 years). However, no such benefit was found in those more recently recruited in the study and tested after a year. The protective effect of beta carotene may be greater for older adults at risk for developing Alzheimer's disease. In a study of 455 healthy older adults, high blood serum levels of beta carotene were associated with lower cognitive decline over 7 years, but only in participants who had the risky e4 allele of the APOE gene.[142]

Another antioxidant is curcumin, found abundantly in South Asian diets and commonly consumed in foods such as curry. In an animal study, a regular diet or a diet high in saturated fat was administered with or without curcumin to rats before a mild experimental brain injury was induced.[143] Curcumin supplementation reduced oxidative damage and normalized levels of BDNF that had been altered after the brain injury. Curcumin also counteracted the cognitive impairment caused by the brain injury. These results are consistent with the evidence we reviewed earlier indicating that oxidative stress acts through the BDNF system to affect synaptic plasticity and cognition. More recently, a synthetic derivative of curcumin called CNB-001 was found to facilitate induction of LTP and to enhance object recognition memory in rats.[144]

Human studies of curcumin have typically been epidemiological, but some clinical trials are underway and more are being planned. One small-scale randomized clinical trial was recently reported.[145] Over 6 months, 27

Alzheimer's patients were given either 1 gram or 4 grams of curcumin or placebo. No significant effects of curcumin on MMSE scores were observed, nor were any effects on change in MMSE score observed over the testing period.

What are the effects of other specific foods, spices, herbs, and micronutrients on overall health and cognition?

The literature on such nutrients is small and still relatively inconclusive. There is some evidence for benefits of antioxidants in the diet, consistent with other evidence for a role for oxidative stress in negative effects of aging.

7.8   Discussion and Closing Comments

The literature on diet, nutrition, and cognition is large and complex. Because of methodological limitations, strong conclusions cannot be drawn from many studies. The strongest statements that can be made about diet concern dietary restriction and excess energy intake—two sides of the same coin. Although there have been few controlled studies of dietary restriction in humans, the animal evidence is strong. Animal studies have shown fairly consistently that DR enhances both neuronal health and cognition, and that such effects are also seen in older animals. The one caveat is that chronic DR in rats was associated in one study with heart dysfunction.[146]

One approach for future research on humans would be to investigate intermittent fasting—for example, once a week or once a month. That dietary restriction is beneficial to serum insulin and glucose even when weight is not lost[147] suggests that the benefits of fasting may be due in part to its stimulation of cellular defense mechanisms. There has been considerable research into cellular responses to starvation—activities in the cell that ensure its survival in the face of nutrient deprivation. Such responses may play a role in the strong effects of DR seen in animals. As DR is never going to be a popular method of forestalling age-related decline in cognition and health, intermittent fasting may be a more acceptable approach.

Most of the evidence on excessive energy intake points to impairment of health and cognition. However, one question that has been little addressed concerns the source of the negative influence of a high-fat diet. One animal study found that, in the absence of weight gain, an 80-

percent-fat ketogenic diet (a high-fat, normal-protein, very-low-carbohydrate diet currently being used to treat refractory epilepsy in children) did not impair hippocampal neurogenesis.[148] This suggests that the culprits in a high-energy diet may be the associated obesity and the accompanying oxidative stress, rather than the fat.

There has been much recent interest in the effects on cognition of resveratrol in animals and of moderate consumption of red wine in humans. Again the animal evidence for the beneficial effects of resveratrol on health and cognition is fairly compelling. Evidence for efficacy in humans is mixed, however, and is based largely on observational studies.

The benefits of polyunsaturated fatty acids in animal studies are clear. However, the evidence as to whether PUFA supplements can enhance cognitive functioning in humans is conflicting. There is good evidence that PUFA in the diet benefits older adults with coronary heart disease and so may reduce the cognitive impairment associated with this condition. However, the evidence also suggests that PUFA must be substituted for saturated fat if its benefits are to be realized.

One issue that is beginning to be addressed with regard to diet is motivation. Most people have a good understanding about an optimal diet, but fail to use that knowledge in daily life. There is increasing research interest in understanding how to motivate people, at least with regard to weight loss. A randomized clinical trial compared (a) an "incentivized" dietary intervention (center-based individualized weight loss counseling on diet and exercise for 2 years) against (b) counseling sessions and (c) regular contact. There was greater weight loss in the intervention group (16.3 pounds, versus 4.4 pounds in the regular-care group) over the 2 years.[149] Research is urgently needed to understand the optimal means of motivating individuals and societies to make lifestyle changes.

Regarding our brain plasticity/cognitive plasticity hypothesis (chapter 4), we argued that the aging brain can be reorganized by cognitive demand, with that reorganization dependent on intact plasticity mechanisms and on factors that enhance neuronal plasticity mechanisms. Of the diet-related factors we have summarized in this chapter, three appear to enhance neuronal plasticity: dietary restriction, resveratrol, and anti-oxidants. Consistent with our hypothesis, these three factors play specific roles in neuronal plasticity, viz., glucose regulation, hippocampal plasticity, oxidative stress, neuroprotection mechanisms, and neurogenesis. In contrast, the benefits from substituting polyunsaturated fatty acids for saturated fat in the diet appear to be exerted more generally and are not limited to effects on neuronal plasticity.

Finally, many studies indicate that the positive effects on health and cognition of dietary restriction, resveratrol, polyunsaturated fatty acids, as well as non-dietary activities such as aerobic exercise, share common mechanisms in the brain.[150] This suggests that optimal health—for heart, brain, and cognition—may follow from a combination of these factors, e.g., aerobic exercise combined with a diet rich in PUFA. We consider such combined effects for ameliorating cognitive aging in more detail in chapter 10.

Summary

Dietary restriction confers significant benefits for cardiovascular health; however, evidence of its impact on human cognition is limited, and its practicality is open to question. There is good evidence that a diet high in calories and in fat is detrimental to both health and cognitive functioning. Several related mechanisms have been implicated in both the benefits of DR and the costs of a high-fat diet. Consumption of foods containing resveratrol may confer benefits on health and cognition that are similar to those of DR. In view of the known dangers of excessive alcohol intake, however, there are risks to encouraging wine consumption. Substitution of polyunsaturated fatty acids for saturated fat in the diet has convincing benefits for animal health and cognition and for human risk of coronary heart disease. The evidence on human cognition remains inconclusive. There is little evidence that B-vitamin supplementation exerts any benefits on the brain or on cognition. Well-controlled studies of the effects of other specific foods, spices, herbs, and micronutrients on overall health and cognition are still few in number, and the results are relatively inconclusive. There is some evidence for benefits of antioxidants in the diet, consistent with other evidence for a role for oxidative stress in negative effects of aging.

8 Estrogen and Other Cognition-Enhancing Drugs

Youth, large, lusty, loving—youth full of grace, force, fascination,
Do you know that Old Age may come after you with equal grace, force,
fascination?
Day full-blown and splendid-day of the immense sun, action, ambition,
laughter,
The Night follows close with millions of suns, and sleep and restoring darkness.
—Walt Whitman, "Youth, Day, Old Age and Night"

Are there drugs that might reduce, forestall, or even prevent cognitive decline late in life? In this chapter we examine two general classes of pharmaceutical substances, estrogen, and certain cognition-enhancing drugs used in the treatment of various neuropsychiatric conditions. The use of either to boost cognitive performance in healthy older individuals raises many ethical issues. Nevertheless, several studies have examined their effects on cognition. We discuss both the scientific evidence and the associated issues surrounding the use of such drugs.

The most effective drug available to improve health and cognition in women in mid-life—estrogen—is also highly controversial because of its potential health risks. Age robs women of estrogen in an abrupt manner in their middle years, usually with negative consequences for health that can range from mild to severe. At best, the loss of physiological levels of estrogen results in thinned bones (prone to breakage) and increased risk of Alzheimer's disease. Depression and cognitive impairment also are common. Of course, estrogen is not the only endogenous factor that declines with age. There are also age-related declines in both sexes in dopaminergic function[1] and in cholinergic function,[2] although the individual differences in those systems are greater than the age-related differences. In contrast, there is no question that estrogen levels decline precipitously in women in mid-life—a decline that has documented consequences for health and for cognition.

During and after menopause, many women experience cognitive change, depression, and bone loss.[3] For a number of years, hormone-replacement therapy (HRT, as the practice was known) was considered a beneficial treatment for those women in whom the consequences of estrogen loss were particularly severe. All that changed suddenly in 2002 when the 5-year results of the Women's Health Initiative (WHI) were released and a gap opened between basic science and clinical science with regard to the risks and benefits of estrogen replacement. In sharp contrast to the previous epidemiological evidence of benefits, the large, randomized WHI trial (the design flaws of which we will discuss below) found instead an increased risk of heart disease, stroke, and breast cancer.[4] As a consequence, the use of HRT suddenly was no longer recommended.

8.1   Use of Estrogen to Slow Cognitive Decline

What is the consensus on the use of estrogen in menopause? In basic science, estrogen has long been known to confer both neuronal[5] and cognitive protection.[6] Estrogen enters the nucleus of the target cell, where it is able to regulate gene transcription, thereby exerting broad effects in the body. Estrogen's target cells are in bone (where it can have profound effects on bone remodeling), in blood vessels (notably coronary arteries), and in the brain (where it has both excitatory and neuroprotection effects). Its effects on brain neurons are rapid. Exposure to estrogen is followed, within minutes, by increased production of choline acetyltransferase (important in cholinergic transmission[7]) and by the formation of new dendritic spines[8] on the apical dendrites of hippocampal CA1 pyramidal neurons.[9] The new spines form synapses largely with preexisting synaptic terminals, thereby increasing the number of postsynaptic spines coupled to a single presynaptic terminal.[10]

These neurochemical and structural alterations are accompanied by functional changes. Effects of estrogen on spine density have been related to memory formation. Estradiol administration has been found to improve memory retention on a delayed match-to-sample task (in this case, remember a location for a period of time when the "sample" or target location is no longer visible).[11] The effect of estrogen on memory formation and retention has been related to spine density in rodents and monkeys.[12] Joanna Spencer and colleagues recently reviewed the evidence on modulation by estrogen of hippocampal function, including neuronal morphology, synaptic plasticity, signaling, and hippocampal-dependent cognitive performance.[13]

Neurogenesis in females is also strongly influenced by estrogen. Neurogenesis—known to occur in the adult hippocampus of rodents,[14] monkeys,[15] and humans[16]—is reduced if the ovaries are removed, but is restored after estrogen replacement in adult animals.[17] Moreover, neurogenesis fluctuates with the estrous (menstrual) cycle, the most neurogenesis occurring during proestrus, when estrogen levels are highest.[18] Both estrogen-induced neurogenesis and estrogen-induced spine formation exert effects on memory performance. The relation between survival of newly formed neurons and learning has also been linked to estrogen. Survival and integration of newborn neurons depends in part on the learning of new information,[19] and better learning is correlated with numbers of new neurons.[20] That female rats showed stronger eyeblink conditioning and retained more new neurons in the hippocampus than males[21] has been attributed to effects of estrogen.

Estrogen's beneficial effects on brain and cognitive functioning have been studied in women. In an fMRI study of healthy young women who were not taking oral contraceptives, greater activation was seen over left frontal brain regions during the high-estrogen phase of the menstrual cycle than during the low-estrogen phase. This was seen during a word-stem-completion task and during a mental rotation task, but not during a simple motor task.[22] There is also evidence of neuroprotective effects of estrogen in the human brain. Epidemiological studies have consistently linked estrogen use after menopause with reduced risk of Alzheimer's disease.[23] Studies have also been conducted longitudinally, typically comparing large groups of women 65 and older who had chosen to take estrogen against women who had chosen not to take it. These studies report that women who chose to use estrogen had higher scores and less decline than non-users on tests of verbal memory,[24] visual memory,[25] and mental status.[26] Moreover, relative to non-users, estrogen users have been found to have less shrinkage of gray matter in prefrontal, parietal, and temporal regions and less shrinkage of white matter in medial temporal regions. The longer the duration of estrogen treatment, the greater the sparing of gray matter.[27]

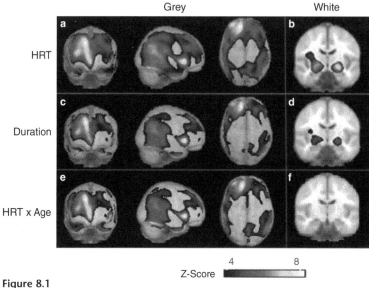

Figure 8.1
Images showing lessened age-related gray-matter shrinkage in older women under-going hormone replacement. Source: K. I. Erickson et al., "Selective sparing of brain tissue in postmenopausal women receiving hormone replacement therapy," *Neurobiology of Aging* 26: 1205–1213. Reprinted with permission of Elsevier (© 2005). This figure appears in color elsewhere in the book.

Even in older women who did not use hormone replacement, endogenous levels of estrogen were related to change in global cognitive functioning and verbal memory over 2 years.[28] The effect of estrogen may be stronger in old age. A longitudinal study of younger women did not observe differences in cognitive functioning related to estrogen use,[29] but a number of studies have found effects in older women. One such study found that the right hippocampal volumes of 56 post-menopausal women (age range 50–74) who were current users of estrogen replacement were larger than those of women who were past users and larger than those of women who had never used estrogen replacement.[30]

Despite the generally positive findings reported by these epidemiological studies, caution should be exercised in their interpretation, as the participants in these studies were self-selected rather than randomly assigned. Women who choose HRT during menopause tend to be younger, better educated, and healthier than women who do not choose it.[31] Although most studies attempt to control for education and age, not all the bias can be removed. This problem can be avoided by taking an experimental approach. By assigning women randomly to either hormone treatment or placebo, problems of self-selection can be eliminated. Such randomized trials are usually conducted in a double-blind manner, with neither the participant nor the experimenter knowing whether a participant is on estrogen or placebo. This minimizes the possibility of bias influencing the results. Previously conducted randomized trials vary in the form of estrogen used (either estradiol, produced by the human ovaries or conjugated equine estrogens, from mare urine), in the manner of estrogen administration (intramuscularly, transdermally, orally), and in the specific cognitive assessment. In contrast to the observational studies, results from randomized trials do not consistently find benefits of estrogen on cognition. Moreover, findings vary with age of participants, with type of estrogen, and with form and route of administration. We consider these points in turn.

Barbara Sherwin conducted an elegant set of studies on the impact of estrogen loss and replacement on memory function in younger women. In women undergoing surgically induced menopause because of benign disease, a crossover design (each woman receiving both treatment conditions serially) allowed the women to be tested cognitively both with and without estrogen replacement. Verbal memory, but not visual memory, was found to decline with the loss of estrogen after ovariectomy, and to be restored to pre-menopausal levels after intra-muscular estrogen replacement.[32] Though these results are impressive in showing clear effects of estrogen on cognition, the women were relatively young—in their thirties and forties. This limits the relevance of the findings to older women.

What about the effects of estrogen on post-menopausal women? Three randomized trials showed a beneficial effect on cognition. In a 2-year randomized trial, 142 women aged 61–87 were randomly assigned to receive either estrogen plus a progesterone or placebo. They were tested cognitively with the California Verbal Learning Test (a test of list learning) before treatment, after a year, and after 2 years. Those women who had been randomly assigned to estrogen and who also scored at or above average on the test at baseline showed less decline over time than their counterparts on placebo. No such effect was seen in women who scored below average on the test.[33]

Another study looked at effects of estrogen on older women who were a number of years after menopause and who had not previously taken estrogen—similar to the WHI participants. Women aged 65 were randomly assigned transdermal patches of either estradiol (19 women, with estradiol levels raised to the level experienced by fertile women) or placebo (18 women) for 3 weeks. Those who received the estradiol improved their performance on verbal learning, memory, and mental rotation (though not on executive functioning); those on placebo didn't improve.[34]

In another study of older women (46 women, mean age 67), 21 days of treatment with conjugated equine estrogens resulted in increased activation of inferior parietal cortex during verbal storage, decreased activation during non-verbal storage, greater activation in right superior frontal gyrus during retrieval, and greater left-hemisphere activation during encoding.[35] Overall, these three randomized trials show that older women experience a cognitive benefit from estrogen, with the evidence strongest for verbal memory.

Another experimental approach using random assignment assessed the effects of estrogen on cognition around the time of menopause. Women (mean age 51) randomly assigned to 12 weeks of transdermal estradiol showed reduced perseverative errors in a verbal recall task relative to a group assigned to 12 weeks of placebo. The estradiol-treated women also showed greater prefrontal activation during verbal and spatial working-memory tasks.[36] In a similar study, 14 women aged 51–64 were administered transdermal estrogen over 3 days, which produced a surge to levels comparable to those at ovulation. Scores on tests of Stroop interference and short-term verbal memory improved during the estrogen treatment over baseline.[37] Overall, these small studies using transdermal estrogen show cognitive benefits in middle-aged and older women.

Despite these reports from randomized trials that older women benefit from estrogen, the largest randomized clinical trial of estrogen treatment— the Women's Health Initiative (WHI) and its sub-study, the Women's Health Initiative Memory Study (WHIMS)—did not find such benefits and furthermore found costs. The WHI recruited 68,132 postmenopausal women between the ages of 50 and 79, who were, on average, 21 years past menopause, and assigned them to several studies. In the WHIMS study, 2,808 women aged 65–79 were randomly assigned to continuous administration of placebo, estrogen only (oral conjugated equine

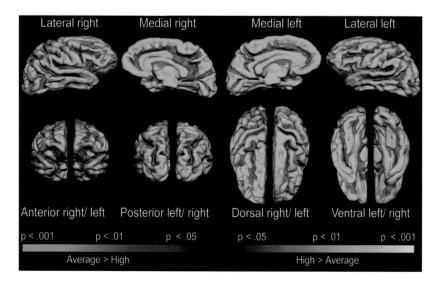

Plate 1 (figure 3.3)
The cortical thickness of old participants with high and normal cognitive (fluid) functioning compared by general linear modeling across the entire cortical mantle. Red and yellow areas indicate where cortices were thicker in the high functioning group. In the right hemisphere, differences were seen in temporal middle gyrus and temporal inferior gyrus, gyrus cuneus, the gyrus and sulcus of the insula, gyrus rectus, the gyrus of the cingulate isthmus, and the posterior cingulate gyrus. Source: Anders M. Fjell et al., "Selective increase of cortical thickness in high-performing elderly—structural indices of optimal cognitive aging," *NeuroImage* 29: 984–994. Reprinted with permission of Elsevier (© 2006).

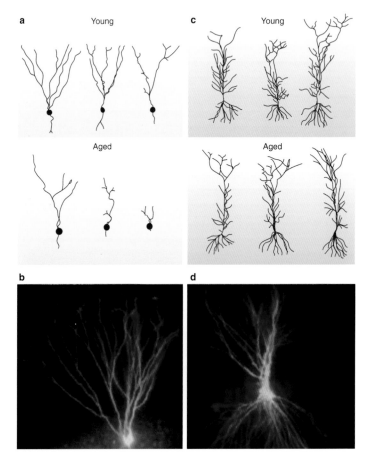

Plate 2 (figure 3.5)

The evidence for brain aging is relatively weak. This composite figure (from S. N. Burke and C. A. Barnes, "Neural plasticity in the ageing brain," *Nature Reviews Neuroscience* 7, no. 1: 30–40) is reprinted with permission of Macmillan Publishers Ltd. (© 2006). (a) Apparent age-related loss of dendritic extent in the dentate gyrus and in CA1 was exaggerated by including demented individuals with healthy aged individuals in the same experimental group and by not using stereological controls. (Original source: M. Scheibel et al., "Progressive dendritic changes in the aging human limbic system," *Experimental Neurology* 53: 420–430.) (b) Two granule cells from the dentate gyrus of an aged 24-month-old rat. Neurons in rat dentate gyrus show no significant age-related change in dendritic extent, but do show significant increases in electrotonic coupling. (Original source: T. C. Foster et al., "Increase in perforant path quantal size in aged F-344 rats," *Neurobiology of Aging* 12: 441–448.) (c) Reconstructions of neurons in hippocampal area CA1 from young rats (2 months) and old rats (24 months) showing no reduction in dendritic branching or length with age in area CA1. (Original source: G. K. Pyapali and D. A. Turner, "Increased dendritic extent in hippocampal CA1 neurons from aged F344 rats," *Neurobiology of Aging* 17: 601–611.) (d) A CA3 neuron filled with 5,6-carboxyfluorescein from a 24-month-old rat. There is no regression of the dendrites, but the aged neurons show a significant increase in the number of gap junctions compared with young neurons. (Original source: G. Rao et al., "Intracellular fluorescent staining with carboxyfluorescein: A rapid and reliable method for quantifying dye-coupling in mammalian central nervous system," *Journal of Neuroscience Methods* 16: 251–263. Reprinted with permission of Elsevier.)

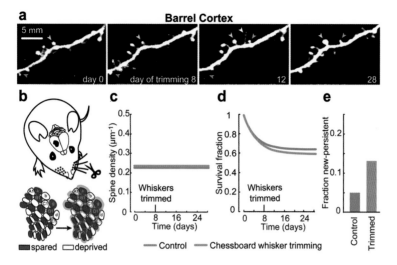

Plate 3 (figure 4.1)
Spine plasticity in adult neocortex associated with experience. (a) Time-lapse image of dendritic spines in barrel cortex before and after "chessboard" whisker trimming at day 9 (indicated in plots c and d). Many spines were "persistent" (yellow arrowhead), but some appeared and disappeared (white arrowhead). After whisker trimming, new persistent spines (orange arrowhead) were more likely to grow and previously persistent spines (green arrowhead) were more likely to disappear. (b) The experimental paradigm. "Chessboard" whisker trimming was associated with changes in the whisker representational map in barrel cortex. (c) Spine density in barrel cortex remained unchanged after whisker trimming. (d) The fraction of surviving spines was slightly decreased after whisker trimming, owing to an increased loss of persistent spines. (e) New persistent spines increased by a factor of about 2.5 after whisker trimming. Source: A. Holtmaat and K. Svoboda, "Experience-dependent structural synaptic plasticity in the mammalian brain," *Nature Reviews Neuroscience* 10: 647–658. Reprinted with permission of Macmillan Publishers Ltd. (© 2009).

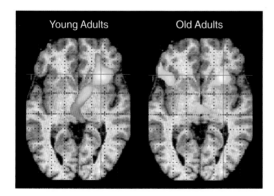

Plate 4 (figure 4.2)
Prefrontal activity during episodic memory retrieval was lateralized to the right hemisphere in younger adults but was bilateral in older adults. Source: R. Cabeza, "Hemispheric asymmetry reduction in older adults: The HAROLD model," *Psychology and Aging* 17, no. 1: 85–100.

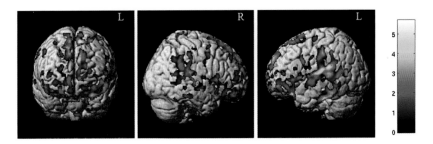

Plate 5 (figure 5.2)
Areas in which gray-matter volume differed significantly between healthy aged groups categorized as having attained high vs. low education. Source: A. Foubert-Samier et al., *Education, Occupation, Leisure Activities, and Brain Reserve: A Population-Based Study*. Reprinted with permission of Elsevier (© 2010).

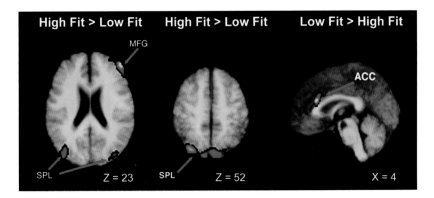

Plate 6 (figure 6.3)
Cardiovascular fitness is associated with different patterns of cortical activation. Source: S. J. Colcombe et al., "Cardiovascular fitness, cortical plasticity, and aging," *Proceedings of the National Academy of Sciences* 10: 3316–3321 (© 2004 National Academy of Sciences). Reprinted with permission.

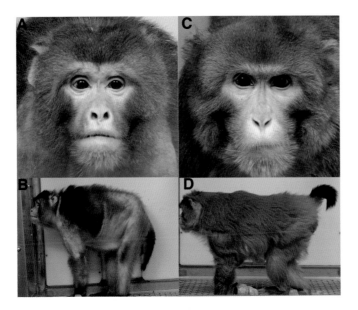

Plate 7 (figure 7.1)
Rhesus monkeys aged 27.6 years (the average life span). Panels A and B show a monkey maintained on a standard diet; panels C and D show a monkey maintained on a DR diet. Source: R. J. Colman et al., "Caloric restriction delays disease onset and mortality in rhesus monkeys," *Science* 325, no. 5937: 201–204. Reprinted with permission of American Association for the Advancement of Science.

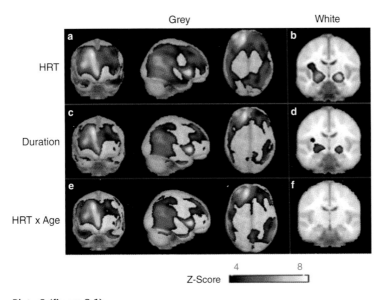

Plate 8 (figure 8.1)
Images showing lessened age-related gray-matter shrinkage in older women undergoing hormone replacement. Source: K. I. Erickson et al., "Selective sparing of brain tissue in postmenopausal women receiving hormone replacement therapy," *Neurobiology of Aging* 26: 1205–1213. Reprinted with permission of Elsevier (© 2005).

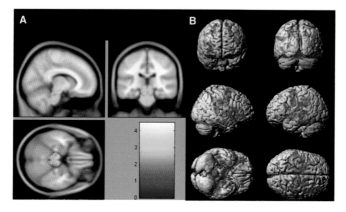

Plate 9 (figure 8.2)

Images showing effects (color-coded *t* statistics) of modafinil on BOLD signal change related to task-switching performance. (A) Increased locus coeruleus activation as a task-related drug effect. (B) Clusters of cortical activation as a task-related drug effect. Source: M. J. Minzenberg et al., "Modafinil shifts human locus coeruleus to low-tonic, high-phasic activity during functional MRI," *Science* 322: 1700–1702. Reprinted with permission of American Association for the Advancement of Science.

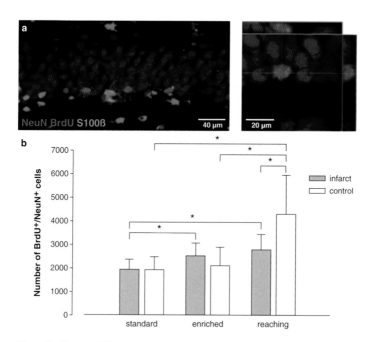

Plate 10 (figure 9.2)

Neurogenesis in the dentate gyrus after an experimental brain lesion (labeled "infarct") or a sham operation (labeled "control") as a function of housing condition (standard, enriched) and skilled reaching training in rats. Source: F. Wurm et al., "Effects of skilled forelimb training on hippocampal neurogenesis and spatial learning after focal cortical infarcts in the adult rat brain," *Stroke* 38: 2833–2840. Reprinted with permission of Wolters Kluwer Health.

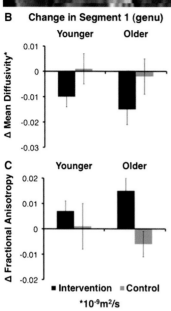

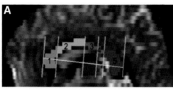

B Change in Segment 1 (genu)

C

Plate 11 (figure 9.5)
Fractional anistrophy in corpus callosum as a function of cognitive training in young and older people. (A) Segmentation of the corpus callosum into regions probably interconnecting the following cortical regions: prefrontal (segment 1: genu), premotor and supplementary motor (segment 2), motor (segment 3), sensory (segment 4), and temporal, parietal, and occipital (segment 5: splenium). (B) Change in mean diffusivity from pre-test to post-test. (C) Fractional anistrophy in segment 1 (genu) of the corpus callosum as a function of age and cognitive intervention group. Source: M. Lövdén et al., "Experience-dependent plasticity of white-matter microstructure extends into old age," *Neuropsychologia* 48: 3878–3883. Reprinted with permission of Elsevier (© 2010).

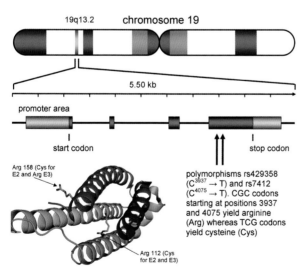

3D-rendering of the apolipoprotein E4 isoform

Plate 12 (figure 12.1)
The APOE gene with the rs 7412 and 429358 SNPs and the ε4 allele. Source: T. Espeseth et al., "Accelerated age-related cortical thinning in healthy carriers of apolipoprotein E epsilon 4," *Neurobiology of Aging* 29, no. 3: 329–340. Reprinted with permission from Elsevier (© 2008).

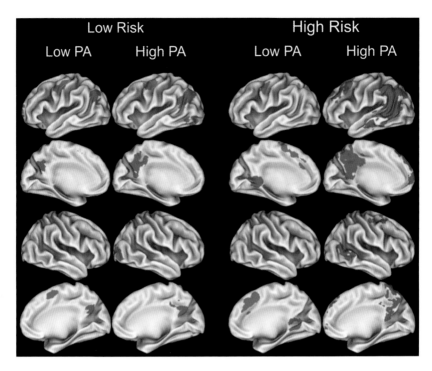

Plate 13 (figure 12.3)
Results of BOLD activation in brain regions during a memory task showing significant differences between conditions as a function of self-reported physical activity (PA) and APOE genotype (low risk = no ε4 alleles; high risk = at least one ε4 allele). Source: J. C. Smith et al., "Interactive effects of physical activity and APOE-ε4 on BOLD semantic memory activation in healthy elders," *NeuroImage* 54, no. 1: 635–644. Reprinted with permission from Elsevier (© 2011).

estrogens), or the same estrogen plus a synthetic progestogen (medroxy-progesterone acetate). The estrogen-only group included only women who had hysterectomies, as they didn't need to take progesterone to reduce the risk of uterine cancer. The WHI trial was stopped in 2002 because of evidence that administration of estrogen plus progestogen was associated with increased risk of heart disease, stroke, and breast cancer. Administration of estrogen only was stopped in 2004 because of an increased risk of stroke and no benefit on cardiovascular measures, although a recent analysis of the data suggests a reduction in the risk of breast cancer.[38] There was also a very slight but statistically significant difference in scores on the Modified Mini-Mental State Exam, a 100-point test of global cognitive function.[39] Women on estrogen plus progestogen scored 94.63 (standard deviation 4.80); those on placebo scored 94.73 (standard deviation 4.48).[40] This is, of course, a tiny difference in women who are scoring near the top of the scale on average. However, the large size of the sample makes it statistically significant, although probably not meaningful. An ancillary study to WHIMS, the WHI Study of Cognitive Aging (WHISCA), looked at longitudinal cognitive change as a function of the same three treatment conditions but used an extensive cognitive battery. The results observed over 3 years were mixed: no change in most measures, including a version of the Mini-Mental State Exam (a screen for dementia); a decrease in verbal memory; an increase in figural memory. However, about half of the women in the estrogen-plus-progestogen group had voluntarily stopped the treatment, and that could have skewed the sample.[41]

Why have smaller, randomized trials observed benefits of estrogen treatment, whereas the WHI study observed either cognitive costs (WHIMS) or mixed effects on cognition (WHISCA)? Several authors have addressed this important question.[42]

First, the route of administration differed between WHI and smaller studies. The smaller studies reporting benefits of estrogen generally used a transdermal patch or intramuscular injection, whereas WHIMS and WHISCA used oral administration. Oral administration of estrogen is known to stimulate induction of clotting, which is consistent with the deep vein thrombosis, heart attacks, and strokes associated with the WHIMS study.[43] There is also some evidence that oral administration results in less estrogen being available to the brain.[44]

Second, the specific hormones differed. The smaller trials used estradiol; WHIMS and WHISCA used conjugated equine estrogens (CEE). Estradiol is the form of estrogen produced by human ovaries; conjugated equine estrogens (from mare urine) are metabolized to estrone sulfate, a biologically weaker estrogen.[45] The above-reviewed benefits of estradiol on dendritic spines have been demonstrated with estradiol, but not with estrone. There is also concern about the progestogen used in WHIMS—medroxyprogesterone acetate. This synthetic progestogen has been shown in vitro to antagonize estradiol-induced neuronal protection and to promote neuron death.[46]

Third, the cognitive assessments differed. Estrogen may be linked specifically with verbal memory,[47] which was assessed in the smaller trials but not in WHIMS. A number of randomized trials of estrogen have reported effects on verbal memory in women.[48] However, verbal memory was not improved in 81-year-old women treated with CEE for 9 months.[49] In contrast to studies that assessed verbal memory, WHIMS assessed cognition with a version of the Mini-Mental State Exam,[50] an omnibus screen for dementia that isn't known for its sensitivity and doesn't assess verbal memory.

Fourth, the age at administration differed. The greatest benefits of estrogen on cognition may be seen only if begun during a narrow window of time around menopause. Susan Resnick and colleagues found that women who continuously used hormone replacement over 6 years around the time of menopause showed no decline in cognitive functioning during that time. In contrast, women who did not take hormone replacement declined over the 6 years. Women who began hormone replacement some years after menopause neither declined nor improved to the level of the women who had begun HRT early in menopause.[51] This and similar work has led to the "critical period" hypothesis,[52] according to which estrogen treatment must be initiated close to the time of menopause in order to be effective.

Perhaps the strongest evidence for the critical-period hypothesis comes from studies of estrogen administration around the time of menopause. Peter Rapp and colleagues tested female rhesus monkeys on a delayed match-to-sample task before and after ovariectomy. In that task, monkeys must remember a stimulus over a short time period (delay). Performance on the task declined after the ovariectomy, but that was reversed by cyclic administration of estrogen (mimicking changes during the menstrual cycle) 30 weeks after surgery. Such reversal was not seen in animals that had been ovariectomized 10 years before the estrogen treatment.[53] In a human study, "early initiators" (women who began estrogen treatment before age 56 or within 5 years of a hysterectomy) performed at a higher cognitive level than "late initiators" (women who started estrogen after age 56). The "never users" also performed better than the "late initiators."[54] Also consistent with the critical-period hypothesis, women who initiated estrogen use around the time of menopause were found to have larger hippocampi than women who initiated estrogen use later.[55]

Does the critical-period hypothesis explain why smaller randomized trials have found benefits of estrogen replacement whereas the larger WHI trials have not? Several recent reviews have concluded that estrogen treatment around the time of menopause confers cognitive benefits with a low health risk.[56] These reviewers agree that estrogen use should not be initiated in women over age 60 or 65. However, there is also some evidence that even short-term estrogen use around the time of menopause may confer lasting benefit. When women who had participated in randomized controlled trials of estradiol and a progestogen were assessed 5–15 years after the initial study, those who had been randomly assigned to 2–3 years of estradiol and progestogen and had then ceased treatment had a 64 percent reduced risk of later cognitive impairment. A similar 66 percent decreased risk of Alzheimer's disease was seen in women who continued the hormone treatment.[57] This suggests that even short-term use of estrogen can have long-term benefits regarding risk of Alzheimer's disease.

Studies measuring brain volume bear on the "critical period" hypothesis as well. Kirk Erickson and colleagues found less age-related shrinkage in prefrontal, parietal, and temporal gray matter and in medial temporal lobe white matter in post-menopausal users of estrogen only (no progestogen) than in "never-users" (overall mean age 68).[58] In a subsequent study, hormone use for more than 16 years was associated with greater shrinkage and worse performance on a complex task of attention and memory (the Wisconsin Card Sort Task). However, in that study about half of the participants used both estrogen and a progestogen, which made comparison with the previous study somewhat difficult. Moreover, other factors can influence regional brain shrinkage. In a cross-sectional study, women who were more physically fit showed better cognitive performance and less shrinkage regardless of hormone use.[59] This is consistent with evidence that effects of estrogen interact with effects of exercise in rodents.[60]

Meta-analyses have been conducted to consider whether hormone replacement in women has a beneficial effect on cognition. Eva Hogervorst and colleagues[61] reviewed both epidemiological studies and randomized trials and concluded that administration of estradiol had a positive effect on paired associate learning, abstract reasoning, and a test of speed and accuracy. In contrast, equine estrogens (e.g., Premarin) did not have a positive effect on cognition. In a meta-analysis, Ronald Zec and Mehul Trivedi calculated effect sizes (the size of a statistically significant effect of estrogen on cognition) and found effects in the medium range for benefits of estrogen on memory in randomized clinical trials.[62] In a more recent meta-analysis, Anne Lethaby and colleagues asked whether randomized clinical trials of estrogen and progesterone treatment prevented cognitive decline in older women and concluded that such treatment did not.[63] However, that meta-analysis did not have sufficient data to consider the age of participants at time of initiation of hormone use and so did not consider the critical-period hypothesis.

In the next few years, results should be forthcoming from three large studies that were initiated to correct the perceived flaws of the WHI study. PREPARE (Preventing Postmenopausal Memory Loss and Alzheimer's with Replacement Estrogens) is a placebo-controlled double-blind trial aimed at determining whether estrogen alone or in combination with progesterone can delay memory loss or Alzheimer's disease in women over age 65 with a family history of the disease. That study is still following participants. Kronos KEEPS[64] is a 5-year randomized placebo-controlled double-blind trial with annual assessments, including a comprehensive cognitive test battery.[65] Women aged 42–58 were randomly assigned to oral CEE or to transdermal estradiol. Both groups are administered a progesterone for 12 days of the month. Therefore, this study compares cognitive effects of equine estrogens (including estrone) with effects of estradiol shown to promote dendritic sprouting. The Early Versus Late Intervention Trial with Estradiol (ELITE), which began in 2004 and will be completed in 2013,[66] is a clinical trial that enrolled 643 women grouped on the basis of years since menopause—either less than 6 years or greater than 10 years. They receive either 17β estradiol or placebo double-blind and are assessed for cognitive change and for signs of early atherogenesis (e.g., thickness of blood vessel lining, coronary artery calcium, coronary lesions). Women with a uterus will be given progesterone the last 10 days of each month. These three studies will provide evidence on early and late administration of estrogen and will avoid the synthetic progesterone that may have been a problem in the WHIMS studies.

What is the consensus on use of estrogen in menopause?

At present there is no consensus. The large, randomized WHI study and the ancillary studies found few benefits of giving estrogen to post-menopausal women and real costs to their health. The costs were sufficiently severe that the WHI study was stopped. The widely reported results of that study led many post-menopausal women to suddenly stop taking estrogen. However, the flaws of the WHI study are hard to ignore. That equine estrogens were used with a synthetic progesterone and were administered orally to women who were 15 or more years past menopause makes the results hard to relate to younger women taking estradiol transdermally around the time of menopause. The WHI study cannot be discounted, but its design flaws mean that there will be little consensus on advice for women experiencing severe symptoms of menopause until the newer clinical trials (KEEPS, ELITE) begin to report results.

The evidence that estrogen has positive effects on numbers of dendritic spines and change in spines density over time is consistent with our brain plasticity/cognitive plasticity hypothesis (chapter 4), which argues that the aging brain can be reorganized by cognitive demand in the presence of factors that support plasticity mechanisms. Exogenous estrogen after menopause influences neuronal plasticity and thus can play a role in compensating for the documented reduction in number of dendritic spines in aged humans.

8.2 Cognition-Enhancing Drugs

As was detailed above, because of the WHI findings physicians currently do not routinely recommend use of estrogen for menopausal women. Also, healthy men would not wish to take estrogen. What about other putative cognition-enhancing drugs? It is well accepted that such drugs should be taken by people with Attention Deficit Hyperactivity Disorder, Mild Cognitive Impairment, Alzheimer's disease, and other cognitive disorders. The use of drugs to minimize age-related cognitive decline in otherwise healthy individuals is more controversial. The use of prescription drugs for cognitive enhancement in healthy aging is not standard practice and raises ethical issues. The question of whether healthy individuals, regardless of age, should use such drugs to improve cognitive performance has been raised in recent years, with arguments pro and con. A proposal in favor of such a practice was recently advanced in a high-profile exchange in the journal *Nature*.[67] This issue has obvious relevance to cognitive aging. There is a small literature on studies in healthy children and adults, but there have been few studies in healthy older adults.

Psychostimulants

Various stimulants have been used since antiquity as cognitive enhancers. Amphetamines occur naturally in several plants, including khat and ephedra sinica, which are widely chewed for their stimulant effects in many parts of Asia and Africa. Amphetamines appear to exert their effects by increasing release of norepinephrine and dopamine. Amphetamines have been controlled substances since 1970 in the United States. In recent years, college students have increasingly turned to Adderal (an amphetamine widely prescribed for ADHD) as a cognitive enhancer.[68] Amphetamines also are among the most commonly abused drugs. Long-distance truck drivers value them for their putative stimulant and fatigue-suppressing properties.

When investigated experimentally, amphetamines do not appear to exert broad beneficial effects on cognition. For example, amphetamines were not found to alter performance of healthy young people on a task of executive function (the Wisconsin Card Sort Task),[69] but they did improve performance on an *n*-back working-memory task, most strongly in people who showed a relatively low working-memory capacity at baseline.[70] Studies using another stimulant used to treat ADHD, methylphenidate, have found somewhat less consistent results, with benefits for performance in young people on spatial working memory (a self-ordered pointing task)[71] and a spatial span task,[72], but not in older people.[73]

Thus, there is only conflicting evidence as to whether amphetamines enhance cognition. In addition, few studies have been conducted with older groups. Even if evidence of amphetamine efficacy in the old were to be found, however, an important drawback to possible use of amphetamines to enhance cognition in older people is their potential to be addictive.

Some attempts to pharmacologically enhance working memory have targeted the dopamine system specifically, on the basis of evidence that working memory is dopaminergically mediated in humans.[74] One recent human study obtained fairly direct evidence of dopamine release during a verbal working-memory task. A PET study revealed reduced D2 receptor availability (based on a dopamine D2 receptor tracer) during task performance. D2 receptor availability was reduced during a working-memory task, but not during a sustained attention task, in ventrolateral frontal cortex bilaterally and in left-hemisphere medial temporal structures. During the working-memory task, greater dopamine release in ventrolateral frontal cortex and in ventral anterior cingulate was associated with better working-memory performance.[75] These results suggest that drugs that increase dopamine levels improve working-memory performance. In a review of such studies, Deanna Barch tentatively concluded that amphetamine and methylphenidate—both non-selective dopamine agonists— have greater effects on people with poorer working-memory performance.[76] Bromocriptine, a selective D2 agonist, appears to have stronger effects at lower doses, consistent with the known inverted "U-shaped" function relating dopamine activity and working memory.[77] Amy Arnsten and colleagues have reported that noradrenergic alpha-2 agonists help to improve working memory in aged non-human primates.[78] However, effects on humans have been much less consistent.

The cholinergic system presents another target for cognition-enhancing drugs. A number of groups have reported that young people's performance on working-memory tasks is improved by drugs that enhance cholinergic function and is impaired by drugs that block cholinergic function.[79] Maura Furey and colleagues have found beneficial effects of the cholinergic agonist physostigmine in young people, but these effects were exerted mainly on attention and not on working memory per se.[80] Of course, cholinesterase inhibitors are routinely prescribed to reduce symptoms of Alzheimer's disease, but these drugs have been little studied in healthy aging. In one study, chronic use of cholinesterase inhibitors improved performance on a flight simulator task in young and middle-aged pilots.[81] In another study, cholinesterase inhibitors were found to heighten the benefits of training in middle-aged pilots (mean age 52).[82]

Another psychostimulant, modafinil (a norepinephrine and dopamine reuptake blocker), appears to enhance cognition and neocortical activation. Danielle Turner and colleagues in Barbara Sahakian's group found that the drug improved neuropsychological task performance in healthy young adults and in a group with ADHD, perhaps by decreasing impulsive responding.[83] In recent work with schizophrenic patients known to have working-memory deficits, modafinil administration before a working-memory task produced greater activation in the anterior cingulate cortex,[84] and the activation was correlated with better cognitive performance.[85] A separate research group, Michael Minzenberg and colleagues, also found that modafinil increased activation in the locus coeruleus and in prefrontal cortex during a task-switching task. The drug was also associated with increased functional connectivity between the two regions.[86] This drug might be particularly suitable for older people, in light of neuroimaging evidence that older individuals come to rely increasingly on top-down processes dependent on prefrontal cortex.[87] (See chapter 3 above.)

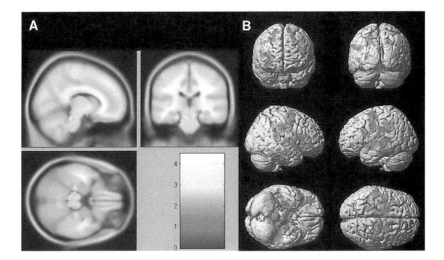

Figure 8.2
Images showing effects (color-coded *t* statistics) of modafinil on BOLD signal change related to task-switching performance. (A) Increased locus coeruleus activation as a task-related drug effect. (B) Clusters of cortical activation as a task-related drug effect. Source: M. J. Minzenberg et al., "Modafinil shifts human locus coeruleus to low-tonic, high-phasic activity during functional MRI," *Science* 322: 1700–1702. Reprinted with permission of American Association for the Advancement of Science. This figure appears in color elsewhere in the book.

A recent review of selective serotonin reuptake uptake inhibitors (SSRIs), which are used as "cognitive enhancers" in treatment of depression, concluded that the few studies that have been carried out have not been able to differentiate the effects on cognition from the effects on mood.[88]

Overall, although some psychostimulants have been found to enhance cognition in young people, little work has been done with healthy older populations.

Caffeine

Caffeine is a widely available legal stimulant that also has been shown to have positive effects on cognition—on vigilance, attention, mood, and arousal.[89] At voluntary dose levels, caffeine produces non-selective antagonism of A_1 and A_{2A} adenosine receptors[90] and increased cortical activity.[91] Very large doses (60 mg/kg per day) administered over 7 days to mice promoted neurogenesis but not long-term survival of the neurons. In contrast, physiologically relevant doses depressed neurogenesis.[92]

Caffeine may influence an important pathology of Alzheimer's disease. Amyloid beta ($A\beta$) is a peptide created when amyloid precursor protein is cleaved by the enzymes beta- and gamma-secretase. $A\beta$ normally regulates synaptic activity, but forms of it are also associated with the amyloid plaques of Alzheimer's disease. In a mouse model of Alzheimer's disease, long-term administration of caffeine protected against cognitive impairment and reduced brain $A\beta$ levels.[93]

In a prospective population study, women (but not men) consuming more than three cups of caffeinated beverages per day declined less in verbal retrieval and visuospatial memory over 4 years than women consuming one cup per day or less.[94] This finding has been confirmed in a different population.[95] However, a recent review has concluded that caffeine is not a "pure" cognitive enhancer, as it also affects general arousal and mood.[96]

Summary

There is little doubt that estrogen protects both the brain and cognitive functioning not only in younger female animals and in women undergoing surgical menopause but also in middle-aged women around the time of natural menopause. However, the WHI study has revealed the health risks of initiating estrogen and progesterone use in women many years after menopause. The disagreement between results of the WHI and results of studies administering estrogen around the time of menopause has raised many questions and has stimulated several new randomized controlled trials of estrogen use. We should await results from these newer, better-designed studies before drawing conclusions about the benefits and costs of estrogen in women.

The effects of other cognition-enhancing drugs on older people have been studied surprising little. In light of the benefits of cholinergic agonists for cognition in young people, it is curious that cholinesterase inhibitors have been so little studied in older people. Perhaps the most promising evidence comes from caffeine, which is already in wide use. There is no compelling evidence that pharmacological agents are useful for ameliorating cognitive aging.

9 Learning, Cognitive Training, and Cognitive Stimulation

Wisdom is the principal thing; therefore get wisdom; and with all thy getting get understanding.
—Proverbs iv:7

Continuous effort—not strength or intelligence—is the key to unlocking our potential.
—Winston Churchill

Retirement from work is an important milestone in the life of an older individual. Retirement brings with it many benefits, among them lowered daily stress and more time to enjoy avocations, to spend with family members, and to learn new skills. However, retirement also tends to reduce opportunities for cognitive stimulation. To the extent that daily cognitive challenges are important in maintaining a high level of cognitive functioning, retirement can be a double-edged sword.

A recent paper by the economists Susann Rohwedder and Robert Willis was titled "Mental retirement," and that title is a good summary of the evidence they report.[1] Rohwedder and Willis examined data from the US government's Health and Retirement Study, which contacts more than 22,000 older people every 2 years, asks about their lives and activities, and gives them a memory test. The memory test presents ten nouns and asks the subject to recall them immediately. Then, after 10 minutes of conversation, the subject is asked to recall the words again. The maximum score is 20 (10 for the immediate recall and 10 for the delayed recall). In 2004, data on retirement ages and on memory functioning from the United States, England, and eleven European countries were obtained from several different public sources and analyzed together. As figure 9.1

shows, the average memory score is highest in the United States, which has a high retirement age (with 65–70 percent of men still working in their early sixties), and lowest in countries where people retire at a much earlier age. Of course, these data are correlational, and people may retire earlier because they are declining cognitively. However, this report also presents data indicating that people are retiring in concert with the age at which they are eligible for a public pension in their country. The authors interpret these data by speculating that people who retire early decline cognitively because cognitive stimulation is reduced in retirement.[2] This interpretation is not new to cognitive psychologists. The notion of "use it or lose it" has been proposed in the literature on cognitive aging by several authors.[3]

In this chapter, we review the evidence from the growing literature on cognitive training. We consider observational studies, but we focus on

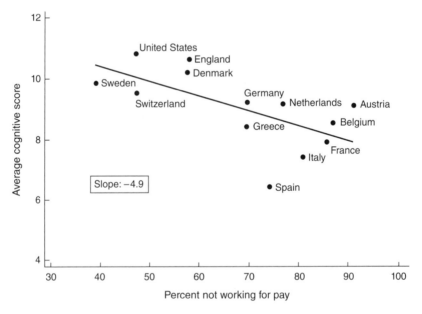

Figure 9.1
Memory performance of 60–64-year-old men and women as a function of percentage of individuals in that age category not working for pay in each of the indicated countries. Source: Rohwedder and Willis, "Mental retirement," *Journal of Economic Perspectives* 24, no. 1: 119–138. Reprinted with permission of the authors and the journal.

studies that use random assignment of people to periods of cognitive training. There are three questions to be addressed. First, do the brain integrity
and the cognitive integrity of older animals benefit from cognitive training? Second, do the brain integrity and the cognitive integrity of older
people benefit from cognitive training? Third, do the benefits of cognitive
training generalize to untrained tasks, and especially to real-world
functioning?

9.1   Cognitive Stimulation in Aged Animals

Do older animals benefit from cognitive stimulation and/or training? As
was reviewed in chapter 2, we know that new neurons are formed daily in
certain brain regions of mammals. To the extent that forming new neurons
and integrating them into the brain's circuitry is beneficial for brain function, factors that promote neurogenesis or promote the survival of the new
neurons would be important. To the extent that cognitive stimulation and
training promote such plasticity mechanisms, that benefit is consistent
with our brain plasticity/cognitive plasticity hypothesis of cognitive aging,
set forth in chapter 4.

Cognitive stimulation improves cognitive performance and neuronal plasticity

The adult brain's responses to cognitive stimulation are a changed cortical
structure and improved cognitive performance. There is a large literature
on brain and cognitive change after environmental enrichment in animals,
some of which we described in chapter 3. Even in old animals, enriched
experiences produce benefits for cognitive performance (e.g., better maze
performance) and changes in cortical structure.[4] There are both macroscopic and microscopic cortical changes related to such training. The
macroscopic changes include increased cortical thickness; the microscopic
changes include increased dendritic extent and synaptic size.[5] It is integral
to the brain plasticity/cognitive plasticity hypothesis of cognitive aging
that benefits of environmental enrichment can occur even in old animals.[6]
Environmental enrichment also increases the birth of new neurons in the
dentate gyrus,[7] and a number of studies have reported a positive correlation
between neurogenesis and learning.[8] In one study, two manipulations
were compared for their effects on the dentate gyrus, a central player in

cognitive and brain aging. Effects of training on skilled reaching were compared with effects of environmental enrichment in rats. Both manipulations increased neurogenesis in the dentate gyrus and improved performance in the Morris water maze. This was seen not only in healthy rats but also in rats that had previously experienced experimentally induced strokes (blood supply to the brain interrupted briefly, causing a focal lesion).[9] Enhanced neurogenesis in the adult dentate gyrus has been associated with improved spatial memory performance.[10]

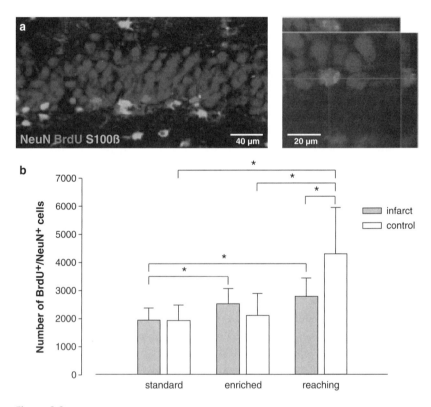

Figure 9.2
Neurogenesis in the dentate gyrus after an experimental brain lesion (labeled "infarct") or a sham operation (labeled "control") as a function of housing condition (standard, enriched) and skilled reaching training in rats. Source: F. Wurm et al., "Effects of skilled forelimb training on hippocampal neurogenesis and spatial learning after focal cortical infarcts in the adult rat brain," *Stroke* 38: 2833–2840. Reprinted with permission of Wolters Kluwer Health. This figure appears in color elsewhere in the book.

Learning can save new neurons from programmed cell death

Although there is evidence that new neurons in adults can become synaptically integrated into the existing neural architecture,[11] many newborn neurons in adults do not survive long.[12] Learning can rescue new neurons from programmed cell death.[13] Programmed cell death (or apoptosis) refers to events in a cell that lead to its death but can have benefits for the organism over the long term. Long-term potentiation has a role in the survival of newborn neurons. LTP is the strengthening of synapses after high-frequency stimulation; as has been mentioned many times, it is thought to be an important neuronal mechanism underlying memory formation. LTP appears to be supported by neurogenesis.[14] Also, there is a direct enhancing effect of LTP on the survival rate of newly formed neurons in the dentate gyrus.[15] Survival of dentate gyrus neurons increased with hippocampal-dependent learning such as that required by the Morris water maze, whereas tasks that did not rely on hippocampal activation had no effect on the survival of new neurons.[16] Overall, the weight of the evidence indicates that learning both promotes neurogenesis and enhances survival of the newborn neurons.

The growing evidence of a relation between neurogenesis and learning has led Gerd Kempermann, a pioneer in the field of neurogenesis, to hypothesize that adult neurogenesis preserves adaptability across the life span. Kempermann argues that ongoing neurogenesis provides a "neurogenic reserve"—a continuous supply of functional units that support new learning by altering and expanding upon the existing architecture of the brain. In this way, neurogenesis sustains functional plasticity beyond youth and into maturity.[17] This theory is consistent with the finding that up to 20 percent of new granule cells remain undifferentiated.[18] Naomi Goodrich-Hunsaker and colleagues[19] and Alexander Garthe and colleagues[20] have argued, from a number of carefully designed animal studies, that adult neurogenesis in the dentate gyrus is critical for integrating novel information when the context of a task has changed.[21] In support of that view, Susanne Diekelmann and Jan Born offers good evidence for a role of the hippocampus in re-organizing and integrating new memories into the network of pre-existing long-term memories.[22] These strands of evidence support Kempermann's proposal that a reserve of new granule cells in the dentate gyrus may be important when an animal must integrate new experiences into existing knowledge.

Neurogenesis and cognitive integrity

One implication of Kempermann's notion that having a pool of undifferentiated neurons in reserve allows an animal to adapt to novel, unanticipated challenges is that presumably a brain experiencing no novel challenges would not need to adapt and therefore would not need to draw on a reserve pool of undifferentiated neurons. Insofar as newborn neurons appear to undergo programmed cell death unless new learning occurs,[23] it could be argued that novel experiences are needed if newborn neurons are to survive to be integrated into the circuitry.[24] Perhaps this argument can be applied to humans, in whom the same regions show ongoing neurogenesis even late in life.[25] We can speculate that older people need novel experiences in order to optimally renew hippocampal circuitry.

Do the brain integrity and the cognitive integrity of older animals benefit from cognitive stimulation and/or training?

Yes, emphatically. Numerous studies have shown that environmental enrichment and motor training improve cognitive performance and promote neurogenesis. Moreover, exposure to novel experiences has been shown to enhance the survival of newborn neurons and may stimulate processes that underlie successful aging.

9.2   Cognitive Stimulation in Older Humans

Are there measurable benefits from challenging the aging human mind and brain with novel demands? Carmi Schooler and colleagues[26] used real-world experiences to evaluate whether self-directed cognitive activity influences cognitive performance late in life. They followed a large group of individuals for 20 years, from mid-life to old age, by assessing cognitive, occupational, and leisure activities. They concluded from their large data set that carrying out self-directed, complex tasks of "substantive complexity," whether for work or leisure, is important for retention of cognitive abilities, even in old age. Observations such as these have led to notions that an "engaged lifestyle" and "cognitive stimulation" are important for maintaining a high level of cognitive functioning late in life. This is referred to as the "use it or lose it" hypothesis.

The acknowledged problem with using observational data to examine this question is one of self-selection: people who choose to be mentally active late in life may be healthier and/or more intelligent than those who do not so choose. Timothy Salthouse carefully reviewed the existing literature on self-reporting of habitual cognitive activities and concluded that there was little evidence in favor of "use it or lose it."[27] Specifically, there was no evidence that the *rate* of age-related decline was different in people who apparently were more mentally active. However, a more rigorous way to test this hypothesis is to randomly assign older people to participate in cognitive experiences varying in complexity and then assess the effects of such experiences on subsequent cognitive performance.

There has long been interest in using cognitive training to help older people maintain functioning late in life. The literature on cognitive training is based largely on the rather simple "use it or lose it" and "disuse" hypotheses mentioned above, which essentially view the brain as a muscle that atrophies without constant, demanding use. If the hypothesis is correct, older people assigned to cognitive training should show better cognitive performance than comparable people who remain untrained over the same interval. And that is generally the finding. A large number of studies show that cognitive training works well, even in older people. Moreover, as we review, cognitive training appears to change the human brain. It has been harder to show that the benefits of cognitive training transfer to untrained tasks or to real-world functioning. We will discuss these matters in the following section, after we review what is known about the effects of training per se.

Because the literature on cognitive training was recently reviewed in a comprehensive way,[28] we will focus mainly on newer work here.

It is well established that older persons can be very successfully trained to achieve substantial improvements in task performance, with the improvements lasting for years.[29] Memory training has been consistently found to be effective in improving memory performance in a task-specific manner. Two meta-analyses have confirmed this.[30] Paul Verhaeghen and colleagues[31] concluded that four variables were important in enhancing performance gains in memory training. Greater gains were seen in younger participants, with shorter sessions, with group sessions rather than individual training, and with "pre-training" involving imagery and motivation. Notably, the type of training employed does not appear to be important.[32] In older people, nearly all studies have found that cognitive training benefits the specific cognitive functions that are trained.[33]

There is some evidence that training not only improves performance but also appears to make the minds and brains of older individuals function differently. We have speculated elsewhere that adopting a more top-down, controlled cognitive strategy dependent on the prefrontal cortex may be the optimal developmental course for efficient cognitive functioning in adulthood.[34] There is evidence for that view. A number of neuroimaging studies have shown that task-related frontal blood-flow activation is unilateral in young adults but bilateral in older adults. Nevertheless, older people do show a unilateral pattern of brain activation under two specific conditions: if they are instructed to use "deep encoding"[36] and if they are "high-functioning" in cognitive performance.[37] Cheryl Grady and colleagues interpret this small literature as revealing greater activity in regions mediating executive activity in the old than in the young.[38] Kirk Erickson and colleagues in Arthur Kramer's lab found that the improvement of older people in dual-task performance after dual-task training was significantly correlated with increased activation in ventro-lateral prefrontal cortex. In contrast, the young showed decreased activation of ventro-lateral prefrontal cortex after training. However, after training, young and old did not differ in left-hemisphere prefrontal activation, although they still differed in right hemisphere activation.[39] This suggests that training not only improves performance but also appears to make older brains function similarly to younger brains in the use of prefrontal cortex.

Consistent with these views of a greater role for prefrontally mediated, controlled processing during successful aging, several groups have found better memory performance among the old to be associated with reduced volume of prefrontal temporal gyri.[40] This suggests that both prefrontal structural remodeling and greater use of the prefrontal cortex during task performance are associated with successful cognitive aging. Interestingly, recent evidence on recovery from depression shows a similar pattern of a shift to a more controlled cognitive approach, as reflected in increased prefrontal activation after either cognitive behavior therapy or antidepressant medication.[41] Also, there is evidence that benefits of training on tasks that require executive functioning and problem solving are more likely to transfer to untrained tasks.

Not only does cognitive training change cognitive performance, it also changes the brain' structure and physiology. Training of working memory changes the physiological response of the brain. Torkel Klingberg and colleagues trained young adults on a working-memory task for 5 weeks to improve their capacity for temporarily storing and keeping things in mind. In one study they found that the training was associated with blood flow activation in middle frontal gyrus and superior and inferior parietal cortices.[42] In another study they found that the same training was associated with changes in the density of cortical dopamine D1 receptors. Dopamine is a neurotransmitter that is important in working memory. Work in monkeys has implicated dopamine D1 receptors specifically in working-memory performance.[43] PET-scan studies of schizophrenics found that cortical D1 receptor availability predicted working-memory performance.[44] Also in humans, release of dopamine from cortex has been observed during the performance of working-memory tasks.[45] A recent study by Klingberg and colleagues linked D1 receptor binding in a PET scan to effects of working-memory training. Fourteen hours of training over 5 weeks was associated with changes in both prefrontal and parietal D1 binding potential. Greater decreases in D1 binding potential were associated with improvements in working memory. D2 binding potential was not affected.[46] This result suggests that training on working-memory tasks has a substantial effect on brain plasticity, specifically in dopamine neurotransmission, which has a known role in working memory.

Another brain change related to training involves white matter. Fast neuronal transmission along an axon depends in part on electrical insulation provided by a myelin layer around the axon. There is evidence that activity in a neuron can induce myelination of that neuron's axon, although the mechanisms are not known. Blockade of action potentials suppressed myelination while increased neuronal firing simulated myelination in vitro and in vivo.[47] These findings support the hypothesis that increased neural activity associated with training could increase myelination and therefore increase white-matter integrity. Recent research has confirmed that hypothesis. Studies by three independent research groups have found that motor and cognitive training alters white-matter integrity, measured in a diffusion tensor imagery (DTI) scan. If the myelin covering the axon is healthy (no areas of thinning, balloons, or blisters[48]), water molecules will move along the length of the axon faster than they will leak out of the axon. Such an axon would be considered anisotrophic—that is, resistant to water diffusing in all directions without impediment. A number of studies have shown that the white matter, which includes both myelinated and unmyelinated axons, begins to degrade in the fifties.[49] DTI scans show that after about age 50 there is increasing isotrophy (less myelin integrity) of the white matter.[50]

It has been reported that young people show rapid increases in white-matter integrity (increased fractional anisotrophy) related to cognitive and motor training. Young people trained on a difficult working-memory task for 25 minutes a day over 2 months showed evidence of increased white-matter integrity (increased fractional anisotrophy) over the course of the training. The increased fractional anisotrophy was seen in white matter underlying the intraparietal sulcus (figures 2.1, 9.3) and in the anterior body of the corpus callosum.[51]

In another study, young people were taught to juggle for 6 weeks, during which they learned a three-ball cascade[52] (figure 9.4). DTI scans were conducted before and after training and also 4 weeks after the new jugglers stopped practicing. The researchers observed increased white-matter integrity (fractional anisotrophy) in the white matter underlying intraparietal sulcus.[53] A third study also found structural changes related to the intraparietal sulcus, but in that study the change was seen in the sulcus itself. Gray matter was observed to increase bilaterally in the intra-parietal sulcus (figure 2.1) and in the right hippocampus after 3 months of intensive cognitive training in German medical students preparing for board exams.[54]

Finally, Ulman Lindenberger's research group has looked at white-matter integrity as a function of training in aged individuals. They administered 100 daily training sessions, each an hour long, to groups of 20 young people and 12 older (65–80 years) individuals. The training involved working memory, episodic memory, and perceptual speed. The researchers' analyses focused on the anterior portion of the corpus callosum, the thick band of myelinated fibers that interconnect prefrontal cortices. In the older group they observed training-related decreases in mean diffusivity and increases in the fractional anisotrophy measure of white-matter integrity. The effects of training were similar in young and old on the mean diffusivity measure but were greater in the older group for perceptual speed and episodic memory. This was the first study to show that older people undergo brain change as a function of cognitive training in which the benefits of training transfer to broader abilities. That the change was seen in white matter, notably in the anterior corpus callosum, indicates

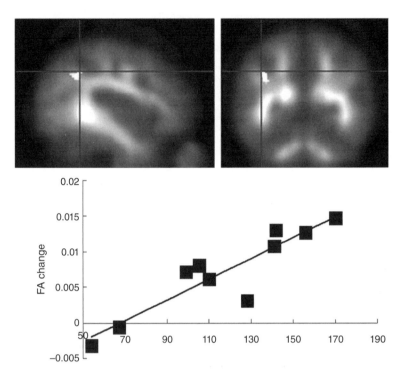

Figure 9.3
Working-memory training affects the integrity of white matter. Fractional anisotrophy in white matter adjacent to intraparietal cortex showed a significant correlation with amount of working-memory training, overlaid on the mean smoothed FA image. The effect of working-memory training at the significant peak voxel in the voxel cluster is highlighted. Source: H. Takeuchi et al., "Training of working memory impacts structural connectivity," *Journal of Neuroscience* 30: 3297–3303. Reprinted with permission of *Journal of Neuroscience*.

continued plasticity regarding axonal integrity in old age.[55] The evidence mentioned above relating action potentials to myelination[56] suggests a possible mechanism by which cognitive training would result in increased myelination.

It is notable that in four of these studies by three different research groups—one training working memory, two training juggling, and the fourth using intensive knowledge acquisition—change was associated with the intraparietal sulcus.[57] Consistent with this evidence of training-related structural change in parietal regions, better recall of items on the California

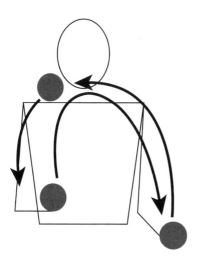

Figure 9.4
Three-ball cascade

Verbal Learning Test over months was associated with thicker regions in frontal and parieto-occipital cortices bilaterally and in left precuneus (see figure 2.1).[58]

Consistent with the brain plasticity/cognitive plasticity hypothesis (chapter 4), it is evident that cognitive training changes the brain—both gray matter and white matter. The precise mechanisms underlying the regional increase in gray matter are not clear at present, but one can speculate from animal studies on environmental enrichment that synaptogenesis and dendritic extension play a role. The mechanisms underlying increased white-matter integrity have been investigated in animals. Experimentally induced demyelination was associated with increased radial diffusivity (water molecules tending to move out of the axon perpendicular to its length), whereas remyelination was associated with decreased radial diffusivity.[59] Relating radial and axial diffusivity to myelin integrity may not be appropriate where there are crossing fibers,[60] but Martin Lövdén and colleagues observed increased radial diffusivity after cognitive training in the anterior corpus callosum, where crossing fibers aren't likely to be a problem.[61] Moreover, as increased neuronal firing has been found to stimulate myelination in vitro and in vivo,[62] there is support for the hypothesis that plasticity changes in myelination play a role in the benefits observed from cognitive training.

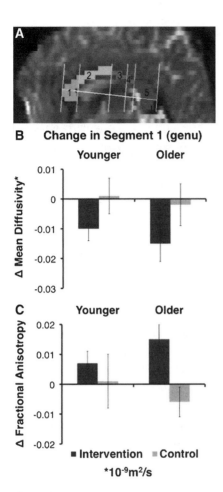

Figure 9.5
Fractional anistrophy in corpus callosum as a function of cognitive training in young and older people. (A) Segmentation of the corpus callosum into regions probably interconnecting the following cortical regions: prefrontal (segment 1: genu), premotor and supplementary motor (segment 2), motor (segment 3), sensory (segment 4), and temporal, parietal, and occipital (segment 5: splenium). (B) Change in mean diffusivity from pre-test to post-test. (C) Fractional anistrophy in segment 1 (genu) of the corpus callosum as a function of age and cognitive intervention group. Source: M. Lövdén et al., "Experience-dependent plasticity of white-matter microstructure extends into old age," *Neuropsychologia* 48: 3878–3883. Reprinted with permission of Elsevier (© 2010). This figure appears in color elsewhere in the book.

Are there measurable benefits from challenging the aging mind and brain with novel demands?

A very large and consistent literature shows that people— even older people—benefit from training. An emerging literature shows that people's brains are altered by both cognitive and motor training, and that the intraparietal sulcus (both gray matter and white matter) is particularly affected. More important, however, is the question of whether benefits of cognitive training transfer to untrained tasks and abilities.

9.3   Transfer of Cognitive Training to Untrained Abilities

The first studies conducted on transfer of training yielded largely negative results. Many studies showed either no transfer[63] or only weak transfer of training to untrained tasks. However, one very large, randomized trial has shown transfer not only to other tasks but also to everyday functioning. Broad transfer was the goal of the ACTIVE (Advanced Cognitive Training for Independent and Vital Elderly) study, a multicenter, randomized controlled trial aimed at assessing the functional consequences of cognitive interventions in older people. Independently living individuals 65 and older were cognitively assessed at entry and annually thereafter. People were randomly assigned to three intervention conditions (Memory, Reasoning, and Speed of Processing) and one control condition. All training involved intensive one-on-one sessions 60–75 minutes long. The memory training involved teaching mnemonic strategies for remembering verbal information. The Reasoning training involved determining logical progressions in a series, such as timetables and schedules (e.g., "a, c, e, g, i, . . ."), with the participant required to supply the next in the series. During the training session, reasoning-type problems were presented, and the participants were trained on relevant strategies, were given feedback, and practiced. Speed of Processing training involved visual search tasks and divided attention tasks. After 2 years, cognitive performance was found to benefit from the training,[64] but there was no benefit on "instrumental activities of daily living" (IADL). Such activities (preparing meals, handling finances, health maintenance, telephone use, shopping, and other daily activities) were assessed by self-rating on a six-point scale of ability to care for oneself. Also assessed were abilities to solve everyday problems.[65] After 5 years of annual "booster" training, participants who received Reasoning training (though not those who received Memory training or Speed of Processing training) showed less functional decline in IADL, with a small effect size (0.29). Although training did not have a general effect on the tasks of everyday functioning (problem solving or speed of processing), additional booster training did lead to better "everyday speed of processing."[66] Thus, although behavior-based interventions may not show transfer to specific untrained tasks, reasoning training has the potential to improve everyday performance in older adults.

The ACTIVE study found evidence that points to the "holy grail" of cognitive training in aging: transfer to everyday functioning. What factors are important in inducing transfer from trained to untrained tasks?

Transfer of training is facilitated by a self-generated strategy

One important factor in whether transfer occurs may be the strategy employed in training. Fergus Craik[67] has long argued that older people typically fail to engage in "self-initiated" processing. David Bissig and Cindy Lustig found that those people who performed well on a word-learning task spent more time on encoding than on retrieval.[68] These tended to be younger participants, who spent twice as much time on encoding as did the older participants. Moreover, when queried after the experiment, people who had performed well reported that they had used strategies such as "Related words to myself, and sometimes to each other" or "Some words combined in sentences." Cindy Lustig and Kristin Flegal[69] explicitly manipulated training strategy by assigning older participants to use one of two explicit strategies: using each word in a sentence or thinking about the meaning of the word by using whatever strategies they preferred. Relative to a group (from a previous study) that was not so forced, both groups benefited from being forced to encode for 14 seconds. However, those allowed to choose their own strategy showed greater transfer to untrained tasks. This finding is consistent with an observation by Anna Derwinger and colleagues in Lars Bäckman's lab[70] that older adults who self-generate successful strategies on a memory task show more durable improvements and greater transfer to untrained tasks than older adults who are assigned a strategy.

Transfer of training is facilitated when trained and untrained tasks share processes

John Jonides has argued that transfer from a trained to an untrained task occurs only if the two tasks share processing components and activate overlapping brain regions.[71] This was tested by Erika Dahlin and colleagues in Bäckman's and Nyberg's research groups, who hypothesized that "updating" in working memory (the term applied to ability to repeatedly change in the information held in working memory) was an important skill that, once trained, would transfer to other tasks that require updating.[72] On the basis of previous work showing that both new learning and updating are dependent on the striatum,[73] Dahlin further hypothesized that the striatum would also be important in training updating.[74] The training task required memory for the last four letters in lists of letters. The lists varied unpredictably in length, so people did not know when the last letter was presented until after it appeared. The transfer task was an "*n*-back" task, requiring retention of the *n*th item (first, second, third, . . .) back in the list from the current item. Therefore, on each trial, the *n*th number has to be changed, or updated in working memory. For example, on a three-back trial in which numbers 5, 2, 6, 9, 3, 7 were presented, the 7 must be compared with the 6, the third number back from the end. Another transfer task used was the Stroop test, which requires people to name the color in which a color word is printed—e.g., "red" printed in green letters. People have trouble ignoring the word itself. In the example above, the strong tendency is to say "red" because reading is so automatic in skilled readers. It takes controlled processing to ignore the word and focus on the color of the letters, and that slows people down. Dahlin et al. predicted that skills learned in the letter list task—which requires updating—would be transferred to the *n*-back task (which also requires updating), but not to the Stroop test (which doesn't require updating). They assessed regional brain activation before and after 5 weeks of "updating" training in a working-memory task. Participants trained for 5 weeks on the letter list task and showed substantial transfer to the letter-memory task, but not to the Stroop test. Moreover, overlapping activation in the striatum was seen between letter memory and *n*-back tasks, but not between letter memory and the Stroop task. These findings support Jonides' hypothesis that transfer occurs best between tasks that share processes (in this case, the updating process) and share dependence on the same brain regions (in this case, the striatum).[75]

In subsequent work investigating whether old people would benefit from transfer of "updating training,"[76] Dahlin et al. randomly assigned 15 young adults and 13 older adults to a training condition or a control condition. Those assigned to training were taught the letter-memory task described above and a task in which they had to keep track of the members of categories. On each trial, they heard 15 words and had to "mentally place" the words into three, four, or five categories (animals, clothes, countries, relatives, sports, professions) shown in boxes at the bottom of the screen. At the end of the trial, they had to type the last word in each category into the box on the screen representing that category. Transfer tasks assessed perceptual speed, working memory, episodic memory, verbal fluency, and reasoning. Participants were trained for 5 weeks and were then tested for transfer. They were tested again for transfer after 18 months. Both young and old people benefited from the training and showed retention of the effects of training at the 18-month testing. The young did show transfer to an *n*-back task—which requires updating—but the old did not show transfer to that task. No transfer was seen on tasks of fluency or reasoning, consistent with the hypothesized importance of shared processing components.[77]

Kramer's group also followed Jonides' views by training and testing on tasks that make similar demands. In contrast to Dahlin et al., they used "dual tasks," in which participants are required to carry out two tasks at the same time. A consistent finding in such experiments is that performance is poorer under dual-task conditions than when each task is performed alone—this is called the "dual-task deficit." However, Kramer's group has found that training reduces the dual-task deficit in older people. They found that training on dual tasks transfers to other untrained dual tasks in older people. In one study, they used computer-based training on a dual task that combined monitoring with an alphabet-arithmetic task. Their older group benefited more from the training than their younger group and the benefit transferred to an untrained dual task. Benefits were retained for 2 months.[78] Even with a different dual task (combined auditory and visual discrimination task), they found their older groups showed greater benefits than their younger group from one day of dual-task training and showed transfer of that training to untrained dual tasks.[79] Kramer et al. have also found transfer of dual-task training to memory tasks.[80] Participants were trained—both separately and together—on a gauge monitoring task and an alphabet-arithmetic task (e.g., K − 3 = ?, where the answer is H). Participants were then transferred to a scheduling task and a running memory task. The transfer tasks were also performed both separately and together. The investigators found that the observed age-related differences on dual-task performance on the transfer tasks were reduced by the dual-task training.

Julia Karbach and Jutta Kray also found broad transfer of training between task switching and broader cognitive functioning.[81] They trained participants on task switching and single-task training and compared transfer to a broad range of abilities in children, young adults, and older adults (mean age 68). The Food task required a judgment of whether a picture showed a fruit or a vegetable; the Size task a required judgment of whether a picture was small or large. These tasks were presented as single tasks or under task-switching conditions (switch tasks after every second trial). Specifically, transfer was tested on fluid intelligence using several reasoning tasks. In letter reasoning, five letters were followed by a question mark (e.g., a c e g i ?) and the missing letter was to be supplied. Transfer was also tested on the Raven's Progressive Matrices, a well-known fluid-reasoning task that presents a set of patterns with the requirement to supply the one that completes the pattern (figure 9.6). There were two pre-test sessions, four training sessions, and two post-test sessions. The investigators found that task switching reduced interference effects on the Stroop Task and improved verbal and spatial working memory, and performance on the Raven's Progressive Matrices test. Single-task training did not alter

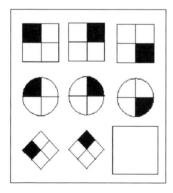

Figure 9.6
Sample problem from Raven's Progressive Matrices

Stroop or working-memory performance. They also found stronger transfer effects in the children and older adults, two groups who typically show task-switching deficits.

Therefore, it appears that training both on task switching and on dual tasks transfers to untrained tasks, including to tasks with no obvious relation to the training tasks. It is less clear that these findings support Jonides' view that transfer from a trained to an untrained task occurs only if the two tasks share processing components. The reasoning tasks Karbach and Kray used to test for transfer by do not obviously require task switching. Nor does it seem that memory and dual-task performance depend on the same skills.

Is transfer of training facilitated when the training is broad-based?

Another approach to cognitive training is to train problem-solving skills. Lesley Tranter and Wilma Koutstaal set out to test the "disuse" hypothesis of cognitive aging.[82] This idea—advanced by Carmi Schooler, Helen Christensen, and others—is that diminished use of problem-solving skills among older people leads to a corresponding reduction in problem-solving abilities. After an extensive cognitive assessment battery assessing fluid intelligence, crystallized intelligence, mental status, and visuospatial ability, Tranter and Koutstaal[83] randomly assigned 44 older participants (mean age 68) to 10–12 weeks of either a series of twelve "mentally stimulating activities" and five social group activities (Experimental group) or two social group activities (Control group). The Experimental group carried out two "home activity" tasks (such as simple mathematics puzzles, creative drawing and modeling, or word-logic puzzles) per week. There were also group sessions in the lab that involved making a marble run, a newspaper tower, and origami figures. Participants' fluid intelligence was then tested by means of an IQ test (Cattell's Culture Fair) and the Block Design subtest of the Wechsler Adult Intelligence Scale (WAIS). "Fluid abilities" (e.g., reasoning and problem solving) are generally thought to decline with age, in contrast with "crystallized abilities" (e.g., vocabulary). The Experimental group showed a greater increase in fluid intelligence than the Control group with an effect size in the small to medium range.

Kramer's group also looked at transfer of problem-solving skills trained in a strategy video game.[84] The training involved 24 hours of an off-the-shelf video game, Rise of Nations, that involves building an empire through military conquest, culture, and/or diplomacy and its attendant requirements of long-range planning, continuous assessment of resources, and tool development. The benefits of that training were found to be extended to various tasks requiring executive functioning—task switching, working memory, short-term memory, and reasoning with medium to large effect sizes. In contrast, benefits were not seen on working-memory capacity or visuospatial attention tasks.[85] This finding that benefits of training on a strategy video game can be transferred to executive functions can be considered in light of Jonides' argument that transfer should occur best when the training and the transfer task share processes. Presumably a strategy video game would place demands on executive processing.

Transfer of training in older people has also been seen with broad-based training. The well-known Berlin Aging Study, founded by Paul Baltes, included a large-scale study of cognitive training that assessed effects of transfer of training to untrained tasks. This is part of the study by Lövdén et al. on effects of training on brain structure described earlier in this chapter.[86] Florian Schmiedek and colleagues trained 101 young and 103 older (65–80) people on twelve tasks for 100 days.[87] The tasks were grouped into perceptual speed, episodic memory, and working memory. Participants were tested cognitively, trained for 100 days, then tested again on 27 tasks from the Berlin Intelligence Structure Test. These tasks were divided into "near transfer" and "far transfer" (meaning whether a task was more similar or less similar to the training task). For working-memory training, the "near transfer" tasks were two spatial working-memory tasks and the "far transfer" tasks were reading span (decide whether sentences were grammatically correct and also remember a letter), counting span (decide whether a group of shapes was odd or even in number and remember the number), and rotation span (decide whether a letter is reversed and remember the direction of a set of arrows presented sequentially). In addition, episodic memory was assessed with noun pairs. In the old people, there was evidence of positive transfer to working memory but not to fluid intelligence or episodic memory. In the young people, evidence of transfer to all three abilities was found.[88]

Elizabeth Stine-Morrow and colleagues also used unstructured and broad-based training.[89] They created an adult version ("Senior Odyssey") of Odyssey of the Mind, a school-based program developed for children and young adults.[90] Like Odyssey of the Mind, Senior Odyssey involves teams working together to solve long-term problems and presenting the solution in a competition. Participants were pre-tested and post-tested on a battery of neuropsychological assessments, then randomly assigned either to Senior Odyssey or to a wait list. The assessments included processing speed, visuospatial processing, working memory (WAIS letter-number sequencing), two tests of inductive reasoning, and divergent thinking (test of fluency). Also assessed was "habitual cognitive engagement" (from tests of Mindfulness, Need for Cognition, and Memory Self-Efficacy). Teams of five to seven people met weekly for about 20 weeks with a coach. At the end of the sessions, two tournaments were held in the context of social events. The same assessment battery was administered to experimental and control participants at post-test. Those randomly assigned to the Senior Odyssey performed significantly better on tests of processing speed, inductive reasoning, and divergent thinking than wait-listed controls. The sizes of the transfer effects were in the small range. In contrast, working memory and visuospatial processing did not change, although a composite measure of cognitive performance showed greater positive change in the experimental group. Measurements of "habitual cognitive engagement" showed only trends toward improvement in the experimental group.

There has also been some success using real-world experience as training.[91] Older African-Americans who live in inner cities and have low education (mean 11.5 years) and low income (less than $15,000 a year) are at higher risk of cognitive decline than either suburban-dwelling counterparts or white older individuals.[92] Michelle Carlson and colleagues devised the Experience Corps program with the aim of creating an incentive for older, cognitively "at risk" individuals to become more cognitively, physically, and socially active. The investigators assessed the effects of volunteering in local elementary schools on older inner-city African-Americans. The participants were generally required to have an Mini-Mental score of 24, although those with scores of 20–23 were allowed to participate if they scored well on parts A and B of the Trail Making Test (28 percent of participants). The Trails B test is widely used to assess executive function. Participants were randomly assigned to be volunteers for 15 hours per week (Intervention group) or to be wait-listed for the next volunteer opportunity (Control group). The Intervention group underwent 2 weeks of training in literacy support, library support, and conflict resolution in an elementary-school setting before being placed with their team and a program coordinator in a school. Volunteers were evaluated initially and 4, 6, and 8 months after the start of their experience. Evaluation consisted of the Mini-Mental State Exam, parts A and B of the Trail Making Test, the Rey-Osterrieth Complex Figure Test (immediate and delayed), and a verbal memory test. The Intervention group improved on Trails B by 8 percent; the control group declined by 13 percent. On the Rey-Osterrieth Figure, the Intervention group showed a 9 percent improvement; the control showed a 10 percent decline. Both tests are considered to assess executive functioning. Likewise, a subgroup of the Intervention group who had shown impairment on Trails B at pre-test showed marked improvement on Trails B (42 percent) and on delayed word list recall (40 percent); impaired Controls showed a 9 percent improvement on Trails B and a 12 percent decline on delayed word list recall. These findings were significant on *t* tests, but only marginally significant in statistical tests that were adjusted for age, education, and exposure duration. Unfortunately the authors did not report the post-intervention Mini-Mental State Exam test scores.

Does cognitive stimulation in older people transfer to untrained abilities?

Yes, with some kinds of training. There is no doubt that older people benefit from cognitive training and that the effects of training are durable, lasting up to 5 years. More important, although transfer of training is not universally seen, transfer of cognitive training to untrained tasks can also occur in older individuals, with small to medium effect sizes. Transfer has been seen with dual-task training, task switching, and reasoning training. However, studies using updating training do not show transfer. Therefore, it is not clear which principles underlie broad transfer of training. Transfer has been seen when there are problem-solving aspects to the training (e.g., reasoning, strategy video games, Senior Odyssey, Experience Corps). However, transfer has also been seen with task switching. It is not clear whether the breadth of training, per se, is a contributory factor in transfer, or simply provides training on a range of cognitive processes, thereby facilitating transfer. It will take more carefully designed studies to determine which specific aspects of training are important for facilitation of transfer of training effects in older people.

Summary

There appears to be good evidence that cognitive training has strong and durable benefits for cognitive functioning in older people. Cognitive training appears to change brain structure and physiology, and neurogenesis may play a role in the benefits of new learning. Under the right conditions, benefits of cognitive training can generalize to untrained tasks and to daily functioning. However, at present data on the type of training that is optimal for maximizing transfer are insufficient. There is some evidence that training on executive functioning may be needed for training to transfer, but that evidence is conflicting.

10 Combined Effects of Interventions and Preventative Actions

. . . we

think the sun may shine someday when we'll

drink wine together and think of what used to

be: until we die we will remember every
single thing, recall every word, love every

loss: then we will, as we must, leave it to
others to love, love that can grow brighter

and deeper till the very end, gaining strength
and getting more precious all the way. . . .

—A. R. Ammons, "In View of the Fact"

Would cognitive decline in old age be minimal if an individual had the best possible diet, exercise, and cognitive stimulation? In the preceding chapters, we reviewed evidence showing that a number of health and lifestyle factors appear to modulate cognitive functioning in animals and humans. However, animal studies reveal growing evidence of overlap between the various factors. This suggests the possibility that even stronger beneficial effects might be observed if the different factors were combined than if they were experienced separately and in isolation.

A recent study provides some supporting evidence for this possibility. The study found that older people who regularly engaged in social dancing performed better than age-matched non-dancers not only on balance and posture assessments but also on tasks of cognition and attention.[1] Although this was an observational study, the finding that a broad-based experience (social dancing) that combines factors of exercise and social interaction was associated with multiple effects on posture, gait, and cognition is consistent with the evidence we discussed in previous chapters. That work points to overlapping mechanisms underlying the effects of resveratrol, dietary restriction, and exercise on cognition. This evidence of overlap suggests that combinations of factors exert additive effects on cognitive and brain integrity. However, the dominant trend in the field has been for different research groups to investigate different factors one at a time, so that there have been relatively few studies comparing multiple factors in a single study. Accordingly, this chapter is one of the shorter ones in this book. Nevertheless, there is a small body of work pointing to the greater efficacy of multiple experiential factors for arresting cognitive decline.

10.1   Animal Studies

Addressing the question of whether combinations of lifestyle factors exert additive effects on cognitive and brain integrity, several studies have compared the effects of two or more lifestyle factors manipulated both separately and in combination. Diet and exercise have been most commonly manipulated together in animals. Studying rats, Fernando Gómez-Pinilla and colleagues—pioneers in this field—examined combined effects of omega-3 fatty acids (specifically, docosahexaenoic acid) and physical exercise (voluntary wheel running) in rats over a period of 12 days. The DHA-enriched diet was associated with improvements in spatial learning, and that effect was heightened in rats also randomly assigned to wheel running. Moreover, rats assigned to both diet and exercise manipulations showed the greatest reductions on a measure of oxidative stress—levels of oxidized proteins in the hippocampus.[2] Measurement of oxidative stress is based on the free radical theory of aging, which posits that reactive oxygen species produce cellular damage (see chapter 7). In aged rats, an antioxidant diet has been found to lead to better learning and retention.[3]

Ideally, effects of diet, cognitive stimulation, and exercise would be compared in one study. Such a study has been carried out, but only in dogs. Dogs may be a better model than rodents for examining combined effects of lifestyle factors on cognitive and brain aging. Like aged humans, aged domesticated dogs show Aβ deposition associated with Alzheimer's disease,[4] increased oxidative damage,[5] and marked individual differences in cognitive aging. Carl Cotman and colleagues have conducted a series of what are essentially randomized clinical trials of diet, exercise, and cognitive enrichment in dogs. In one of these studies, young and old dogs were randomly assigned to either a standard diet or an "antioxidant" diet (a standard diet supplemented with vitamins, "spinach flakes, tomato pomace, grape pomace, carrot granules and citrus pulp," and "mitochondrial cofactors"). The aged dogs with the enriched diet learned a landmark-discrimination task faster and were more likely to learn a harder version than the aged dogs on a standard diet. The young dogs did not show any effect of diet.[6]

Having shown a benefit of diet, the same group then looked at separate and combined effects of dietary enrichment and "behavioral" enrichment. The behavioral enrichment manipulation involved physical exercise (two walks a week), toys in the cage, and training on landmark-discrimination and oddity-discrimination problems. When tested on the size-discrimination and reversal task after a year of these manipulations, the aged dogs that experienced the combined dietary enrichment and behavioral enrichment showed the fewest errors (figure 10.1). The same group randomly assigned young and old dogs to a control diet (standard) or an antioxidant-enriched diet (vitamins C and E plus fruits and vegetables); or to control housing (not described) or enriched housing (with kennelmates, toys, two outdoor walks per week, and 20 minutes of cognitive testing a day); or to both an antioxidant-fortified diet and enriched housing. In this case the intervention lasted 2.7 years. In both these studies, the dogs assigned to the "behaviorally enriched" condition were also walked more, causing a confound so that the separate contribution of exercise and cognitive testing cannot be determined. Though both manipulations separately improved performance of the older dogs on a size-discrimination and reversal learning task relative to the controls, cognitive performance of the old dogs was best in the group which experienced the two manipulations together, indicating additive effects.[7]

10.2 Human Studies

The literature on combined effects in humans is even smaller than that on combined effects in animals. Nevertheless, there is some evidence, some from observational studies and some from randomized designs, of additive

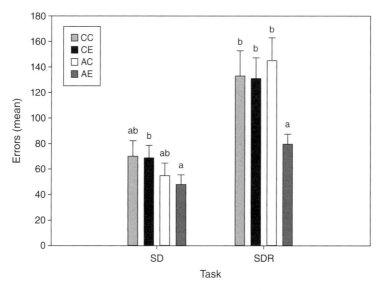

Figure 10.1

Mean total errors in acquisition of a size-discrimination task (SD) and a size-discrimination and reversal task (SDR) in dogs as a function of diet and enriched living conditions (described in text). CC (control diet/no enrichment); CE (control diet/ behavioral enrichment); AC (antioxidant food/no enrichment); AE (antioxidant food/behavioral enrichment). Source: N. Milgram et al., "Long-term treatment with antioxidants and a program of behavioral enrichment reduces age-dependent impairment in discrimination and reversal learning in beagle dogs," *Experimental Gerontology* 39: 753–765. Reprinted with permission of Elsevier (© 2004).

effects of lifestyle factors in humans. In an observational study, 1,880 community-dwelling older people were assessed for effects on risk of Alzheimer's disease of two lifestyle factors: adherence to a Mediterranean -type diet (low in red meat and saturated fat, high in fruits, vegetables, cereals, and fish) and amount of physical exercise. The results showed that both stricter adherence to a Mediterranean diet and more exercise were independently associated with lower risk of developing Alzheimer's disease over 15 years.[8] Therefore, similar to the dog studies, effects of diet and exercise appear to be additive.

There are also a few human studies using random assignment, although these have very small samples. Gary Small and colleagues conducted such a study,[9] randomly assigning 17 screened middle-aged adults to either an intervention group (n = 8) or a control (n = 9) group. The intervention group underwent 14 days of training that involved memory training, exercise (brisk walks), relaxation techniques, and diet (diet plan provided with emphasis on fruits and vegetables, omega-3 fats, complex carbohydrates). The training was not well described in the paper but seems to have been provided to the participants in a notebook giving instructions for each day. There was no direct supervision. The paper refers the reader to Small's mass-market book,[10] which describes a similar scheme as a "memory prescription" and gives detailed instructions—including pictures—for each day for memory, diet, exercise, stress reduction. For example, on Monday, people are advised to take a 5-minute walk in the morning (this is increased to 30 minutes by Friday), eat yogurt and raisins for a mid-morning snack, write their first name with the non-dominant hand, then write their first name with both hands simultaneously, have a tuna sandwich on whole wheat bread and an apple for lunch, and so on. To assess cognitive and brain change, cognitive testing and PET scans were administered before and after the 14 days of training. The controls were asked to carry out their usual routines over the 2 weeks. The only objective measure of cognitive functioning that changed over the 2 weeks was fluency. The intervention group showed increased fluency; the fluency of the control group did not change. On the PET-based cerebral metabolism measure, the intervention group showed a 5 percent decrease in activity in the dorsolateral prefrontal cortex; the control group showed no change. Although these are not strong results, it is impressive that both cognitive and brain changes were seen after only 2 weeks of a rather weak, poorly monitored intervention (participants had phone access to a "research nurse," whose role was not explained in the paper). Ideally there would have been separate groups for each intervention.

Another small-scale study did have separate groups for each intervention and also a fairly rigorously controlled intervention. Claudine Fabre and colleagues randomly assigned older people to aerobic training, cognitive training, or both ($n = 8$ in each group).[11] Participants underwent extensive physical exams and fitness assessments. The aerobic training consisted of two supervised 45-minute sessions of interval training (brisk walking and/or jogging) per week; the cognitive training was 90 minutes per week of broad-based memory and attention training aimed at learning strategies. The memory quotient from the Wechsler Memory Scale showed improvement from pre- to post-training in all three experimental groups (aerobic,

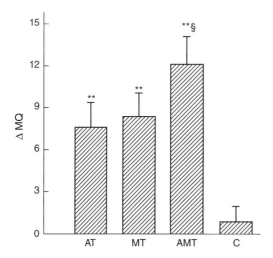

Figure 10.2
Memory performance was better in the group experiencing combined aerobic training and memory training group than in the groups receiving either alone. Source:
C. Fabre et al., "Improvement of cognitive function by mental and/or individualized
aerobic training in healthy elderly subjects," *International Journal of Sports Medicine*
23, no. 6: 415-421. Reprinted with permission of Georg Thieme Verlag KG.

cognitive, and combined groups); the controls showed no improvement.
Improvement was greatest in the combined groups (figure 10.2). Thus, in
humans, as in dogs, there is evidence of additive effects of combined factors.

Summary

Only a few studies have examined the effects of an optimal lifestyle (involving diet, exercise, and cognitive stimulation) on cognitive decline in old
age. The general conclusion is that lifestyle factors have greater beneficial
effects on cognitive aging when they are jointly experienced than when
individually experienced. Overall, human observational studies and randomized animal and human studies conducted by different research groups
consistently find additive effects of diet, exercise, and cognitive training,
although there are confounds in some of the studies. The literature is small,
the manipulations are not powerful, and the effects are not strong. Nevertheless, the studies suggest real benefits from combined interventions.
Clearly this area of research would benefit from additional investigations
comparing different combinations of experiential factors for effectiveness.

11 Modifying the Work Environment and the Home Environment

There is a great satisfaction in building good tools for other people to use.
—Freeman Dyson

Can we change the environments in which people work and live to better accommodate the limitations of the aged? In previous chapters we have considered a number of ways in which older individuals can adapt themselves to the effects of brain and cognitive aging by engaging in various lifestyle-related activities. A complementary method to having people pursue such activities would be to "adapt" not the person, but the environment (e.g., home and work environments), so as to support optimal cognitive functioning in old age. Sometimes such changes will involve the introduction of some form of technology, but simple physical modifications of the home (e.g., adding handrails to stairs, or widening hallways to allow use of a walker) would also belong to this category.

Any such changes to the environment, or any implementation of new devices, must be carefully designed. The new environments or technologies must also be tested with appropriate users so that the modifications are indeed found to be useful and safe.

Designing technologies and environments so that they can be used effectively and safely falls within the discipline of human factors and ergonomics, also known as cognitive engineering.[1] In this chapter, we describe some important features of this discipline before discussing how environmental modifications and technology can support cognitive vitality in older adults. A number of demonstration projects have been undertaken, most of them aimed at helping older people "age in place" (that is, remain in their own homes as they age despite physical health problems). This has real benefits, since people often prefer to remain in their own homes. There are also considerable financial benefits to avoiding "assisted living" or a nursing home. Examples of such demonstration projects include "virtual assistants"—computer-controlled systems of sensors and interaction devices (keyboard, telephone) that issue prompts and monitor the status of an older person. This field, still in its early stages, promises future benefits as more scientific evaluations of the efficacy of such systems are conducted.

11.1   Limitations on Coverage

Our coverage of the broad area of human factors and ergonomic design for older adults in this book is limited to a few topics, including health-care technologies aimed at older adults and assistive technologies for the home. A number of other aspects of human factors and age-sensitive design, such as the design of input and output devices, computer inter-faces, webpages, instructional and training materials, and transportation systems, are beyond the scope of this book. Readers wishing to understand the basic research issues in human factors and aging are referred to the *Handbook of Human Factors and the Older Adult*.[2] Several other sources can be consulted for coverage of products, technologies, and design guidelines.[3]

Another area that we do not cover in this chapter is display or inter-face changes that are made to accommodate age-related reductions in sensory-perceptual functioning. It is well known that old age is accom-panied by a loss of acuity in vision, audition, taste, and smell.[4] Such sensory changes can lead to difficulties with everyday activities, such as reading a medication label that is in very small type or understanding a spoken command that isn't quite loud enough. Guidelines for modifying visual displays (by improving lighting, reducing glare, enhancing contrast, and so on) have been published,[5] as have descriptions of how speech and other auditory displays can be changed to compensate for age-related hearing loss.[6] Such sensory "environmental" modifications are an impor-tant adjunct to other human-factors interventions aimed at reducing the impact of age-related cognitive decline on everyday functioning. Sensory factors are also important to consider because they may interact with cognitive factors to further exacerbate the effects of age-related cognitive decline.[7]

11.2 Human Factors and Cognitive Engineering

The field of human factors and cognitive engineering seeks to take both human capabilities and human limitations into account when designing products, technologies, and environments. The goal is to match the demands that products and new technologies impose on people with the sensory, perceptual, cognitive, and physical characteristics of the population of users.

Human factors is a both a basic research field and a profession concerned with applications. The field has its roots in the post-World War II period, during which the military witnessed many events and accidents that led to injuries or fatalities among soldiers and workers operating increasingly complex military equipment. The incidents were not attributable to carelessness or poor motivation on the part of soldiers and other military personnel, but rather to poor equipment design, inadequate training, or both. The field of human factors expanded considerably beyond military applications in the 1960s and the 1970s with the rapid development and introduction of advanced technologies in the workplace, and more recently with the personal-computer revolution and the rise of the Internet.

Research in experimental and cognitive psychology has provided a number of general principles (and even "laws") that describe human characteristics and that, if applied judiciously, can inform the design of products that are usable, safe, and even enjoyable, or least not frustrating to operate. An early example is the principle of stimulus-response compatibility. This principle states that responses to visual displays are made more speedily and with less error when the spatial configuration of the displays with respect to the response devices to which they are linked are consistent, intuitive, and easy to understand. This principle has been successfully used to design the layout of various display-response devices in homes (e.g., the knobs on a stove), in cars (e.g., turn signals), and in several work environments.[8] Another principle is Fitts' Law,[9] according to which the how long it takes a person to reach a target area (e.g., to use a mouse to move a cursor to some position on a computer screen) depends on the distance to be traveled and the size of the target. This law has been applied to determine the layout of controls and dials in such settings as cockpits of airplanes and the control rooms of power plants. Cognitive constructs such as working-memory capacity (and its limits) are similarly relevant to the design of such things as how many choices should be presented in a command menu in a computer interface. Wickens and Hollands describe various other general principles guiding human performance that are relevant to designing products and technologies for effective use.[10]

Unfortunately, the principles and guidelines of human factors are often not followed in the initial design and manufacture of products; typically they are considered after the fact, when problems of usability arise, or if the products are found to compromise safety. Human-factors professionals may then be asked to redesign a product, or to institute new training programs in its use. Though useful, such "retrofit" solutions are not ideal, and the cognitive engineering profession is constantly striving to influence systems engineers and manufacturers to consider human capabilities and limitations in the early stages of design.

Human-factors research has achieved some notable successes. The graphical user interface (exemplified by Microsoft Windows or the Macintosh Operating System) and the computer mouse both arose from the research efforts of human-factors scientists at the Xerox Corporation's Palo Alto Research Center the 1970s. In particular, relative to the command-line interfaces that existed before that time (e.g., entering the DOS command "delete" followed by a file name), graphical interfaces made it more intuitive and much easier for users to select and perform operations on computer-based objects such as files (e.g., by simply dragging the icon for a file to the icon of a trash can). Such research and development efforts were first brought to market by Apple. Millions of computer users, young and old, now enjoy the benefits of ease of use and personal productivity that these well-designed products have conferred.

However, many products enter the market with little or no attention to human factors, and as a result people often experience frustration and difficulty in using them. For example, one survey of adults across a wide range of ages (18–90 years) found that more than 70 percent of those surveyed reported usability problems with many everyday household products, such as toiletries, health-care products, and medications.[11] In such cases, the consequence of being forced to use poorly designed products is frustration. In other instances, the products may pose a danger to safety. In an analysis of several non-prescription household products available to older consumers, Lorraine Gardner and colleagues found that the majority of the products were inadequately designed to perform the tasks for which they were intended, and that some were potentially hazardous and could cause injury.[12]

Insofar as older adults exhibit changes in different aspects of sensory, perceptual, and cognitive functioning, good human-factors design may be even more important for them than for the young. Indeed, some cognitive engineers argue that by designing to accommodate the needs of older adults, one is designing for all—the concept of *universal design*.[13] The profession of cognitive engineering has a number of methods and tools to help the design process. In addition to the general psychological principles described previously, *task analysis* can be used to define the component processes involved in a user's interaction with a device or a product. Task analysis involves decomposing the elementary sub-tasks involved in carrying out a complex task (for example, using a word processor), from the user's top-level goal (e.g., composing a letter), to the component tasks involved (entering the word processing program, selecting a name and address, and so on), to the elementary physical operations required to carry out the tasks and achieve the goal (pressing a key on the keyboard, pressing the space bar, etc.). One of the most widely used methods of task analysis is the Goals–Operators–Methods–Selection rules (GOMS) model.[14] The use of GOMS or other methods of task analysis can lead to a better understanding of the requirements for using a product effectively and safely.

In addition to task analysis, there are several methods that fall under the general rubric of *usability testing*. In this area of human factors, researchers observe and quantify the behavior of users interacting with prototypes of products in order to pinpoint areas of difficulty. Usability methods range from simple questionnaires and "think aloud" protocols to eye-movement recording and physiological recording during use. A number of other methods and tools have also been used by human-factors professionals.[15]

11.3   Self-Care, Assistive Technologies, and "Aging in Place"

A number of assistive technologies are being developed to help older adults to "age in place"—that is, continue to live independently in their homes. This is important because more than 90 percent of older adults continue to live in their own homes, or with relatives, or in independent-living communities. Even those who may be experiencing the beginnings of physical or cognitive decline may fear losing their independence and may wish to remain at home.

The concept of "aging in place" is also consistent with the movement toward "self-care." It has been claimed that more than half of health care is provided by the individual or the family, with no intervention by a physician.[16] Such self-care includes disease prevention, treatment of illnesses, management of chronic diseases, and rehabilitation.[17] Since the need for health care increases with age, it makes sense to support and encourage the efforts of individuals, families, and communities to promote self-care. In many countries, increasing life expectancies and low birth rates combine to generate concerns about the number of health workers that will be available to provide services to the aged in the future. "Aging in place" has become a movement of sorts, with older people banding together to organize self-help "villages" that screen service providers (e.g., plumbers) and that offer direct services (e.g., meal delivery) to dues-paying members, typically older people living in their own homes. Such efforts can help older people stay out of nursing homes.

Assistive technologies and changes in the home can help greater numbers of older individuals to remain at home. However, any such changes must take human factors into account. In particular, the methods and tools described above must be applied to ensure that there is no mismatch between the demands imposed by the new technologies and the capabilities of older adults. Even fit and healthy people can experience difficulty and frustration when using poorly designed devices, and the difficulties can be exacerbated by declining physical and cognitive abilities.

Several advanced technologies are being developed to improve the daily lives and maintain the health of older people living independently. These technologies are aimed at various problems that older adults may face in their daily activities—for example, managing multiple medication regimens, and summoning help if incapacitated. Some systems go beyond those more basic needs by using classification techniques such as "fuzzy logic" to determine if there is a longer-term change in behavior patterns that might arise from illness or depression.[18]

Researchers at the Georgia Institute of Technology have been in the forefront of efforts to introduce "intelligent" technologies to help older adults age in place. They have developed the Aware Home, a conventional-looking house with considerable sensing and computing infrastructure that was designed to, among other things, assess technological innovations aimed at improving the lives of older individuals. The Aware Home has within it several component technologies designed to deal with particular aspects of perceptual or cognitive functioning. For example, the "Digital Family Portrait" (figure 11.1) provides information that can be viewed remotely. The frame has icons that are updated regularly. As the older person moves about the home, motion detectors are activated. The butterfly icons vary in size depending on how much the person moves in each room of the house. A butterfly would be small if the person sat quietly reading much of the day, but would be large if the person was painting the walls or exercising. In addition, other information (about, for example, the temperature, the lights, opening of the refrigerator door, and opening of the front door) can be sent to a relative or a neighbor. By this means, a person at a remote location can check to see that all is well with the older individual.

More generally, the market is beginning to respond to the need for home-based technologies to monitor and support older people. Several computer systems intended to provide support to older people and their caregivers are under development. One often-cited example is Honeywell's

Figure 11.1
Georgia Tech's "digital family portrait." The butterfly icons in the frame are associated with various rooms and change in size with the amount of activity in those rooms. Courtesy of Aware Home Research Initiative.

ILSA (Independent LifeStyle Assistant), an ambitious system that originally was aimed at supporting an independently living older person with extensive monitoring and management (including monitoring of temperature, blood pressure, and heart rate) and with the ability to remotely control lights, power, a thermostat, door locks, and water flow. These plans were scaled back when ILSA was implemented for a real-world trial. Computers and sensors were linked so that a family member could monitor an older relative's activity in his or her own home, and receive alerts if necessary, and so that the older adult could receive information (e.g., appointments) and reminders (e.g., of medication needed). The equipment included "interaction devices," including a security panel, a Web browser, a television, speakers, and a telephone. Sensors were placed in each room except the bathroom, so that family members would be able to remotely monitor participants' activity and to intervene if activity was reduced by 50 percent over 24 hours.

A prototype of ILSA was field-tested in the homes of eleven older individuals in 2003.[19] One problem that arose was an excess of "no motion" alerts. Some of the erroneous alerts were system-generated, but in one case a client forgot to turn ILSA off before leaving her home for several hours. Another source of dissatisfaction was that users expected more sophistication than the system was designed to provide—for example, one woman complained that ILSA failed to recognize that she had made a change in her pattern of medication. There was also a perceived working-memory load arising from the requirement to press "1" or "2" on the phone keypad. For phones with the keypad integrated into the receiver, this requires the client to move the receiver to look at it, remember which key to press for the choice, then move the receiver back to continue the phone call. Other problems were related to users' social perceptions of the system. Apparently the users came to dread phone calls reminding them to take their medication. Users also disliked the recorded message, which was perceived as "rude." Empirical research shows that computer systems judged "polite" are viewed as more reliable than those judged "impolite."[20]

Other companies are beginning to get involved in providing home-based technological support to older people. A company called Grandcare[21] sells a system that is similar to ILSA in that it uses sensors that can be monitored remotely. Such sensors send information about activity in the home—when room doors and cupboard doors were opened, when the stove is on, when a person got into and out of bed, whether a person has been moving around a room for certain time period. In addition, information about physiological measures such as weight and blood pressure can be obtained. This information is then sent out via e-mail, text messaging, or voice mail. There are also interactive features that allow the older person being monitored to receive messages and photos and to play games with a grandchild. A television program on the Discovery Channel recently featured a 13-minute segment about one family's experiences using Grand-care.[22] Other similar systems are now offered commercially.[23]

The use of ILSA and other advanced technologies that are being developed to address the needs of older adults and allow them to remain in their homes longer also raise important issues of overreliance and complacency, since no system is 100 percent reliable. When older adults interact with advanced assistive devices, their decisions about when to rely

Figure 11.2
A GrandCare interface in the home of a woman who is remotely monitored by a son who lives in a distant state. Joseph Hilliard photo (© 2011).

on automation may not be optimal, and they may be more susceptible to automation errors than younger adults are. Geoffrey Ho and colleagues, working in Charles Scialfa's laboratory, suggested that this was the case because older people typically have less experience with computer systems.[24] As a result, older people may exhibit greater complacency in interaction with powerful but imperfect automated decision aiding systems. Here "complacency" refers to how trusting people are of automated systems and how that trust changes through experience in interacting with the automation.[25] If an automated system works reliably and provides correct information to the user, trusting the system will not have any adverse consequences. However, overreliance on imperfect automated systems can lead to problems on occasions when they provide incorrect information or wrong advice. In a recent review of the literature, Raja Parasuraman and Dietrich Manzey noted that complacency is typically seen when users of highly reliable but imperfect automated systems have other tasks that compete for their attention.[26] Complacency therefore reflects a redirection of attention away from the automation to other manual tasks. A widely cited example of an accident in which automation complacency was a contributing factor was the grounding of the cruise ship *Royal Majesty* off the coast of Nantucket in 1995.[27] That ship had been fitted with an automatic radar plotting aid (ARPA) that depended on a GPS signal to work correctly. The crew had to monitor the automated system while engaged in other duties. Because of a loss in the GPS signal due to a frayed cable from the antenna, the ARPA system reverted to "dead reckoning" mode and didn't correct for the prevailing tides and winds, so the ship was gradually steered toward a sand bank in shallow waters. The National Transportation Safety Board's report on the incident cited the crew's overreliance on the ARPA system and complacency associated with insufficient monitoring of other sources of navigational information, such as another radar and visual lookout.[28]

Complacency is typically associated with individuals who tend not to verify the accuracy of information or recommendations provided by automation. In view of the link between complacency and attention,[29] an increased tendency in older adults to exhibit complacency is consistent with the finding that they are typically less adept at attending to multiple tasks than the young. In fact, greater "automation complacency" in older adults has been reported in two studies comparing multiple-task performance in young and older adults, but only under high task loads.[30]

Related studies on differences between young and older adults in the degree of reliance on automated systems were carried out in Wendy Rogers' laboratory by Julian Sanchez et al. and by Neta Ezer et al.[31,32] In the latter study, young and older participants performed a counting task that required them to count the number of circles on a display with assistance from an automated decision aid. The advice the decision aid provided, though mostly reliable, wasn't always correct. Participants earned points for correct answers and lost points for errors. They could also verify the advice of the automation regarding the number of circles present in the display. For example, if there were six circles in the display but the aid said there were only five, participants could either go along with that advice or check the decision aid, the latter involved a "cost" of losing a point for every 2 seconds spent verifying the automation. The percentage of trials in which participants agreed with the decision aid and did not perform the task manually was recorded as reliance. Ezer et al. found that all participants increased their reliance with a lower cost of verification but older adults were less responsive to changes in costs, indicating greater overreliance.

Geoffrey Ho and colleagues further investigated complacency in older people using a simulated medication-management system.[33] The automated system provided either correct or incorrect advice on taking medications, at varying levels of reliability. In the first experiment, older and younger participants performed a simple mathematical task at the same time as a simulated medication-management task. The older participants trusted the aid more than the young and were less confident in their performance. In a second experiment, the rates of errors of omission and errors of commission made by the automation were varied. Older adults were again more reliant on the decision aid and were less sensitive to automation failures. Since complacency and other forms of bias related to automation usage may not only degrade users' performance but also sometimes compromise safety[34] (as evidenced by the grounding of the *Royal Majesty*), the finding of increased complacency in an older group is troubling. However, Elin Bahner and colleagues in Dietrich Manzey's group showed that training with exposure to actual automation failures reduced complacency in a group of young adults.[35] These results indicate that any (imperfect) decision-support system for older adults to use at home should be supplemented by appropriate training—for example, exposure to instances of automation failures.

In addition to helping healthy older adults manage everyday tasks at home, assistive technologies have also been developed for independently living people with chronic diseases that affect cognitive function. One example is home diabetes management. Many older adults suffer from Type 2 diabetes, a chronic disease in which disordered metabolism affects glucose control and results in abnormally high blood sugar. Diabetes can have very serious consequences, including blindness, amputation, and kidney failure. Particularly relevant to the topic of amelioration of cognitive aging is the fact that elevated blood glucose contributes to age-related memory decline.[36] Despite the potential seriousness of diabetes, Type 2 (the more common type) can often be controlled well with medication, diet, and exercise—sometimes even with the latter two alone. In light of the known effect of blood glucose on memory, effective diabetes self-care is particularly important for optimal cognitive health in older diabetics. However, most diabetics do not receive the recommended care. Russell Glasgow and Lisa Strycker reported that fewer than 40 percent of diabetes patients received care that met recommended guidelines.[37] Aside from substandard care by physicians, patients' "medical adherence"—the extent to which a patient follows medical advice for a condition, such as diet, exercise, and medication—is reported to be only about 50 percent.[38] "Disease management" methods have been developed to help physicians and patients achieve optimal, evidence-based care for various conditions. This is sometimes implemented by health-care providers who act as personal coaches, but sometimes also by commercial software.

An important aspect of the management of diabetes is support of self-care, particularly regular monitoring of blood sugar. To support diabetes self-care, Olivier Blanson Henkemans and colleagues developed a prototype of a "virtual" health-care assistant, called Health-Pal, that monitored the patient's medical plan, medical records, and somatic complaints and determined the patient's care plan, including self-care tasks.[39] To study the effectiveness of this approach, a study was conducted in which healthy participants cared for a virtual patient using the Health-Pal. This was done either autonomously (in which case Health-Pal decided on the ideal care plan and required a participant to implement that plan for the virtual patient) or cooperatively (in which case Health-Pal reviewed the care plan and suggested care tasks to improve the patient's health). Participants could deviate from Health-Pal's plans to make changes to improve the patient's health. Participants were responsible for caring for their virtual patient for 2 weeks, instructing the patient to carry out various tasks—e.g., exercise, take a medication, or play a game with Health-Pal. This was done either under the autonomous condition or under the cooperative condition. There was a strong preference for the cooperative version, but there was no difference in performance between the two versions. It will be important to test this approach with real patients.

There are also devices aimed at supporting the ability of older people to "age in place." One such device, developed by the Georgia Tech Aware Home group, is Cook's Collage, a video display that guides users through a recipe.[40] In a demonstration study, it was judged useful by one younger participant and one older participant. It was particularly useful to the older person, who tended to misread recipes and forget ingredients. Both participants used it as a memory aid, although the technology didn't work perfectly and made some errors. It helped with the missed ingredients, but not with the misreading.

Most such devices generally exist only in prototype form, and to our knowledge none have been formally assessed. Full development, including rigorous testing, will require long-term study of in-home use by members of the target group.

Summary

Modifying the home and work environments is a complementary approach to lifestyle activities that can be used to support optimal cognitive functioning in older adults. Designing technologies and environments so that they can be used effectively and safely requires the use of methods from the field of human factors or cognitive engineering. Human-factors methods involve taking human capabilities into account while minimizing the effects of cognitive limitations associated with aging. In addition, task analysis and usability analysis are required to ensure the safety and ease of use of assistive technologies for older adults. Several demonstration projects aimed at helping older people "age in place" have been described as self-health-care technologies. Evaluation of the effectiveness of these technologies requires that overreliance and complacency be addressed. The field, still in its infancy, promises future benefits as more scientific evaluations of the efficacy of such systems are conducted.

The retirement of the first baby boomers is likely to fuel the development of assistive technologies aimed at older people who plan to "age in place." There is evidence that older people are increasingly knowledgeable about and accepting of technology. According to a recent report from Pew Research Center, 76 percent of people aged 56–64 and 58 percent of people aged 65–73 are "online."[41] One hopes that products for these age groups will be designed soon, and with attention to human factors. Indeed, a website called Aging in Place Technology Watch is aimed at entrepreneurs designing products for people aging at home.[42]

12 Nurture via Nature: Genetics, Environment, and Cognition

Genetics and experiential factors shape the biological and behavioral manifestations of human life, but they do not suffice to account for the totality of human nature. Man also enjoys a great degree of freedom in making decisions; he is *par excellence* the creature that can choose, eliminate, organize, and thereby create.

—René Dubos, *So Human an Animal*

In previous chapters we described the several ways in which the older brain and mind can be nurtured so as to limit cognitive decline. Regular physical exercise, careful attention to proper diet and nutrition, and engaging in opportunities for new learning are all activities that can exploit the power of brain plasticity in order to slow or arrest cognitive decline in late life. In chapter 11 we described how modifications to the home environment or the work environment can supplement such self-initiated activities so as to enhance cognition in older adults. Our general view is that all older individuals can benefit from such activities and interventions. But is it the case that some individuals may profit more than others? Do some individuals benefit little or not at all? If the latter, are such differences between people related to genetic factors?

12.1 Nature and Nurture

The relative roles of nature and nurture in biological and psychological function have been a topic of investigation and debate for more than 100 years. Long pitted against one another, nature and nurture are now seen as working in concert to shape the developing organism. Further knowledge of the mechanisms of the interplay between nature and nurture will, no doubt, help in understanding how the ameliorative effects of various experiential and lifestyle factors on age-related cognitive decline vary between individuals.[1] To better understand this issue, let us first consider the association between genes and cognition.

12.2   Molecular Genetics and Cognition

The rapidly expanding field of molecular genetics has revolutionized our understanding of how DNA—nature's blueprint—sets the human body and brain in motion on their long journey through life, and of how, together with the environment, it shapes our capabilities and limitations. The breakthrough was provided by the success of the Human Genome Project in decoding the entire human genome.[2] That achievement, coupled with evidence from cognitive neuroscience studies linking cognitive functions to the activation of specific networks in the human brain, set the stage for a molecular genetics of cognition.

How do genes affect cognition? Behavioral geneticists have traditionally addressed this question by comparing the performance of identical and fraternal twins to assess the heritability of a cognitive function. By this method, investigators can clearly determine the existence and the degree of genetic influence for a particular function. For example, twin studies have established that human intelligence is highly heritable, with a heritability estimate of about 70 percent in adults.[3] Heritability estimates of general intelligence have been found to increase with age, beginning in childhood. For example, a recent study of more than 8,700 pairs of twins found that the heritability of general cognitive ability increased from 23 percent in children 2–4 years old to 62 percent in those 7–10 years old.[4] Heritability estimates of general intelligence are higher in adults than in children, and heritability remains high throughout adulthood. One study found high heritability estimates for different measures of cognitive ability in a sample of pairs of twins over the age of 80.[5]

Molecular genetics provides a complementary approach to behavioral genetics for understanding the genetic basis of cognition. The particular advantage that molecular methods provide is that specific genes can be associated with individual variations in cognition. Many studies use the *allelic association* method. Genes are found in chromosomes. Some but not all genes come in different forms, called alleles, which represent variations in the component DNA sequence at a single location or at multiple locations within the string of several thousand DNA nucleotides that make up the gene. Since a chromosome has two strands, there are two possible alleles at the critical location, each inherited from one parent or the other. The functional consequence of such normal allelic variation, if there is any, can be examined. The allelic association method can thus be used to relate specific genes to cognitive performance measures in healthy persons.[6] In some recent studies, the method has been applied to individual differences in cognition in healthy individuals.[7] These studies have provided evidence of modulation of different aspects of cognition by specific genes, individual variation in cognitive functioning being linked to allelic variation in a particular gene.[8] More recently, this method has been extended to examine the associations and interactions between multiple specific genes and cognition.[9]

Genome-wide association studies

The single-gene and multiple-gene approaches have recently been extended (and in some cases replaced) by studies in which an exhaustive search is conducted through the entire genome. Such *genome-wide association studies* (GWAS) have been made possible by progressive reductions in the cost of genetic assays. Variations in large numbers of genes now can be related to diseases or conditions.[10] Technological and database developments now make it possible to examine from 300,000 to more than a million single-nucleotide polymorphisms (SNPs) over the whole genome, covering or tagging most human genes.

There are now several published GWAS studies of neuropsychiatric diseases, including Alzheimer's disease, schizophrenia, and autism. The results have generally identified novel disease genes, most with very small effects. GWAS studies of cognition are still relatively rare. In an important GWAS study of the heritability of general cognitive ability (intelligence) in a large adult sample, Davies, Deary, and colleagues found that none of the more than 500,000 SNPs reached statistical significance. However, they found, when the proportion of variation explained by all the SNPs together was considered, that 40–50 percent of the variation in intelligence among individuals could be explained by common SNPs in linkage disequilibrium with causal variants. This suggests that, as with other complex phenotypes, differences in intelligence among people are due to small effects of many genes.[11] GWAS studies of Alzheimer's disease are also relevant to cognitive aging. A recent example of this approach is a study in which Jean-Charles Lambert and colleagues looked at genes associated with Alzheimer's disease in 16,000 people and identified two genes not previously known to be related to the disease.[12]

The GWAS approach is unbiased in that no specific gene is given greater consideration *a priori*. However, it is also atheoretical, which can lead to problems in interpreting results when associations are found for multiple genes. That is true in the case of the previously mentioned study by Lambert et al.[13] The functions of the two new genes linked to Alzheimer's disease have not been well studied and were not previously strong candidates. In the candidate-gene approach, a single gene is selected on a theoretical basis because of its hypothesized effects on brain mechanisms mediating the cognitive function in question. If an association is found, then there is, in principle, a clear theoretical basis for explaining the association.

Moreover, with regard to Alzheimer's disease at least, GWAS studies may have reached their limit of usefulness. A recent GWAS study of Alzheimer's disease by Sudha Seshadri and colleagues included more than 35,000 people, 8,371 of them with Alzheimer's disease. Seshadri et al. confirmed associations found in previous studies (including the well-documented apolipoprotein E gene as well as the newer genes CLU and PICALM) and discovered two SNPs not previously identified. They replicated those new loci in an independent Spanish sample. When they added those newly identified SNPs to a model that predicted incident Alzheimer's disease on the basis of age, sex, and the APOE gene, there was no improvement in prediction of Alzheimer's. Nor did the addition of CLU and PICALM genes to the model add to clinically useful predictability.[14] By way of interpretation, Seshadri et al. point to recent polygenic models of schizophrenia that assume tens of thousands of alleles, each contributing about 5 percent increased risk. In schizophrenia, such models fit the existing data better than models that assume a small number of alleles, each contributing moderate risk.[15]

In an editorial accompanying the paper by Seshadri et al., the well-known population geneticist Nancy Pedersen speculated that Alzheimer's disease is not only polygenic, but may arise from an interaction between risk alleles and environmental factors.[16] Thus, some of these genes may be important, but only under specific environmental conditions. Thus, Pedersen points "back to the environment." The Alzheimer's disease concordance rate—that is, how often members of a pair of identical twins are both diagnosed with Alzheimer's disease—is around 60 percent,[17] which reminds us of the importance of environmental factors in determining whether an individual at increased genetic risk of Alzheimer's disease develops the disease. One important goal of this book is to describe what is known of environmental and lifestyle factors related to cognitive aging.

Candidate-gene studies

In contrast to the unbiased, atheoretical GWAS approach, allelic or candidate-gene studies attempt to use theories to guide associations between genes and cognition. In an article published in 2004, we outlined a general theoretical framework for allelic association studies of cognition.[18] First, candidate genes are identified. These are genes that, owing to the functional role of each gene's protein product in the brain, are likely to influence a certain cognitive ability or trait. An exhaustive search through the genome typically isn't necessary, although the decreasing cost of genetic assays has led to many genome-wide association studies.

Variations in the DNA sequence that define a candidate gene occur in different ways. Many are due to substitution of one of the four nucleotides in the DNA "alphabet"—adenine (A), guanine (G), cytosine (C), and thymine (T), where A and G form a complementary base pair (bp) (A/G) and C and T another (C/T)—with its complement. Such substitutions are referred to as single-nucleotide polymorphisms. For example, a gene having as part of its complete DNA sequence the series ACATAGA could have a variant in which the G is substituted for an A, resulting in the sequence ACACAAA. Other variations in DNA gene sequences include insertions, deletions, and repetitions of nucleotides.

The candidate-gene approach has produced some interesting discoveries, although it has acknowledged limits. Typically, SNPs of candidate genes are selected for further study if they are likely to influence neurotransmitter function or to affect neurotrophic activity in the brain. Examples of the former type are genes that influence the action of the neurotransmitter dopamine, such as COMT.[19] Examples of the latter type are the APOE and BDNF genes.[20] One also can review neuroscience research on the cognitive function in question to identify the brain networks that mediate that function. Pharmacological and neurophysiological studies in animals can be particularly useful in this regard, for they may provide information on the neurotransmitters that are active in these networks. These multiple approaches can be used to narrow down potential candidates for a genetic association study of cognition.

12.3   The Apolipoprotein E Gene, Aging, and Alzheimer's Disease

The apolipoprotein E (APOE) gene is among the genes studied most widely in research on aging. Its importance lies in the discovery (made in the 1990s by Allen Roses and colleagues) that a particular allelic variation in APOE, known as ε4, is a major risk factor for the development of Alzheimer's disease.[21] A large body of research since that time has therefore examined the role of APOE not only in the development of this disease but also in normal, age-related cognitive decline.

Apolipoprotein E is a lipoprotein (a protein that binds to lipids) that is important in lipid storage, transport, and metabolism.[22] It has also been found to play a role in neuronal repair and axonal outgrowth.[23] The gene that codes for this protein is found on chromosome 19 and has three variants, ε2, ε3, and ε4, which occur in the US population with respective frequencies of about 8 percent, 78 percent, and 14 percent.[24] Figure 12.1 shows a representation of the APOE gene. Risk of Alzheimer's disease increases with inheritance of one ε4 allele and is higher still for those with two ε4 alleles.[25] Possession of the ε4 allele also lowers the age of onset of the disease.[26]

If APOE-ε4 is a risk factor for Alzheimer's disease, can the effects of this gene be observed in healthy older adults before clinical diagnosis of Alzheimer's? Several studies have provided a positive answer to this question. APOE-ε4 has both broad and domain-specific effects on cognition in healthy adults, the cognitive abilities most affected being episodic memory and attention.[27] For the cognitive domain of visuospatial attention, Greenwood and colleagues also found that the pattern of attentional changes in healthy APOE-ε4 carriers was qualitatively similar to that seen in clinically diagnosed Alzheimer's patients, except that the deficit in Alzheimer's disease was quantitatively higher.[28]

Though these results are informative regarding the development of Alzheimer's disease, they also carry implications for our understanding of normal cognitive aging and of factors that may ameliorate age-related cognitive decline. In chapters 2 and 3 we summarized the evidence that

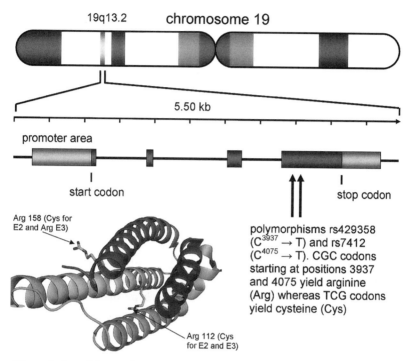

19q13.2   chromosome 19

5.50 kb

promoter area

start codon

stop codon

Arg 158 (Cys for E2 and Arg E3)

Arg 112 (Cys for E2 and E3)

polymorphisms rs429358 ($C^{3937} \rightarrow$ T) and rs7412 ($C^{4075} \rightarrow$ T). CGC codons starting at positions 3937 and 4075 yield arginine (Arg) whereas TCG codons yield cysteine (Cys)

3D rendering of the apolipoprotein E4 isoform

Figure 12.1
The APOE gene with the rs 7412 and 429358 SNPs and the ε4 allele. Source: T. Espeseth et al., "Accelerated age-related cortical thinning in healthy carriers of apolipoprotein E epsilon 4," *Neurobiology of Aging* 29, no. 3: 329–340. Reprinted with permission from Elsevier (© 2008). This figure appears in color elsewhere in the book.

not all older adults show age-related decline in cognitive function or in associated measures of brain function. We also suggested that the individuals who show deficits may be in poor vascular health. Another possibility is that age-related cognitive decline may occur only in those who possess one or two copies of the APOE-ε4 allele—about 14 percent of the US population. Can such individuals, despite their genotype, benefit from the ameliorative factors that we have discussed? Because only a few studies have systematically examined the cognitive effects of lifestyle and experiential factors in individuals differing in APOE or other genes, this question can't be answered in full yet.

12.4   The Molecular Genetics of Individual Differences in Cognition: Neurotransmitter Genes

Several genes other than APOE have recently been examined for their association with individual differences in cognition. Most of the candidate genes that have been examined are involved in modulation of different brain neurotransmitter systems. A large body of work has focused on the COMT gene, which is involved in the availability of dopamine in the synaptic cleft and is widely expressed in the prefrontal cortex.[29] A specific variation in the COMT gene, rs4680, has been associated with working-memory capacity and with activation of prefrontal cortex during tasks requiring working memory and executive control.[30] Genes that influence activity in other neurotransmission systems—acetylcholine, norepinephrine, and serotonin—have also been associated with different aspects of cognition in both healthy adults and various neuropsychiatric groups.[31]

Here we will describe one example of the candidate-gene approach outlined in our 2004 article[32] to show how molecular genetics can be combined with cognitive neuroscience research to establish links between genes and specific aspects of cognition. The example involves the brain's cholinergic system and the CHNRA4 gene.

Selective attention is known to involve a distributed network of brain regions, with the parietal cortex as a major locus.[33] This cortical region is richly innervated by acetylcholine. Anticholinergic agents such as scopolamine, when administered directly to the intraparietal cortex in monkeys, impair the speed of reorienting visuospatial attention.[34] Alzheimer's patients who have reduced metabolic activity in posterior parietal cortex (revealed by positron-emission tomography) are similarly slowed in disengaging attention from a cued spatial location.[35] The parietal cortex is known to have cholinergic receptors that modulate neuronal function[36] there. Attentional orienting is also modulated by nicotine administration in rats[37] and in humans who smoke.[38] Other pharmacological studies in animals also point to an important role for nicotinic receptors in attention.[39] All these lines of evidence indicate that genes that modulate nicotinic acetycholine receptors may be good potential candidates genes for association with selective attention. The most widely distributed nicotinic receptor in the central nervous system is composed of alpha-4 and beta-2 subunits assembled together.[40] Parasuraman and colleagues[41] therefore examined a polymorphism in this subunit receptor gene, involving a common C-to-T substitution at position 1545 (CHRNA4 C1545T rs#1044396), for its potential role in modulating the efficiency of visuo-spatial attention.

Eighty-nine healthy adults with a mean age of 35 years were genotyped for the CHRNA4 C1545T (rs1044396) polymorphism.[42] Participants were administered a cued letter-discrimination task modeled after the orienting task introduced by Michael Posner.[43] An arrow cue indicated which of two locations to the left or to the right of fixation would contain a letter target. After a cue-target delay of 200–2,000 milliseconds, the target letter appeared. Participants were required to make a speeded decision as to whether the target was a consonant or a vowel. Cue validity (valid, invalid, neutral) was varied so that both benefits (neutral cue RT—valid cue RT) and costs (invalid cue RT—neutral cue RT) of cueing could be obtained. Both RT benefits of valid cues and RT costs of invalid cues on letter discrimination varied in a systematic manner with CHRNA4 genotype. With an increased "gene dose" of the C allele (from no C alleles to one to two), RT benefits increased progressively, whereas RT costs decreased, also in a similarly progressive manner. Thus, the T allele was associated with greater costs of invalid cues, i.e., slowed discrimination of events outside the attended region. These systematic results provided the first evidence for an association between a nicotinic receptor gene, CHRNA4, and individual differences in the efficiency of shifting spatial attention in response to location cues. Subsequent studies by Thomas Espeseth and colleagues showing that the T allele is associated with ability to focus at the cued location and ignore distractors have replicated and extended these findings.[44] In addition to finding that variation in CHRNA4 modulates ability to maintain the focus of attention, Greenwood and colleagues also found that the SNP modulates the ability to adjust the size or scale of the attentional focus. Such modulation appeared to be limited to the CHRNA4 SNP and was not seen with a noradrenergic SNP.[45] This suggests some degree of specificity in neurotransmission gene effects, as predicted.

Parasuraman and colleagues also found that the CHNRA4 gene was not associated with individual variation in working memory, which was found to be linked to a dopaminergic/noradrenergic gene: dopamine beta hydroxylase (DBH).[46] The DBH gene is involved in converting dopamine to norepinephrine in adrenergic vesicles.[47] A single-nucleotide polymorphism in the DBH gene involving a G-to-A substitution at 444 location in exon 2 (G444A, rs#1108580) has been linked to changes in the dopamine to noradrenaline ratio in brain[48] and to attention deficits in children.[49] Parasuraman et al. genotyped 103 participants and subdivided them into three groups on the basis of number of G alleles: none (AA genotype), one (AG genotype), or two (GG genotype).[50] The working-memory task involved maintaining a representation of up to three spatial locations over a period of 3 seconds. After a fixation period, participants were shown target dots at one to three locations for 500 milliseconds. Simultaneous with the offset of the dot display, the fixation cross reappeared and was followed by a 3-second delay, at the end of which a single red test dot appeared alone, either at the same location as one of the target dot(s) (match) or at a different location (non-match). Participants had 2 seconds to decide whether the test dot location matched or didn't match one of the target dots. Accuracy was equivalent for all three genotypes at the lowest memory load, but increased with higher gene dose of the G allele, particularly for the highest load (three targets), as confirmed by a simple effects analysis. Memory accuracy for the GG genotype (G gene dose = 2) was significantly greater than that for both the AG (G gene dose = 1) and AA genotypes (G gene dose = 0). These findings point to a significant association between the DBH gene and individual differences in working memory.

The DBH G444A variation is a so-called functional SNP (that is, it has a direct effect on protein and amino acid production), and the product, the DβH enzyme, can be measured in cerebral spinal fluid and in blood. Joseph Cubells and colleagues first reported that the G444A polymorphism of the DBH gene influences levels of the DβH enzyme in plasma.[51] In a more recent study, however, Cubells and Zabetian found that another SNP which is located upstream of the G444A SNP, −1021 C/T, has a much greater effect on plasma levels of DβH; specifically, the CC allele combination of this SNP was associated with a tenfold increase in plasma DβH levels.[52] This indicates that the CC genotype would be associated with greater conversion of synaptic dopamine to norepinephrine than would the TC or the TT genotype. Because PFC dopamine levels have been linked to working memory, therefore, the CC genotype should be associated with poorer and the TT genotype with superior working-memory capacity, respectively. This prediction was confirmed using the same spatial working-memory task in another sample of participants genotyped for both DBH SNPs.[53]

Effects of these SNPs can interact. Based on our previous reports (reviewed above) that the CHRNA4 rs#1044396 nicotinic receptor SNP affected visuo-spatial attention but not working memory, and the DBH rs#1108580 nor-adrenergic enzyme SNP affected working memory but not attention, we predicted that the two systems would interact when working memory was manipulated by attention. In a separate study, we found that memory performance was modulated by both SNPs when we manipulated the scale of visuospatial attention deployed around a working-memory target. However, the CHRNA4 SNP exerted a stronger effect than the DBH SNP on memory performance when visuospatial attention was manipulated.[54]

Several studies have also found that the effects of neurotransmitter genes such as COMT, CHRNA4, and DBH can vary with age.[55] For example, Ming-Kuan Lin and colleagues in Raja Parasuraman's lab recently observed that a well-studied polymorphism in the COMT gene affects memory per-formance more strongly in older than in younger individuals.[56] The obser-vation that healthy aging modulates the effects of genes suggests a role for environmental and experiential factors.

12.5   The Interplay of Nature and Nurture

Thus far we have discussed the effects of genes on cognition in isolation. Yet it is widely appreciated that various environmental factors can modify the genetic influence on cognition. The interplay between nature and nurture includes effects of gene-environment interaction, in which the influence of particular genes on a particular psychological function is observed to arise only when certain environmental conditions are encoun-tered. Many psychiatric disorders are known to be partly heritable, but, as in the case of schizophrenia, the genetic disposition results in mental illness only in the presence of certain stressful life events.[57] Gene-environ-ment interaction may also play a role in variability in the ameliorative effect of the various experiential and lifestyle factors we have discussed on age-related cognitive decline.

Nature and nurture also interact in the phenomenon of epigenetics— that is, heritable functions that do not involve changes to the underlying DNA sequence. (The Greek prefix "epi" means "in addition to"—hence, epigenetic traits are those that occur in addition to the molecular genetic basis for inheritance.) Epigenetic effects involve changes in the way a certain gene affects biological or psychological function without alteration in the gene itself, but with alteration in its expression. Such changes can occur in response to various environmental events and may influence age-related cognitive decline. The environmental drivers of epigenetic change can occur at any age, from infancy[58] to old age.[59] Currently we know only a little about what role such epigenetic effects play in moderating the effects of experiential and lifestyle factors on age-related cognitive decline.

12.6   Interactive Effects of APOE Genotype and Exercise

Only a few studies have examined gene-environment interactions in the context of cognitive aging. The relevant studies have primarily looked at the interaction of genes and lifestyle factors such as exercise and cognitive training in relation to cardiovascular disease and the incidence of dementia, age-related cognitive decline, and Alzheimer's disease. A much smaller number of studies have focused on the interaction of genes and lifestyle factors in ameliorating cognitive changes in healthy older adults. Again, such studies have largely focused on genetic interactions with the effects of exercise and cognitive stimulation, although some work on interactive effects with dietary factors, notably dietary restriction and polyunsaturated fats, has also been reported.[60]

The APOE gene is the best-studied gene in cognitive aging, largely because of its association with Alzheimer's disease, and is also the main gene that has been examined in these studies on the interaction between genes and lifestyle factor. However, although the ε4 allele of the APOE gene is the strongest known genetic risk factor for Alzheimer's disease, only about half of ε4 homozygotes with the riskiest genotype eventually develop Alzheimer's disease.[61] Many Alzheimer's patients have not inherited even one ε4 allele. Therefore, some other factor or factors must be important.

There is recent evidence that lifestyle exerts an effect. The age-adjusted prevalence of dementia is very low in India and in sub-Saharan Africa.[62] Further, APOE genotype has little influence on the incidence of Alzheimer's disease in Africa, but does influence it in African-Americans living in the United States.[63] According to a recent report, people with hereditary amyloid angiopathy, which is characterized by amyloid deposition in brain arteries, have undergone a markedly shortened life span in the very brief span of time of 20 years. Astridur Palsdottir and colleagues attribute this to modern dietary changes exacerbating the effects of the genetic mutation linked to the disease.[64]

One important lifestyle factor, physical exercise, can positively influence cognitive aging. However, its influence appears to be modulated by APOE genotype. Evidence bearing on this comes from studies that use as measures incidence of dementia and/or Alzheimer's disease and cognitive change short of dementia. A longitudinal study conducted in Finland by Suvi Rovio and colleagues showed that APOE genotype modulates the link between mid-life physical activity and later dementia or Alzheimer's disease.[65] Capitalizing on the existence of two large population-based studies of cardiovascular risk factors, Rovio et al. obtained data on their leisure-time physical activities of people initially assessed at mid-life (mean age 50.6) in the 1970s and the 1980s. At the mid-life assessment they were asked "How often do you participate in leisure-time physical activity that lasts at least 20–30 minutes and causes breathlessness and sweating?" Answers were rated on a range from 1 (meaning daily) to 6 (meaning not at all). These people were then invited for re-assessment in 1998, when their mean age was 71.6. At an average follow-up interval of 21 years, 1,449 people were re-assessed for dementia and Alzheimer's disease. Those who were physically active at least twice a week in mid-life had approximately 60 percent lower odds of later developing Alzheimer's disease than sedentary individuals. This effect was modulated by APOE genotype. The association was significant among APOE ε4 carriers in all models tested but was not significant among APOE ε4 non-carriers.

An even larger study asked the same question but in a somewhat different way. Laura Jean Podewils and colleagues assessed effects of physical activity in old age rather than in mid-life.[66] They followed a large sample of 3,375 healthy adults over the age of 65 for 5 years, and found that the incidence of dementia in the sample after 5 years was significantly lower in those with high levels of self-reported physical activity (e.g., walking, household chores, mowing, raking, gardening, hiking, jogging, biking, exercise cycling, dancing, aerobics, bowling, golfing, general exercise, and swimming) than in those who engaged in no or little activity. In direct contradiction to the study by Rovio et al., this association was seen only in those who did *not* possess the APOE-ε4 allele and was absent in ε4 carriers.

These two studies agree that exercise is protective against late-life dementia, but disagree on how that effect is modulated by APOE genotype. The difference may lie in the age at which the physical activity occurred. Despite its association with Alzheimer's disease, the APOE ε4 allele is known to exert protective effects in young individuals. The ε4 allele is under-represented among spontaneously aborted fetusus[67] and is more common among people who attained higher education and less common in those who left school early.[68] It protects against cognitive deficits in impoverished children with chronic diarrheal disease.[69] Better memory has also been claimed for young adult APOE-ε4 carriers,[70] as has greater activation in the hippocampus.[71] However, the latter finding occurs in older ε4 carriers too and is generally interpreted as reflecting compensation.[72] In view of the conflicting results of the studies by Rovio et al. and by Podewils et al., it may be that the ε4 allele confers a benefit from exercise on cognitive integrity that is strongest when the exercise is carried out at younger ages.

Unfortunately, no neuropsychological or other cognitive test data were reported in either of those studies for those individuals (the majority of the samples) who did not progress to dementia, so the interactive effects of APOE genotype and exercise on healthy cognitive aging could not be determined. However, in a more recent study, Jennifer Etnier and colleagues obtained the neuropsychological test scores of 90 healthy women of mean age 62.[73] The tests included the auditory verbal learning test (AVLT), the Complex Figures Test (CFT), the Wisconsin Card Sort Task, a block design test, and the Paced Auditory Serial Addition Task (PASAT). An important methodological strength of this study was that self-reports of physical activity were supplemented by a physical test of aerobic fitness (VO_2 max). Etnier et al. found that aerobic fitness was associated with significantly better performance on the AVLT, CFT, and PASAT tests in APOE-ε4 homozygotes, but not in heterozygotes or in non-carriers. They suggested that beneficial effects of exercise are more likely to be found in those individuals whose "cognitive reserve" is likely to fall below a threshold (i.e., the homozygotes) than in others.

Sean Deeny and colleagues in Bradley Hatfield's laboratory[74] also measured cognitive performance in mid-life, and found broadly similar effects. They examined how APOE genotype affected memory performance on a Sternberg task and how it affected cortical activation during the memory task (measured with magnetoencephalographic recordings from the scalp). The participants were divided by APOE genotype, and were categorized as sedentary or active on the basis of the Yale Physical Activity Survey for

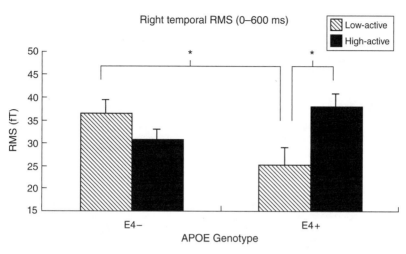

Figure 12.2

Magnetoencephalographic response over the right temporal region during a memory task as a function of APOE genotype and amount of physical activity (high vs. low). Source: S. P. Deeny et al., "Exercise, APOE, and working memory: MEG and behavioral evidence for benefit of exercise in ε4 carriers," *Biological Psychology* 78, no. 2: 179–187. Reprinted with permission from Elsevier (© 2008).

Older Adults.[75] The Yale survey yields a measure of total minutes of activity and was used to estimate total calories expended in a week. Consistent with the results of Rovio and Etnier and their colleagues, Deeny et al. found that on the memory task highly active ε4 carriers were faster to decide whether a letter was included in a "memory set." Highly active ε4 carriers also had greater right temporal lobe activation on matching probe trials than low-active ε4 carriers (figure 12.2). Physical activity levels of non-carriers did not affect behavioral performance or cortical activation.

Interactive effects of APOE genotype and exercise have also been reported for brain activation patterns associated with memory performance in older adults. One recent fMRI study examined changes in regional brain activation during a test of semantic memory in participants (aged 65–85) with and without the APOE ε4 allele.[76] Greater activation in several brain regions typically associated with semantic processing was observed in those with the ε4 allele who reported higher levels of physical activity than in less active ε4 carriers.[77]

Considered together, then, many studies agree that the benefits of exercise appear to be greater in APOE-ε4 carriers. The results of such studies

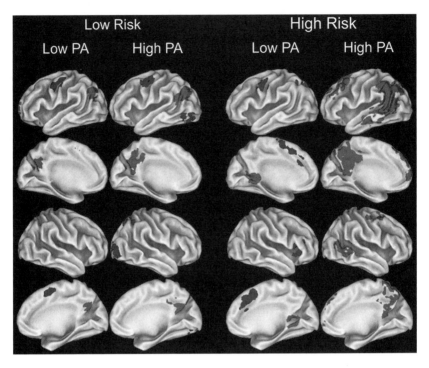

Figure 12.3
Results of BOLD activation in brain regions during a memory task showing significant differences between conditions as a function of self-reported physical activity (PA) and APOE genotype (low risk = no ε4 alleles; high risk = at least one ε4 allele). Source: J. C. Smith et al., "Interactive effects of physical activity and APOE-ε4 on BOLD semantic memory activation in healthy elders," *NeuroImage* 54, no. 1: 635-644. Reprinted with permission from Elsevier (© 2011). This figure appears in color elsewhere in the book.

are somewhat at odds with the meta-analytic study by Colcombe and Kramer discussed in chapter 6, which found broad-based benefits of aerobic exercise on performance in many cognitive tasks in healthy older adults as a group, not only in the small percentage (about 5 percent) who are likely to be APOE-ε4 homozygotes.[78] Nevertheless, these studies are valuable in indicating that even inheritance of two ε4 alleles, which substantially increases the risk of Alzheimer's disease, does not prevent older adults from benefiting cognitively from physical exercise.

As we discussed in chapter 6, most studies examining the relationship of physical exercise to cognition in older adults have used self-report measures and observational methods, and only a few studies have used the randomized trial method (in which participants are randomly assigned to either an exercise group or a control group). Hence, a recent report by Nicola Lautenschlager and colleagues on interactive effects of APOE genotype and exercise using such a method is noteworthy.[79] Lautenschlager et al. randomly assigned older adults (mean age 68) who had self-reported memory problems but did not meet the clinical criteria for dementia to either a 6-month home-based program of physical activity or a "usual care" program. At the end of the intervention period, the activity group had significantly higher scores than controls on the ADAS-Cog, a general measure of cognitive function, and modestly higher scores on word list recall. When the two groups were subdivided according to presence or absence of ε4, ADAS-Cog scores were higher in the ε4 non-carriers who had participated in physical activity than in any of the other subgroups.

More evidence bearing on this question comes from studies assessing the effect of physical activity on the risk of cardiovascular disease—itself a contributor to cognitive decline late in life.[80] The ε4 allele of the APOE gene has long been associated with atherosclerosis.[81] Lipoproteins (namely chylomicrons, VLDL, IDL, LDL, and HDL) transport dietary lipids through the bloodstream from the intestines to other regions of the body. A mature chylomicron contains four apolipoproteins. Apolipoproteins—APOA, APOB, APOC, and APOE—are proteins that bind to lipid (fat) and carry lipids, including cholesterol, in the blood. Population studies have reported that people without the ε4 allele (genotypes ε2/2, ε2/3, and ε3/3) have lower levels of plasma LDL ("bad") cholesterol and lower risk of heart disease. People who inherited one or more ε4 alleles (genotypes ε3/4, ε4/4) have higher levels of plasma LDL and higher risk of heart disease. This conclusion is consistent with a recent meta-analysis of 82 studies of lipid levels and 121 studies of coronary outcomes that concluded that APOE-ε4 carriers have a "slightly higher" risk of coronary heart disease.[82]

Does variation in the APOE gene alter the benefits of exercise on risk of cardiovascular disease related to serum HDL and LDL? Both aerobic training[83] and strength training[84] exercise reduce risk of cardiovascular disease by raising the level of serum HDLs ("good" cholesterol). HDL acts by ferrying cholesterol to the liver for excretion or re-use. However, there is substantial variation in the extent to which individuals experience positive effects on HDL levels from exercise. A meta-analysis of 59 studies found an average increase of only 2 mg/dL in HDL levels after exercise.[85] Several groups have hypothesized that one source of that variation is the APOE gene. James Hagberg and colleagues assessed serum lipids in initially sedentary overweight men (mean age 59) before and after 9 months of supervised cycling, walking, and jogging, eventually reaching 3 days a week for 45 minutes. Those with an APOE-ε2 allele (n = 6) experienced a greater increase in serum HDL after training than ε3 or ε4 carriers.[86] Arthur Leon and colleagues compared the effects on black participants and white participants (mean age 35) of training on stationary bicycles three times a week for 20 weeks from 30 to 50 minutes per week. They observed a greater increase in serum HDL in those with genotypes ε2/3 (n = 37) and ε3/3 (n = 144) than in those with ε4/4 genotype (n = 3). However, this was seen only in the white females.[87] (These were very small groups.) In another study, the same investigators found that after 24 weeks of supervised endurance exercise training in sedentary older adults (mean age 58), black participants with no ε4 allele (genotypes ε2/3 and ε3/3) had larger HDL particles (and thus a better ability to ferry cholesterol) and higher HDL serum levels than other groups with an ε4 allele.[88] (These sample sizes were somewhat larger: eighteen ε4 non-carriers and nine ε4 carriers.)

In a larger study that did not consider race, Paul Thompson and colleagues equated APOE genotype groups in middle-aged individuals (mean age 39). They found that 6 months of supervised aerobic training (using treadmills, stationary cycles, etc.) was associated with greater training-related reductions in triglycerides and other cardiovascular risk measures (LDL/HDL ratios) in the ε2/3 and ε3/3 groups (n = 16 and 20, respectively) but increased risk measures in the ε3/4 group (n = 17). However, the attendant increase in exercise capacity reflected in maximum oxygen consumption (VO$_2$ max) was actually less in the ε3/3 group than in the ε2/3 and ε3/4 groups.[89] Hagberg et al. also observed that ε4 carriers benefited less than non-carriers in LDL levels but more in VO$_2$ max from exercise.[90] Those carrying the APOE ε4 allele (n = 12) experienced the largest mean increase in VO$_2$ max (25.9 percent), whereas those with the APOE ε2 allele experienced a mean increase of 11.1 percent (n = 6). There was an intermediate increase (17.9 percent) in those with an APOE ε3 allele (n = 33). Despite the small sample sizes in some of these studies, these results on effects of physical exercise on measures of cardiovascular health are fairly consistent in showing that aerobic exercise has a greater benefit on blood lipid levels in those without an ε4 allele. However, the two studies that measured VO_2 max found more improvement in exercise capacity in ε4 carriers than in non-carriers. This suggests that APOE genotype exerts differential effects on the benefits of exercise on risk factors for cardiovascular disease and on capacity for exercise.

12.7   Interactive Effects of APOE Genotype and Cognitive Activity

There is increasing evidence that living a life of intellectual engagement, including education, reduces the risks of Alzheimer's disease late in life. This topic was reviewed in chapter 9, but we now ask whether APOE genotype modulates that relationship.

Increasing years of education have been associated with a lower risk of Alzheimer's disease in several studies.[91] However, that effect appears to interact with APOE. Several groups have found that the well-documented negative effects of the APOE ε4 allele on cognitive decline late in life[92] and on risk of dementia[93] are greater in people with low educational attainment. Apart from education, individuals vary in the amount of cognitive stimulation they habitually seek. Retrospective studies have shown a relationship between choosing cognitively stimulating activities and Alzheimer's disease. A retrospective study asks people to remember previous behavior. For example, people who claimed high participation in cognitively stimulating activities in mid-life were less likely to become demented later in life.[94] A better way to ask this question is to do so prospectively—that is, to ask people about their current cognitive activities and then follow them over time to see if the level of cognitive activity they reported affects the likelihood of later development of dementia.

The first prospective study to ask the question this way assessed the role of APOE. Robert Wilson and colleagues recruited 801 Catholic clergy (nuns, priests, brothers) to a longitudinal study.[95] The participants were administered extensive assessments, which included their rated frequency of common activities that require "information processing" (watching television, reading books, going to museums, playing cards, checkers, and so on). During a follow-up period averaging 4.5 years, 111 of the subjects developed Alzheimer's disease. Analysis showed that a one-point increase in cognitive activity was associated with a 33 percent reduction in risk of being diagnosed with Alzheimer's over that time. However, APOE genotype did not alter this relationship.

Study of twins provides another way to assess the effect of mid-life cognitive activity on later development of Alzheimer's disease. (The advantage of a twin design is that identical twins share the same genes and typically the same environment into adulthood.) Data from the Swedish Twin Registry show that if one identical (monozygotic) twin has Alzheimer's disease, in 59 percent of instances the other twin will have it too.[96] This indicates a strong role for genetics in determining whether a person at genetic risk of Alzheimer's actually develops the disease. However, the time between the diagnoses of monozygotic (identical) twin pairs can differ substantially, ranging in a recent study from 4 years to 18 years.[97] This also indicates a substantial influence of environmental and experiential factors. Consistent with that, mid-life cognitive activity influenced discordance between female twins but not between male twins.[98]

Michelle Carlson and colleagues asked whether APOE genotype modulated the effect of mid-life cognitive activity on risk of Alzheimer's disease among twins.[99] In 1967, questionnaires on health and physical and cognitive activities were mailed to 15,924 twin veterans of World War II who were identified by the National Academy of Sciences' Twin Registry. Their average age was 44.7 years at that point. Between 1990 and 2005, 147 twin pairs were identified who had completed the 1967 questionnaire and were also discordant for Alzheimer's disease—that is, one of the pair had developed Alzheimer's disease and other had not for at least 3 years. They were assessed for dementia 20–40 years after the activity questionnaire. Suspected dementia was assessed in an in-home formal assessment by a psychometrician and a nurse. Using the 1967 questionnaire, the reported cognitive activities were categorized according to the degree of "novel information processing" required. Activities judged as having high novelty included reading and studying for courses. Tasks judged as having low novelty included watching television and listening to radio. Social activities were judged intermediate. Overall, greater cognitive activity in mid-life was associated with a 26 percent reduction of risk for developing dementia, but the effect was stronger in APOE-ε4 carriers. Among monozygotic APOE-ε4 allele carriers (n = 46), cognitive activity in mid-life was associated with a 30 percent reduction in risk for developing dementia first. Moreover, novel activities were most strongly associated with reduced dementia risk.

No effect was seen from physical activity. Thus, in this prospective twin study, the APOE-ε4 allele had protective effects when combined with a relatively active mental life in mid-life.

In a similar study conducted in Singapore over a shorter period, Matthew Niti and colleagues identified 1,635 older adults in the Singapore Longitudinal Aging Study (a population-based sample identified from census records) who were of mean age 66 (range 55–93) at baseline assessment.[100] The Mini-Mental State Examination and participation in physical, social, and "productive" activities (including reading, music, computing, painting, gardening, shopping, and community work) were assessed at baseline. The MMSE was re-assessed 1.5 years later on average. The benefits on the MMSE of "high leisure activity" were greater in the APOE ε4 carriers than in the non-carriers, but the association was stronger for "productive" activities than for physical and social activities. This is consistent with Carlson's finding on risk of dementia.

What can we conclude from studies of genetic modulation of the effects of cognitive experience on cognitive aging?

Although the literature is relatively small, it appears that APOE genotype may differentially influence the beneficial effects of cognitive activity on cognitive functioning late in life. In three of four independent studies, heightened cognitive activity had a greater effect on risk of dementia in APOE ε4 carriers than in non-carriers. This may be somehow related to increased brain activity found in ε4 carriers in several different studies during memory encoding. In contrast, greater physical activity conferred a stronger benefit on cognition (including avoidance of dementia) late in life in ε4 non-carriers than in ε4 carriers in two of three studies. Consistent with that, more physical exercise resulted in greater improvements in blood chemistry in ε4 non-carriers than in ε4 carriers. However, it was the ε4 carriers who benefited the most in exercise capacity (VO2 max).

Summary

A growing body of research has identified genes that are associated with normal variation in various cognitive functions. Only a few studies have examined how such genes may interact with the lifestyle factors considered in this book—exercise, diet and nutrition, cognitive stimulation—to modify age-related cognitive decline. Most studies have looked at the APOE gene. The weight of the evidence suggests that the APOE-ε4 allele—a known risk factor for Alzheimer's disease and other neurodegenerative diseases—interacts with lifestyle factors. Notably, carriers of the ε4 allele obtain a greater benefit from exercise than non-carriers for late-life cognitive functioning, but most strongly when the exercise is carried out in mid-life. Moreover, cognitive experience also confers stronger benefits on APOE ε4 carriers.

13 What Can and What Should Be Done to Support Cognitive Vitality in Older Adults?

In a substantial proportion of the healthy older population—perhaps 40–50 percent, according to estimates from several investigators[1]—there may be little or no decline in cognitive functioning. However, cognitive aging is a real phenomenon with tangible consequences for the earnings, the health, and the quality of life of the rest of the older population. Moreover, older people who have not begun to decline undoubtedly have a strong interest in maintaining their functioning—especially if they decide, or are required, to stay in the workforce.

Our goals in writing this book were twofold. We wanted to describe what is known about the effect of aging on healthy minds and brains. We also wanted to describe what is known about factors that can modulate age-related decline in minds and brains, focusing on the best available scientific evidence. We have reviewed the scientific evidence on what *can* be done. We must also examine what *should* be done.

Not everything that can be done to maintain cognitive functioning in late life is feasible or even advisable. Although some practices, including advanced education and avoiding excess weight gain, are advisable early in life, it is not likely that young people can be induced to adopt such behaviors with the goal of avoiding cognitive decline when they are older. The people with the strongest motivation to prevent or delay cognitive decline late in life probably are middle-aged or older. Fortunately, there appear to be some practices that older people can adopt to maximize their cognitive integrity.

In the remainder of this chapter, we summarize the evidence described in detail in the previous chapters and identify the specific lifestyle practices that can support neuronal plasticity and thereby help prepare individuals to sustain cognitive functioning during old age.

13.1   What Can Be Done? Adapt the Person

As should be evident from our discussion in chapter 3, most brain changes in otherwise healthy older individuals are subtle and selective,[2] reversible,[3] and not universal.[4] Such findings suggest that cognitive decline can be avoided or slowed in many individuals. The brain plasticity/cognitive plasticity hypothesis (discussed in chapter 4) argues that the aging brain can be reorganized by cognitive demand, but only in the presence of intact plasticity mechanisms that are supported and enhanced by specific lifestyle factors.

Exercise aerobically

Aerobic exercise can counteract the negative effects of a high-fat diet[5] and can promote glucose regulation. In light of the important finding that poor glucose control has direct negative effects on hippocampal and memory function,[6] any steps taken to improve glucose control will have beneficial effects on the hippocampus and thereby on memory formation. Exercise promotes neurogenesis in animals, and in humans exercise is associated with preserved brain volume. Exercise is also easier and less unpleasant than dietary restriction. Even moderate aerobic exercise (brisk walking three times a week) has benefits for cognition.

Substitute polyunsaturated fats for saturated fats

A diet low in saturated fats (found in red meat and in butter) and high in monounsaturated fats (found in olive oil) and polyunsaturated fats (found in fish oil, walnuts, and flax seed) is easily achieved and carries no known risk. It appears from several studies that simply adding polyunsaturated fat (PUFA) supplements to the diet is not sufficient. As was noted in chapter 7, a growing literature indicates that benefits are obtained only when PUFAs are substituted for saturated fats in the diet, not when they are merely added as supplements.[7] Several hypotheses have been advanced to explain the mechanisms by which omega-3 PUFA, notably the specific components DHA and EPA, affects neuronal function. Two mechanisms that may be involved are diffusion of proteins and diffusion of signaling molecules across the membranes of neuronal cells.[8]

Restricting dietary fat is also a way to achieve and maintain optimal weight. Randomized trials conducted to determine the optimal diet for weight control find that no one type of diet is necessarily better than others for weight loss. One recent study compared four types of diets, some low in fat (20 percent) and some high in fat (40 percent), and found that for weight loss the most important factor was calories consumed, not proportion of fat.[9] However, as we noted in chapter 7, there is animal and human evidence indicating that a diet high in fats (especially saturated fats) may be detrimental to brain health and specifically to memory functioning. In sum, the literature argues for a diet that is low in saturated fats and in which polyunsaturated fats are substituted for saturated fats.

Restrict calories, or at least avoid excess calories

There is a substantial literature showing increased health and longevity associated with dietary restriction (DR) in animals. The smaller literature on DR in humans is consistent in showing benefits of DR for cardiovascular disease and for blood insulin levels and other markers of longevity.[10] Evidence of cognitive benefits is weaker. Only a few studies have looked for cognitive effects of DR, and only one study has found such a benefit.[11] Moreover, it is not clear whether DR would be well accepted by most people. In one study of alternate-day fasting, participants reported constant hunger.[12] Another study looked at chronic calorie restriction for 3 months and found not only that memory improved but also that the treatment approach was well tolerated.[13] Nevertheless, it seems unlikely that dietary restriction will ever be popular.

Avoiding obesity appears to be beneficial, as was reviewed in chapter 7. Avoiding excess calorie intake without fasting is something that many people in the world do regularly. Restricting calories to achieve and/or maintain an optimal weight also carries the important benefit of reducing risk of diabetes, which has detrimental effects on cognition and especially on memory. Future research might look into whether periodic fasting (perhaps weekly or monthly) confers detectable benefits on health. Episodic fasting might be tolerated better than daily fasting. Even if fasting doesn't directly benefit cognition, it may benefit hippocampal function by improving glucose regulation.[14]

Learn new skills and acquire new knowledge

Both formal education and cognitive training—learning new skills and integrating new knowledge—appear to heighten brain integrity and enhance cognition in old age. Neurogenesis continues even late in life,[15] and the survival of newborn neurons is heightened by learning.[16] New neurons may have a role in sustaining the ability of the dentate gyrus to integrate new information.[17] To the extent that learning promotes the survival and integration of new neurons, it may be important to continue to learn new material and skills throughout life. That both skill learning[18] and knowledge acquisition[19] result in regional enlargement of specific brain regions in young people is consistent with that speculation. Though training-related enlargement of gray matter has not been investigated in older people, an increase in white-matter integrity with cognitive training has been seen.[20] Three randomized trials in older people have found that reasoning training,[21] video-game training, and combined memory and speed training[22] resulted in improved cognitive functioning that trans-ferred broadly to untrained cognitive functions. Such broad transfer is, of course, the ultimate goal of cognitive training. Improvements were seen in everyday functioning,[23] in working memory (but not working-memory capacity),[24] and in task switching, memory, and reasoning.[25] Thus, cogni-tive training can change older minds and brains.

More work will be needed to determine the optimal form of cognitive training. It is important to ask not only what skills improve with training but also what types of training generalize to untrained abilities. The three studies mentioned above are important because they show that large samples of older people benefit from training effects that transfer. However, other studies have observed transfer only in young people.[26] We don't yet know what factors govern transfer of training, but there is increasing evidence that transfer can occur broadly to untrained abilities in old people.

Use estrogen? (for women)

The four lifestyle factors we have discussed thus far appear to provide benefits and carry little or no cost. In contrast, estrogen "replacement" confers benefits for post-menopausal women but also carries potential costs. The review of evidence we presented in chapter 8 and other recent reviews[27] conclude that estrogen can benefit cognition in older women when initiated around the time of menopause, but may be harmful when initiated later. The optimal duration of treatment is not known. In the next few years, several well-designed randomized trials of estrogen administration will report results. Until then, although estrogen does promote neu-ronal plasticity, the relation of health risks to benefits remains unclear.

Take cognition-enhancing drugs?

Several prominent neuroscientists recently argued for wide use of cogni-tion-enhancing drugs by both adults and children. They called for research into the costs and benefits of such drugs, and for the development of ethical guidelines to govern their use.[28] Although those scientists did not specifi-cally address the issue of prescribing drugs to ameliorate age-related cogni-tive decline, its remediation would be an obvious target of such drugs. As we discussed in chapter 8, research on the cognitive effects of such drugs in healthy individuals has been sparse, and only a few studies have looked at effects on older people. Moreover, stimulant drugs such amphetamines have the potential to be addictive. The literature on the cognition-enhancing effects of selective serotonin reuptake inhibitors (SSRIs) is limited by a confound with effects on mood. Work on caffeine is a little more prom-ising, with both animal and human studies reporting cognitive benefits. However, caffeine is already in wide use. In general, at present, there is little basis for a recommendation to use cognition-enhancing drugs.

Which "lifestyle treatment" is best?

We have summarized the evidence regarding the efficacy of several lifestyle factors in ameliorating cognitive decline. The next step would be to compare these manipulations as "treatments" to prevent or slow cognitive aging in longer-term clinical trials with older human participants. As was noted in chapter 10, studies of dogs indicate that combinations of diet, exercise, and cognitive stimulation yield stronger effects than any of these three factors in isolation.[29] Comparable studies have not yet been carried out in humans. One approach could be modeled on the ACTIVE project, in which healthy older people are followed annually after being randomly assigned to a treatment that is also administered annually. The goal would be to assess rate of cognitive decline as a function of treatment type over several years. A large sample would be needed to assess combinations of treatments with reasonable numbers in each treatment group—for example, exercise alone, cognitive training alone, and exercise plus cognitive train-ing. Of course diet is also important, but diet is harder to control in humans using random assignment. Another approach would be to assess the effects of various treatments in people at increased risk of progressing to Alzheimer's disease in the near future. People diagnosed with Mild Cognitive Impairment show some cognitive deficits but do not meet clinical criteria for Alzheimer's disease.[30] However, they do progress to Alzheimer's at a relatively high rate (10–15 percent a year)—especially those who are carriers of the APOE-ε4 allele, a known genetic risk factor for Alzheimer's. A treatment that slowed the conversion rate would be very valuable.

Until human studies are conducted, the evidence from animals predicts that a combined approach would produce the best results. There is certainly no harm in combining a diet that substitutes polyunsaturated fats for saturated fats with regular aerobic exercise and deliberate exposure to novel learning experiences. Such an approach is likely to confer benefits with little or no cost.

13.2 What Can Be Done? Adapt the Environment

Thus far we have described the many ways in which individuals can adapt themselves to the sensory, motor, and cognitive losses associated with old age. In addition, an aged individual's physical surroundings and the devices he or she uses at home and at work can be adapted in various ways to accommodate reduced cognitive capabilities as well as reduced physical capabilities. However, any such environmental adaptation must be designed with sufficient attention to human factors—that is, it must take into consideration the capabilities and limitations of older users.

The relatively new field of human factors and technology for the older adult provides a complementary approach to adapting the individual. A number of technologies have been proposed, but evaluations of their efficacy are relatively rare. A number of the proposals are aimed at supporting older people's desire to "age in place" (that is, in their own homes) with automated or semi-automated systems supporting medication use, diabetes management, access to emergency services, monitoring by relatives, and even cooking. Some of the experimental research on these systems has shown that issues such as perceived politeness and complacency should be considered when designing automation for older people. Memory load also should be considered. And there is concern that poorly designed automation could have the perverse effect of limiting cognitive stimulation. Well-designed automation should not rob older people of their autonomy and their need for cognitive challenge. With the baby boom generation reaching retirement age, there will be a market for well-designed automation aimed at helping older people "age in place."

13.3 What Should Be Done?

Not everything that *can* be done to ameliorate cognitive aging *should* be done. The weight of the evidence from human and animal studies suggests that substitution of polyunsaturated for saturated fats, regular aerobic exercise, and cognitive training can improve cognitive performance in older individuals. The first two are easily achieved, although motivation is always an issue. There is increasing knowledge about the best way to motivate people regarding diet,[31] and research is needed to determine whether the motivating factors differ among lifestyle factors. Cognitive training (that is, learning new skills and acquiring knowledge) appears to confer benefits with no costs other than monetary ones. At present, though, the optimal form and means of delivery of cognitive training remain unknown, and how cognitive training could be conducted on a large scale is not clear. Although software for cognitive training is becoming available, we are not aware of any comparisons of the effectiveness and appeal of such software. Arthur Kramer, who has done pioneering work on the effect of exercise on cognition in older people, suggests "walking book clubs." Many universities are offering older people housing on their campuses as a way to give them greater access to university services and programs. To an extent, older people are developing some of their own solutions for increasing cognitive stimulation. At Lifelong Learning Institutes, self-organized groups of older people arrange classes for their members, typically in affiliation with a university or a community college.[32] This is the sort of community-wide approach that is needed if cognitive training is to become widely available and affordable.

In addition to adapting the individual to old age, the world can be adapted to the aged individual. That approach has promise if used thoughtfully so as to avoid reducing cognitive stimulation; however, the field is in its infancy, with mainly demonstration projects and with little rigorous scientific evaluation of the benefits of the various technologies that have been developed. Automation (such as Honeywell's ILSA) that is aimed at helping people "age in place" is intuitively appealing, but its development will require rigorous experimental work. An optimal approach to automation for older people may entail finding a way to assist in daily life without reducing cognitive stimulation. It is hoped that the marketplace will generate progress in this area.

However, not everything that can be done should be done. Other potential lifestyle changes appear to have some benefits for cognitive aging but cannot at present be recommended. Dietary restriction, as has already been noted, is probably never going to be very popular. Nonetheless, it should studied further, not only to confirm earlier findings of beneficial effects on cognition in humans but also to determine what method (e.g., daily calorie reduction, alternate-day fasting, or weekly fasting) is most tolerable. The strength of DR's effects in animals suggests that clinical trials might be warranted in humans who are at increased risk of cognitive decline, such as those with diabetes and Mild Cognitive Impairment (MCI). Resveratrol may be found to have a benefit similar to that provided by DR, but health risks arise after two drinks per day. B-vitamin supplementation, beyond that currently mandated in flour, does not appear to benefit cognitive functioning in old age and may increase the risk of cancer. Nor can estrogen use by women be recommended without further research. Although beneficial for cognition, estrogen may carry too great a health risk to be used widely and for more than a few years. The costs and the benefits of cognitive-enhancing drugs have not been sufficiently investigated to warrant recommendations.

The first of the baby boomers, born in 1946, have now reached the age at which age-related cognitive decline is detectable. The retirement age is being raised in many Western nations, and the recent economic recession has reduced many people's retirement savings markedly. Older people will have to work longer, even after retirement, to support themselves, and they will need to have their wits about them to do so. The time has come to begin large-scale clinical trials in older people who appear to be "at risk" of decline (for example, those with MCI and diabetes)—trials comparing various lifestyle manipulations that have been shown to heighten cognitive and brain functioning in healthy older people (diet, cognitive training, exercise). Reaching a consensus on the optimal ways to limit decline in healthy older people as well as in those "at risk" will require a cognitive science approach to research. In the meantime, aging individuals can undertake lifestyle changes with the goal of enhancing brain and cognitive plasticity and sustaining cognitive functioning.

Glossary

Aβ (amyloid beta) A peptide created when the amyloid precursor protein (which straddles the neuron membrane) is cleaved by the enzymes beta- and gamma-secretase. Aβ normally regulates synaptic activity, but forms of it are also associated with the amyloid plaques of *Alzheimer's disease*.

Acetylcholine A chemical that acts as a *neurotransmitter* in the brain and the spinal cord.

ADHD Attention Deficit Hyperactivity Disorder.

Ad libitum eating Eating "at one's pleasure," with food continuously available.

Allele A variant form of a gene. Alleles differ in one or more amino acids. The differences may or may not affect the protein produced by the gene.

Alternate-day fasting A form of dietary restriction in which the individual eats only every other day.

Alzheimer's disease A neurological disease involving progressive loss of cognitive function. It is associated with both genetic and environmental risk factors.

AMPA α-amino-3-hydroxyl-5-methyl-4-isoxazole-propionate, a compound that is an agonist for AMPA *receptors*. It mimics effects of the important excitatory *neurotransmitter* glutamate.

APOE An abbreviation used to refer both to the apolipoprotein (lipid-carrying protein) and to the gene that controls the apolipoprotein. The substance has broad effects in the brain, notably in axonal repair and *plasticity*. The ε4 allele of the APOE gene is the strongest known genetic risk factor for *Alzheimer's disease*.

Axon A long projection from a neuron that carries neuronal signals from one neuron to another.

Axonal sprouting Growth of an *axon* from either cut or intact fibers over some distance.

Barrel cortex   Regions of layer IV of somatosensory *cortex* of rodents that represent whiskers. Inputs from the thalamus carrying information from each whisker terminate in discrete areas of layer IV.

BDNF   Brain-derived neurotrophic factor. One of the *neurotrophins*, naturally occurring secreted proteins that act as growth factors in the brain and the blood.

BOLD signal   Blood-oxygen-level-dependent signal. Obtained with functional magnetic-resonance imaging (*fMRI*) of the brain; used to measure changes in blood flow related to neural activity.

BrdU   Bromodeoxyuridine, a thymidine analog used to label proliferating cells such as cancer cells or neurons.

Brodmann's Areas   A scheme of categorizing brain regions. First published in 1909 and still in wide use, it is based on the cytoarchitectonic organization of neurons.

CA fields   The four cornu Ammonis (CA) fields in the *hippocampus*. These fields have tightly packed pyramidal cells that look similar to *neocortex*.

Caudate nucleus   A nucleus of neurons in the basal ganglia that is important in learning and in memory.

Cholinesterase inhibitors   Chemicals that inhibit the cholinesterase enzyme from breaking down *acetylcholine* in a *synapse*, thereby increasing the level and duration of the effect of the *neurotransmitter* acetylcholine.

CNS   The central nervous system, consisting of the brain and the spinal cord.

COMT   Catechol-o-methyltransferase, one of a group of enzymes that break down catecholamines (*dopamine*, epinephrine, norepinephrine) in a *synapse*. The COMT gene controls production of the enzyme.

Constraint-induced movement therapy (CIMT)   A therapy developed by Edward Taub for people in whom unilateral brain damage has caused paresis of one or both of the limbs on one side of the body. The unaffected limb is restrained for much of the waking day. Daily therapy is instituted to shape desired movements of the more affected limb.

Corpus callosum   A thick band of *white matter* (myelinated axons) that interconnects the two cerebral cortices.

Cortex   The outer "skin" of the cerebral hemispheres, containing neurons. It surrounds the deeper *white matter*, which contains the axons of neurons. Neocortex has six layers; allocortex, as in the hippocampus, has fewer.

Covert attention   Visuospatial attention that can be deployed with or without eye movements.

CREB A protein with well-established links to formation of long-term memory and neuronal *plasticity*. The abbreviation stands for cAMP response element-binding.

CSF Cerebrospinal fluid, a clear fluid in which the brain and the spinal cord are bathed and rinsed of metabolic waste. It also provides cushioning for the brain and the spinal cord. Its chemical composition can affect the brain.

Dendrites, dendritic branches, dendritic spines Dendrites are branched projections of a neuron that *synapse* with an *axon* of an adjacent neuron to conduct neuronal signals. The dendrites and their branches are referred to collectively as the dendritic arbor. Spines protrude from dendrites to provide points of synaptic contact with other neurons.

Dentate gyrus A dense layer of small granule cells that curve around the end of the *hippocampus* proper.

DHA Docosahexaenoic acid, a nutritionally important omega-3 fatty acid found in cold-water fish oil and in a few types of nuts and seeds.

Dietary restriction (DR) Restriction of calories below the *ad libitum* level.

Dopamine A major *neurotransmitter* in the brain. Thought to mediate *working memory*, it binds to several different *receptors*, termed D1 to D5.

DTI (diffusion tensor imaging) An *MRI* technique that measures the diffusion of water in *axons* and in other tissues.

Effect size The size of a statistically significant treatment effect, for example, between group means.

Electrotonic coupling Electrical *synapses*.

Entorhinal cortex A cortical region in the medial temporal lobe that lies close to the *hippocampus* and, along with that structure, is known to play an important role in memory formation.

Environmental enrichment Housing animals in large group cages with toys, tunnels, hidden food, and so on.

EPA Eicosapentaenoic acid, a nutritionally important omega-3 fatty acid found in cold-water fish oil and in a few types of nuts and seeds.

Episodic memory Memory for events that occurred previously in specific places or at specific times—those with a "time-stamped" quality.

EPSP Excitatory postsynaptic potential—the graded voltage change caused by movement of positively charged ions into the postsynaptic neuron, which makes the neuron more likely to fire.

Executive attention, executive functioning A somewhat loosely defined set of functions characterized by ability to form and carry out plans, resolve conflicts, and so on.

Filopodia Protrusions from dendrites that are thought to develop into dendritic spines.

fMRI Functional magnetic-resonance imagery, which measures the change in blood flow (termed "hemodynamic response") related to neural activity in the brain. Neurons that are active use oxygen, which results in changes in the ratio of oxygenated to deoxygenated blood in an activated region of the brain.

Fractional anisotrophy (FA) Measures the extent to which the flow of water molecules is restricted to one direction by the composition of different tissues, such as cell membranes and *myelin*.

Glial cells Support cells for neurons. More numerous than neurons, they support neurons by producing *myelin*, providing oxygen, and removing dead neurons.

Glucose regulation Levels of glucose (sugar) in the blood are monitored by cells in the pancreas that release hormones to increase or decrease blood sugar.

GWAS Genome-wide association study—a study of all or most of the genes (the genome) of a species to determine genetic variation among individuals of the species.

Hemiparesis Weakness in a limb that occurs when the sensorimotor cortex controlling that limb is damaged. It is typically seen contralateral to the hemisphere containing the lesion; for example, a lesion in the right-hemisphere motor strip in the hand area would result in weakness of the left hand.

Hippocampus The "hippocampus proper" consists of the four *CA* fields. The "hippocampal formation" consists of the hippocampus proper, the *dentate gyrus*, and the subiculum.

Insulin resistance A condition in which the body produces insulin but does not use it effectively.

In vitro In glass—that is, outside of the living organism, as when neurons are studied outside of the brain.

In vivo In the organism, as when neurons are studied as part of the living brain.

Ischemia A state of insufficient blood supply.

LTP Long-term potentiation—enhanced synaptic transmission between neurons after repeated stimulation of one by the other. It is considered an important cellular mechanism underlying learning and memory.

MCI Mild Cognitive Impairment.

Meta-analysis A "study of studies" looking at all the relevant papers on a topic and formally determining how convincing the effects are overall.

Mini-Mental State Exam (MMSE) A 30-item test of orientation and memory.

Modafinil An analeptic drug manufactured by Cephalon and approved by the US Food and Drug Administration for treatment of sleep disorders. The precise mode of action is not understood.

Morris water maze A standard test of spatial memory in rodents, illustrated above in figure 3.6.

MRI Magnetic-resonance imagery, an imaging method used to visualize structures inside the brain and other portions of the body by means of nuclear magnetic resonance. A strong magnetic field is used to align the magnetization of hydrogen atoms in the body. Radio frequency fields are then used to alter the alignment of the induced magnetic field, producing a rotating magnetic field that can be detected by the scanner.

Myelin An electrically insulating material produced by specialized glial cells in the brain and the spinal cord. Myelin is wrapped around neuronal *axons* much like the insulation of an electrical cord. Myelinated axons conduct more rapidly than non-myelinated ones.

Neocortex See *Cortex*.

Neurogenesis The process by which new neurons are formed. This occurs in development, but it can also occur in certain brain regions of adults.

Neuroimaging Various techniques used to image the brain, including *MRI* and *PET*.

Neuro-muscular junction The synapse between motor neurons and muscle fibers.

Neurotransmitters Chemical substances found within the nervous system that transmit neural information across *synapses*.

Neurotrophins Naturally occurring secreted proteins that act as growth factors. Found in the brain and in blood, they have long been known to play a major role in neuronal development. Nerve growth factor (NGF), brain-derived neurotrophic factor (*BDNF*), neurotrophin-3 (NT-3), and neurotrophin-4 (NT-4) are structurally related to one another. In general, they have a role in preventing neurons from initiating programmed cell death (apoptosis); they also promote formation of progenitor cells during neurogenesis.

NMDA receptors Glutamate *receptors* that play a major part in the molecular process by which synaptic *plasticity* and memory formation are processed in the brain.

Oxidative stress   Balance between the production of peroxides and free radicals from oxidation ("reactive oxygen species") during aerobic metabolism and the ability of the body to repair the resulting damage to neurons.

Paresis   Impaired movement, typically of limbs, usually attributable to a brain lesion.

Parietal cortex   See figure 2.1.

Parkinson's disease   A neurological disorder involving progressive loss of motor functions. It involves the gradual depletion of *dopamine* from the substantia nigra in the midbrain.

PET   Positron-emission tomography, a method of imaging brain regions that are active during task performance. Gamma-ray detectors are used to measure positrons emitted from a radioactive tracer (e.g., labeled water or glucose) injected into arterial blood. Regions of the brain that are active during a task take up the radioactive substance preferentially.

PFC   Prefrontal cortex—the anterior part of the frontal lobes, anterior to the motor and premotor regions. See panels a and b of figure 2.1.

Plasticity   The process by which the brain and neuronal structures can change as a result of experience.

Polyunsaturated fatty acids (PUFAs)   Fatty acids are an integral part of phospholipid cell membranes. Polyunsaturated fatty acids have two or more double bonds, whereas saturated fatty acids have no double bonds.

Posterior cingulate   See figure 2.1.

Precuneus   See figure 2.1.

Prefrontal cortex (PFC)   See figure 2.1.

Progenitor cells   Cells in the process of becoming neurons during *neurogenesis*.

Radioligand   A substance that normally binds to a *receptor* but can be made radioactive so that the binding of the molecule to the receptor can be measured.

Randomized assignment   A form of experimental design in which participants are assigned to an experimental condition by a random number. For example, participants in a study comparing effects of a dietary manipulation would be randomly assigned to one type of diet. This approach is in contrast to an "observational" design, in which people choose a treatment.

Receptor   A protein molecule on a post-synaptic neuron that receives and responds to a *neurotransmitter* ("binding") or some other substance.

SIRT1   One of a family of sirtuin genes in mammals. Sir is the ortholog in yeast. Sirtuins are hypothesized to play a role in the response of an animal to stress. SIRT1 is the first SIRT gene to be discovered; sirt1 is the enzyme controlled by the gene. (SIR is an acronym for Silent Information Regulator.)

Somatosensory cortex   A region of cortex, just posterior to the central sulcus, that contains a sensory map of the body such that each region of skin has a region of neurons in somatosensory cortex which respond to touch of that skin.

Stroop task   A task that measures the ability to name the color of a word that is printed in ink of a color that differs from the meaning of the word. There is a strong tendency to say the word rather than name the color.

Synapse   The active junction between neurons across which a chemical or electrical signal is transmitted.

Synaptogenesis   The formation of *synapses*, which occurs throughout life.

TMS   Transcranial magnetic stimulation, a non-invasive method for assessing cortical function by inducing a weak current in targeted brain areas through the intact scalp.

Upregulation, downregulation   Increase or decrease of a cellular structure, respectively.

Variance   A measure of how far a set of numbers are spread out from the mean of the set of numbers.

VO_2 max   A commonly used measure of physical fitness that is used to index aerobic capacity (the maximum capacity of an individual's body to transport and use oxygen).

Wechsler Memory Scale   A widely used neuropsychological test of memory published by PsychCorp. The latest edition is the fourth.

White matter   An important component of the brain, composed of myelinated and unmyelinated axons.

Working memory   The ability to hold and manipulate information in a short-term memory store.

Notes

Chapter 1

1. US Administration on Aging, "Older Population as a Percentage of the Total Population: 1900 to 2050," Excel spreadsheet, 2009.

2. See, e.g., Cockerham 1991; Kinsella and Phillips 2005; Restrepo and Rozental 1994; Smith and Kington 1997. For additional examples, see http://GlobalAging.org.

3. Andersson 2004.

4. *Source:* http://www.adb.org.

5. Farrell 2008.

6. Park and Schwarz 2000.

7. Craik and Salthouse 2007.

8. Schaie 2005.

9. Baltes and Mayer 1999.

10. Park et al. 2002; Park and Schwarz 2000.

11. For a recent version of the ongoing debate, see Salthouse 2009 and Schaie 2009.

12. For reviews, see Greenwood 2000 and Raz 2000.

13. Shakespeare, *As You Like It,* act ii, scene VII, lines 139–166.

14. Farrell 2008.

15. Alzheimer's Association 2010.

16. Morris and Becker 2004.

17. Morris and Kopelman 1986.

18. McKhann et al. 1984.

19. Baddeley et al. 1991; Parasuraman and Haxby 1993; Parasuraman and Martin 1994.

20. Greenwood and Parasuraman 1994; Greenwood et al. 1993.

21. Parasuraman et al. 1992.

22. Posner 1980.

23. Greenwood and Parasuraman 1994; Greenwood et al. 1993; Hartley et al. 1990.

24. Parasuraman et al. 1992.

25. Ibid.

26. For a review, see Buckner 2004.

27. Parasuraman and Greenwood 2004.

28. Rebeck et al. 1994; Schoenhofen et al. 2006.

29. Folstein et al. 1975.

30. Sliwinski et al. 1996.

31. Bäckman et al. 2002.

Chapter 2

1. Gallagher et al. 1993; Lee et al. 1994; Pawlowski et al. 2009.

2. Rapp and Amaral 1991.

3. Voytko et al. 2001.

4. Willis and Schaie 1986.

5. Salthouse 2006.

6. Raz et al. 2005.

7. Salat et al. 2002; Van Petten et al. 2004.

8. Rodrigue and Raz 2004.

9. Hof and Morrison 2004.

10. Rakic 1985.

11. Das and Altman 1971.

12. Ibid.

13. Nottebohm 2002.

14. Kaplan and Hinds 1977.

15. Rakic 1985.

16. Specter 2001.

17. Ibid.

18. Eriksson et al. 1998.

19. Milner 1958.

20. Leonard et al. 2009.

21. Curtis et al. 2007.

22. Knoth et al. 2010.

23. Srivareerat et al. 2011.

24. Rogers et al. 1998.

25. McEwen et al. 2001.

26. Cotman et al. 2007.

27. Galimberti et al. 2006.

28. Park et al. 2002. One of the most widely referred to papers in the field of cognitive aging, this is often cited for its systematic demonstration of the almost linear decline in different cognitive functions with adult aging.

29. Schaie 1994.

30. For opposing views on the debate, see Salthouse 2009 and Schaie 2009.

31. Gallagher et al. 1993, 2003; Morse 1993; Schupf et al. 2004; Wilson, Beckett, et al. 2002.

32. Schaie 2009.

33. Wilson, Beckett, et al. 2002.

34. Schupf et al. 2004.

35. Raz et al. 2005.

36. Persson et al. 2006.

37. Borghesani et al. 2011.

38. Raz et al. 2007.

39. Wu, Brickman, et al. 2008.

40. Carson et al. 2003.

41. Hasher et al. 1991.

42. Healey et al. 2008.

43. Campbell et al. 2010; Kim, Hasher, and Zacks 2007.

44. For example, Park et al. 2002.

45. Taylor et al. 2007.

Chapter 3

1. Giedd et al. 1999.

2. Huttenlocher 1984.

3. Sowell et al. 2001.

4. Jernigan et al. 2001.

5. Allen et al. 2005.

6. Ibid.; Jernigan et al. 2001; Tisserand et al. 2002.

7. Raz et al. 2005.

8. Raz et al. 2004.

9. Piguet et al. 2009.

10. Reviewed in Van Petten et al. 2004.

11. Rapp and Gallagher 1996.

12. Rodrigue and Raz 2004; Salat et al. 2002.

13. Van Petten 2004.

14. Chantome et al. 1999; Foster et al. 1999.

15. Van Petten et al. 2004.

16. Salat et al. 2002.

17. Sowell et al. 2001.

18. Rodrigue and Raz 2004.

19. Brickman et al. 2011.

20. Van Petten 2004.

21. Salat et al. 2002.

22. Braak and Braak 1995.

23. S. A. Small et al. 2002.

24. Braak and Braak 1995; Ghebremedhin et al. 1998.

25. S. A. Small et al. 2002.

26. Rodrigue and Raz 2004.

27. Killiany et al. 2000; Rodrigue and Raz 2004.

28. Raz et al. 2007.

29. Shaw et al. 2006.

30. Sowell et al. 2004.

31. Shaw et al. 2006.

32. Fjell et al. 2006; Salat et al. 2002.

33. Fjell et al. 2006.

34. Fjell et al. 2007.

35. Espeseth et al. 2008.

36. Henderson et al. 1995.

37. Espeseth et al., forthcoming.

38. Fjell et al. 2006.

39. Fjell et al. 2007.

40. Buckner 2004.

41. Walhovd et al. 2006.

42. Schaie 1989, 1994, 2005.

43. Giedd et al. 1999.

44. Fjell et al. 2006.

45. Walhovd et al. 2006.

46. Cabeza 2002.

47. Cabeza et al. 2002.

48. Morcom et al. 2003.

49. Grady et al. 2003.

50. Greenwood 2007.

51. Kleim et al. 2003.

52. Sterr et al. 1998.

53. Elbert et al. 1995.

54. Kalisch et al. 2009.

55. Draganski et al. 2004.

56. Kaas et al. 1984; Merzenich et al. 1984.

57. Pons et al. 1991.

58. Merzenich et al. 1984.

59. Jenkins et al. 1990.

60. Carlson 1991.

61. Pons et al. 1991.

62. Ramachandran and Hirstein 1998.

63. Taub et al. 1993, 2002.

64. Taub et al. 2002.

65. Kunkel et al. 1999; Taub et al. 1993; Wittenberg et al. 2003.

66. Bütefisch et al. 1995.

67. Frost et al. 2003.

68. Liepert et al. 1998.

69. On the TMS technique, see Hallett 2000 and Walsh and Cowey 2000.

70. Liepert et al. 1998.

71. Ro et al. 2006.

72. Leary 1991.

73. For a review, see Nudo et al. 2001.

74. Dancause et al. 2005; Stroemer et al. 1995.

75. Dancause et al. 2005.

76. Nudo et al. 1996.

77. Hihara et al. 2006. For supporting evidence, see Quallo et al. 2009.

78. Xu et al. 2009.

79. Sterr et al. 1998.

80. Maguire et al. 2000.

81. Liepert et al. 1998; Ro et al. 2006.

82. Draganski et al. 2004.

83. Hihara et al. 2006.

84. Xu et al. 2009.

85. Draganski et al. 2006.

86. Erickson et al. 2007.

87. Raz et al. 2007.

88. Fjell et al. 2006; Salat et al. 2002; Shaw et al. 2006.

89. Cabeza et al. 2002; Morcom et al. 2003.

90. Huttenlocher 1984.

91. Picciotto and Zoli 2002; Rogers et al. 1998.

92. Hao et al. 2006; Morrison et al. 2006.

93. Rosenzweig and Bennett 1996.

94. Smith et al. 1999.

95. Molteni et al. 2004; Vaynman et al. 2004.

96. Draganski et al. 2004, 2006; Sterr et al. 1998.

97. Ro et al. 2006.

98. Burke and Barnes 2006.

99. Ibid.

100. Ball 1977, 1978; Brody 1955.

101. Brody 1955.

102. Hof and Morrison 2004; Morrison and Hof 1997.

103. Smith et al. 2004.

104. Masliah et al. 1993.

105. Scheff et al. 2007.

106. Burke and Barnes 2006.

107. Buell and Coleman 1979.

108. Flood 1993; Hanks and Flood 1991.

109. Flood 1993.

110. Pyapali and Turner 1996; Turner and Deupree 1991.

111. Markham et al. 2005.

112. de Brabander et al. 1998; Uylings and de Brabander 2002.

113. de Brabander et al. 1998.

114. Lee et al. 2006.

115. Valdez et al. 2010.

116. Williams and Matthysse 1986.

117. Curcio and Hinds 1983.

118. Duan et al. 2003.

119. See, e.g., Potier et al. 1993; Turner and Deupree 1991.

120. Barnes and McNaughton 1980.

121. Kerr et al. 1989; Moyer et al. 1992; Potier et al. 1993.

122. Wilson et al. 2005.

123. Diana et al. 1994; Dieguez and Barea-Rodriguez 2004; Landfield et al. 1978.

124. Chang et al. 2005.

125. Ibid.

126. Morris 1984.

127. Tombaugh et al. 2002.

128. Moser et al. 2008.

129. Wilson and McNaughton 1993.

130. Barnes et al. 1997; Wilson et al. 2003, 2005.

131. Nicholson et al. 2004.

132. Pawlowski et al. 2009.

133. Newman and Kaszniak 2000.

134. Rapp et al. 1997.

135. Barnes 1979; Gallagher and Rapp 1997.

136. Maguire et al. 2000.

137. Walhovd et al. 2004.

138. Maguire et al. 2006.

139. Alkire et al. 1998.

140. Abrous et al. 2005; Gould et al. 1999.

141. Braak and Braak 1991b.

142. Awad et al. 2004; Convit et al. 1997.

143. Convit et al. 2003.

144. Wu, Brickman, et al. 2008.

145. Convit et al. 2003.

146. McNay et al. 2000.

147. McNay and Gold 2001.

148. Pereira et al. 2007.

149. Good et al. 2001.

150. Courchesne et al. 2000; Jernigan et al. 2001.

151. Giedd et al. 1999.

152. Allen et al. 2005; Walhovd et al. 2005.

153. Bartzokis 2004; Walhovd et al. 2005.

154. Piguet et al. 2009.

155. Bartzokis 2004.

156. Allen et al. 2005; Walhovd et al. 2005.

157. Peters et al. 2000.

158. Xi et al. 1999.

159. Peters and Sethares 2002.

160. Beaulieu 2002.

161. Schmierer et al. 2007.

162. Beaulieu 2002.

163. Beaulieu 2002; Wozniak and Lim 2006.

164. Benedetti et al. 2006; Charlton et al. 2006; Madden et al. 2004; O'Sullivan et al. 2001; Walhovd et al. 2005.

165. Salat et al. 2005.

166. Madden et al. 2004.

167. Fjell et al. 2008.

168. Wang et al. 2009.

169. Charlton et al. 2006.

170. Deary et al. 2006; Grieve et al. 2007.

171. Bucur et al. 2008; Sullivan et al. 2006.

172. Vernooij et al. 2009.

173. Ziegler et al. 2010.

174. Fjell et al. 2006.

175. Scholz et al. 2009.

176. Takeuchi et al. 2010.

177. Lövdén et al. 2010.

178. Persson et al. 2006.

Chapter 4

1. See, e.g., Raz et al. 2005.

2. Bäckman et al. 2000.

3. Gallagher et al. 1993; Glisky et al. 2001; Lee et al. 1994; Rapp and Amaral 1991; Walhovd et al. 2006; Willis and Schaie 1986; Persson et al. 2006; Borghesani et al. 2011.

4. Nicholson et al. 2004; Pawlowski et al. 2009; Walhovd et al. 2004; Wilson et al. 2003.

5. Greenwood and Parasuraman 2010.

6. Greenwood 2007.

7. Chklovskii et al. 2004.

8. Lee et al. 2006; Stepanyants et al. 2004.

9. For a recent review, see Holtmaat and Svoboda 2009.

10. Denk et al. 1990.

11. Holtmaat and Svoboda 2009.

12. Xu et al. 2009.

13. Ambrogini et al. 2010.

14. Rainer and Miller 2000.

15. Olesen et al. 2004.

16. McNab et al. 2009.

17. Abi-Dargham et al. 2002.

18. Pons et al. 1991.

19. Ibid.

20. Merzenich et al. 1984.

21. Voss et al. 2006.

22. For a review of research on phantom limbs, see Ramachandran 2005.

23. See Florence et al. 1998; Pons et al. 1991.

24. Taub et al. 2002.

25. Ro et al. 2006.

26. Wolf et al. 2010.

27. Nudo and Friel 1999.

28. Dancause et al. 2005.

29. Hihara et al. 2006.

30. van Praag et al. 1999.

31. van Praag et al. 2005.

32. Curtis et al. 2007; Curtis et al. 2003; Eriksson et al. 1998.

33. Das and Altman 1971.

34. Rakic 1985.

35. Bhardwaj et al. 2006; Gould and Gross 2002; Rakic 2002.

36. Englund et al. 2002; Kokoeva et al. 2005; Song et al. 2002.

37. Toni et al. 2008.

38. Curtis et al. 2003; Eriksson et al. 1998; Kempermann et al. 1998, 2002; van Praag et al. 2005.

39. Knoth et al. 2010.

40. Faherty et al. 2003; Greenough et al. 1985.

41. Arendash et al. 2004; Frick and Fernandez 2003.

42. Shaw et al. 2006.

43. Fjell et al. 2006.

44. Walhovd et al. 2006.

45. Sterr et al. 1998.

46. Elbert et al. 1995.

47. Draganski et al. 2004.

48. Draganski et al. 2006.

49. Hasher et al. 1991.

50. Healey et al. 2008.

51. Kim, Hasher, and Zacks 2007.

52. Campbell et al. 2010.

53. Gigerenzer and Gaissmaier 2011.

54. Cabeza 2002; Cabeza et al. 2002; Gutchess et al. 2005; Morcom et al. 2003; Reuter-Lorenz et al. 2000; Rugg et al. 2002.

55. Cabeza et al. 1997.

56. For reviews, see Cabeza et al. 2002 and Greenwood 2007.

57. Cabeza et al. 1997.

58. Davis et al. 2008; Grady et al. 1994.

59. Reuter-Lorenz et al. 2000.

60. Morcom et al. 2003.

61. Springer et al. 2005.

62. Chein and Fiez 2001.

63. Cooke et al. 2002.

64. See, e.g., Paulus et al. 2002.

65. Rodrigue and Raz 2004; Tisserand et al. 2004.

66. For a review, see Greenwood 2007.

67. Grady et al. 2003.

68. Rypma et al. 2005.

69. Duckworth and Seligman 2005; Tangney et al. 2004.

70. Shamosh et al. 2008.

71. Gray et al. 2003.

72. Shamosh et al. 2008.

73. Greenwood 2007.

74. Carlson et al. 1995; Hasher et al. 1991.

75. Logan et al. 2002.

76. Cabeza 2002.

77. Drapeau et al. 2003.

78. Farioli-Vecchioli et al. 2008; Snyder et al. 2005.

79. Bruel-Jungerman et al. 2006; Dupret et al. 2008; Waddell and Shors 2008.

80. Dobrossy et al. 2003; Leuner et al. 2004; Waddell and Shors 2008.

81. Kempermann 2008.

82. Kempermann and Gage 1999.

83. Mouret et al. 2008; Waddell and Shors 2008.

84. Straube et al. 2003.

85. Chen, forthcoming.

86. Reviewed in Grossman et al. 2003.

87. Kempermann and Gage 1999.

88. Colsher and Wallace 1991.

89. Yu et al. 1989.

90. Teipel et al. 2009.

91. Carlson et al. 2008.

92. Springer et al. 2005.

93. Reviewed in Hillman et al. 2008.

94. van Praag, Kempermann, and Gage 1999.

95. Naylor et al. 2008.

96. Kronenberg et al. 2006.

97. Farmer et al. 2004.

98. Pereira et al. 2007.

99. Convit et al. 2003.

100. Reviewed in Baur and Sinclair 2006.

101. Greenwood and Winocur 1996.

102. Tschop and Thomas 2006.

103. Convit et al. 2003.

104. Henderson et al. 2000; Matthews et al. 1999.

105. Hogervorst et al. 1999; Resnick et al. 1997.

106. Tanapat et al. 1999.

107. Brinton 2009.

108. Mumenthaler et al. 2003; Yesavage et al. 2002.

109. de Saint Hilaire et al. 2001.

110. Randall et al. 2004.

111. Turner et al. 2003a.

112. Gill et al. 2006.

113. See, e.g., Takasawa et al. 2002.

114. Leuner et al. 2004.

115. Bruel-Jungerman et al. 2006.

116. Kempermann and Gage 2002; Kempermann et al. 1997.

117. Gould et al. 1999; Shors et al. 2001.

118. For human examples, see Cabeza et al. 2002 and Erickson et al. 2007.

119. Cabeza 2002.

Chapter 5

1. Jennifer Gonzalez, "High-school dropout rate is cited as a key barrier to Obama's college-completion goal," *Chronicle of Higher Education* Web site (http://chronicle.com), May 25, 2020.

2. Hebb 1947.

3. Rosenzweig and Bennett 1972.

4. For a review, see Grossman et al. 2002.

5. Rampon et al. 2000.

6. Chapillon et al. 1999.

7. Faherty et al. 2003; Greenough et al. 1985.

8. Kempermann et al. 2002.

9. Cracchiolo et al. 2007.

10. van Praag et al. 2005.

11. Curtis et al. 2007; Curtis et al. 2003; Eriksson et al. 1998.

12. Eriksson et al. 1998.

13. Gould et al. 1998.

14. Kempermann et al. 1998.

15. Eriksson et al. 1998.

16. Bhardwaj et al. 2006; Curtis et al. 2007; Knoth et al. 2010.

17. Gould et al. 1998.

18. Kempermann et al. 1998; Kuhn et al. 1996.

19. van Praag, Kempermann, and Gage 1999; Kronenberg et al. 2006.

20. Kempermann et al. 1998.

21. Ambrogini et al. 2010; Hastings and Gould 1999; van Praag et al. 2002.

22. Englund et al. 2002; Kokoeva et al. 2005; Song et al. 2002.

23. Toni et al. 2008.

24. See, e.g., Urakawa et al. 2007.

25. Bruel-Jungerman et al. 2005.

26. Meshi et al. 2006.

27. Goodrich-Hunsaker et al. 2008.

28. Garthe et al. 2009.

29. Goodrich-Hunsaker et al. 2008.

30. Teipel et al. 2009.

31. Jacobs et al. 1993.

32. Katzman et al. 1997; Yu et al. 1989.

33. Breitner et al. 1999; Katzman 1993; Lindsay et al. 2002.

34. Colsher and Wallace 1991.

35. Starr et al. 1997.

36. Christensen et al. 1997.

37. Stern 2002.

38. Ibid.

39. Foubert-Samier et al., forthcoming.

40. Springer et al. 2005.

41. See, e.g., Rypma and D'Esposito 2000; E. E. Smith et al. 2001.

42. Gray et al. 2003.

43. Westerberg and Klingberg 2007.

44. Cabeza et al. 2002; Morcom et al. 2003. But see also Colcombe et al. 2005.

45. Reviewed in Cabeza 2002.

46. Fotenos et al. 2008.

47. Cao et al. 1999.

48. Greenwood 2007.

Chapter 6

1. CDC 2008.

2. Yeung et al. 2010.

3. Colditz 1999.

4. Physical Activity Guidelines for Americans, 2008 (http://www.health.gov).

5. Cotman and Berchtold 2002.

6. Hillman et al. 2008.

7. van Praag et al. 2005.

8. Nichol et al. 2007; van Praag et al. 2005; Vaynman et al. 2004.

9. van Praag et al. 2005.

10. Molteni et al. 2004.

11. For a review, see Cotman et al. 2007.

12. Hayes et al. 2008.

13. Ding et al. 2006.

14. Luo et al. 2007.

15. Rhyu et al. 2010.

16. Colcombe and Kramer 2003; Etnier et al. 2006; Heyn et al. 2004.

17. Colcombe and Kramer 2003.

18. Erickson et al. 2009.

19. Cardiovascular Health Study-Cognition Study, 2009 (https://biolincc.nhlbi.nih.gov).

20. Erickson, Raji et al. 2010.

21. For a review, see Buckner 2004.

22. Colcombe et al. 2004.

23. Ibid.

24. Smiley-Oyen et al. 2008.

25. Pontifex et al. 2009.

26. Colcombe et al. 2004.

27. Ibid.

28. Colcombe et al. 2006.

29. Erickson et al. 2011.

30. Lautenschlager et al. 2008.

31. Rolland et al. 2007; Teri et al. 2003.

32. Larson et al. 2006.

33. Milner 1958.

34. Gould et al. 1999; van Praag, Kempermann, and Gage 1999.

35. Lou et al. 2008; van Praag, Kempermann, and Gage 1999.

36. Kronenberg et al. 2006.

37. Ang et al. 2003.

38. Trejo et al. 2001; van Praag, Kempermann, and Gage 1999; van Praag, Christie, Sejnowski, and Gage 1999.

39. van Praag et al. 2005.

40. Kim et al. 2007.

41. Kronenberg et al. 2006.

42. Luo et al. 2007.

43. Silva et al. 1998.

44. Eadie et al. 2005.

45. Valdez et al. 2010.

46. Lee et al. 2006.

47. Whitlock et al. 2006.

48. Farmer et al. 2004.

49. Ibid.

50. Berchtold et al. 2005.

51. Ang et al. 2003.

52. Trejo et al. 2001.

53. Fabel et al. 2003; Lou et al. 2008.

54. Reviewed in Poo 2001.

55. Lafenetre et al. 2010.

56. Rasmussen et al. 2009.

57. Knaepen et al. 2010.

58. G. Li et al. 2009.

59. Erickson et al. 2009.

60. Erickson, Raji et al. 2010.

61. Erickson, Prakash et al. 2010.

62. Li et al. 2009.

63. Amenta et al. 1995; Jucker et al. 1990.

64. Pereira et al. 2007.

65. Wu et al. 2008.

66. Ibid.; Convit et al. 1997.

67. Pereira et al. 2007.

Chapter 7

1. Calabrese and Baldwin 1998.

2. Rattan 2004.

3. Mine et al. 1990.

4. Mattson et al. 2002.

5. Walford et al. 1999.

6. Roth et al. 2004.

7. For reviews, see Heilbronn and Ravussin 2003 and Mattson 2000.

8. Elbaz et al. 2009; Ritz and Costello 2006; Tanner et al. 2009.

9. Maswood et al. 2004.

10. Lane 2000.

11. Colman et al. 2009.

12. Ingram et al. 1987.

13. Eckles-Smith et al. 2000.

14. Carter et al. 2009.

15. Yanai et al. 2004.

16. For an example, see the Web site of the Calorie Restriction Society International (http://www.crsociety.org).

17. Fontana and Klein 2007; Fontana et al. 2004; Walford et al. 1999.

18. Walford et al. 1999.

19. Heilbronn et al. 2006; Allard et al. 2008.

20. Green and Rogers 1998.

21. Bryan and Tiggemann 2001.

22. Martin et al. 2007.

23. Witte et al. 2009.

24. W. Wu, Brickman, et al. 2008.

25. Witte et al. 2009.

26. Stote et al. 2007.

27. For reviews, see Mattson et al. 2003 and Varady and Hellerstein 2007.

28. Anson et al. 2003.

29. Johnson et al. 2007.

30. Anson et al. 2003.

31. Ahmet et al. 2010.

32. Heilbronn et al. 2005.

33. For example, Sabia et al. 2009.

34. Gunstad et al. 2007.

35. Kuo et al. 2006.

36. Fotuhi et al. 2009.

37. Farr et al. 2008.

38. Laitinen et al. 2006.

39. Barberger-Gateau et al. 2002; Morris et al. 2003, 2004; Solfrizzi et al. 2006.

40. Mawuenyega et al., 2010.

41. Bayer-Carter et al. 2011.

42. C. E. Greenwood and Winocur 1996.

43. Granholm et al. 2008.

44. Pistell et al. 2010.

45. C. E. Greenwood and Winocur 2005.

46. W. Wu, Brickman, et al. 2008.

47. Murray et al. 2009.

48. Tozuka et al. 2009; Walker et al. 2008.

49. White et al. 2009.

50. Convit et al. 2003.

51. Anson et al. 2003; Heilbronn et al. 2005; Walford et al. 2002.

52. Anson et al. 2003.

53. Fontan-Lozano et al. 2007.

54. Li and Tsien 2009.

55. Bliss and Collingridge 1993.

56. Eckles-Smith et al. 2000; Shi et al. 2007.

57. Lee and Shacter 1999.

58. White et al. 2009.

59. Ibid.

60. Arumugam et al. 2010.

61. Wu et al. 2008.

62. Johnson et al. 2007.

63. Roberge et al. 2008.

64. Arumugam et al. 2010.

65. Ibid.

66. Bramham and Messaoudi 2005; Mattson et al. 2004; Pang and Lu 2004.

67. Stranahan et al. 2008.

68. Lindqvist et al. 2006.

69. Mattson et al. 2004.

70. Park et al. 2010.

71. Molteni et al. 2004.

72. Pfluger et al. 2008.

73. For a review of the evidence, see Baur and Sinclair 2006.

74. Ibid.

75. Pearson et al. 2008.

76. Oomen et al. 2009.

77. Barger et al. 2008.

78. Arunachalam et al. 2010.

79. Feng et al. 2009.

80. Bournival et al. 2009.

81. See http://clinicaltrials.gov/ct2/results?term=resveratrol.

82. Browner et al. 2004; Burnett et al. 2011; Viswanathan et al. 2011; Lombard et al. 2011.

83. Yamamoto et al. 2007.

84. Guarente and Picard 2005.

85. D. Kim et al. 2007.

86. Pearson et al. 2008.

87. Renaud and de Lorgeril 1992.

88. Baur and Sinclair 2006.

89. Cleophas 1999.

90. See, e.g., Naimi et al. 2005; Shaper et al. 1988.

91. Fillmore et al. 2007.

92. Lakshman et al. 2010.

93. Baur and Sinclair 2006.

94. Simopoulos 2002.

95. Mensink et al. 2003.

96. Lewington et al. 2007.

97. Crawford et al. 1999.

98. Jump 2002a.

99. Duplus et al. 2000; Puskas et al. 2003.

100. Jump 2002a 2002b.

101. McGahon et al. 1999.

102. Stillwell et al. 2005.

103. Gómez-Pinilla 2008.

104. A. Wu et al. 2008.

105. Fernandes et al. 2008.

106. Hashimoto et al. 2009.

107. Jiang et al. 2009.

108. Holguin et al., 2008.

109. For a review, see Oster and Pillot 2010.

110. Virtanen et al. 2008.

111. Vercambre et al. 2010.

112. Cherubini et al. 2007.

113. van de Rest et al. 2009.

114. Cole et al. 2009.

115. Muldoon et al. 2010.

116. van Gelder et al. 2007.

117. McNamara et al. 2007.

118. Lukiw et al. 2005.

119. Lim et al. 2006.

120. Yurko-Mauro et al. 2010.

121. Dangour et al. 2010.

122. van de Rest et al. 2008.

123. Ibid.

124. Mozaffarian et al. 2010.

125. Ma et al. 2007.

126. Reviewed in Stillwell et al. 2005.

127. Troen and Rosenberg 2005.

128. Seshadri et al. 2002.

129. Bunce et al. 2004.

130. Honein et al. 2001.

131. Jacques et al. 1999.

132. Cole et al. 2007.

133. For example, Lonn et al. 2006.

134. Kang et al. 2008.

135. Aisen et al. 2008.

136. Smith et al. 2010.

137. Kang et al. 2008.

138. Malouf and Grimley Evans 2008.

139. Gómez-Pinilla 2008.

140. Gray et al. 2003; Zandi et al. 2004; Dai et al. 2006.

141. Grodstein et al. 2007.

142. Hu et al. 2006.

143. A. Wu et al. 2006.

144. Maher et al. 2010.

145. Baum et al. 2008.

146. Ahmet et al. 2010.

147. Anson et al. 2003.

148. Strandberg et al. 2008.

149. Rock et al. 2010.

150. See also Gómez-Pinella 2008.

Chapter 8

1. Bäckman et al. 2000; Volkow et al. 1998.

2. Voytko et al. 2001.

3. Freeman et al. 2007.

4. Shumaker et al. 2003.

5. Bishop and Simpkins 1994; Luine 1985.

6. Singh et al. 1994.

7. Luine et al. 1980.

8. Morrison et al. 2006.

9. Woolley and McEwen 1994.

10. Woolley 1999.

11. Sandstrom and Williams 2001.

12. Hao et al. 2006; Liu et al. 2008; Murphy et al. 1998; Woolley and McEwen 1994.

13. Spencer et al. 2008.

14. Tanapat et al. 1999.

15. Gould et al. 1997, 2001.

16. Eriksson et al. 1998.

17. Tanapat et al. 1999.

18. Ibid.

19. Shors et al. 2001.

20. Dalla et al. 2009.

21. Ibid.

22. Dietrich et al. 2001.

23. Henderson et al. 2000; Yaffe et al. 1998. For a review, see Sherwin 2002.

24. Hogervorst et al. 1999.

25. Resnick et al. 1997.

26. Matthews et al. 1999.

27. Erickson et al. 2005.

28. Yaffe et al. 2007.

29. de Moraes et al. 2001.

30. Lord et al. 2008.

31. Matthews et al. 1996.

32. Sherwin 1988.

33. Tierney et al. 2009.

34. Duka et al. 2000.

35. Shaywitz et al. 1999.

36. Joffe et al. 2006.

37. Krug et al. 2006.

38. Ragaz et al. 2010.

39. Teng and Chui 1987.

40. Espeland et al. 2004.

41. Resnick et al. 2006.

42. Brinton 2009; Sherwin and Henry 2008; Simpkins and Singh 2008.

43. Reviewed in Simpkins and Singh 2008.

44. Sherwin and Henry 2008.

45. Kuiper et al. 1997.

46. Nilsen and Brinton 2003.

47. Sherwin and Henry 2008.

48. Joffe et al. 2006; Krug et al. 2006; Sherwin 1988.

49. Binder et al. 2001.

50. Folstein et al. 1975.

51. Resnick et al. 1997.

52. Gibbs and Gabor 2003.

53. Rapp et al. 2003.

54. MacLennan et al. 2006.

55. Erickson, Voss, et al. 2010.

56. Farquhar et al. 2009; Sherwin and Henry 2008; Simpkins and Singh 2008.

57. Bagger et al. 2005.

58. Erickson et al. 2005.

59. Erickson, Colcombe, Elavsky, et al. 2007.

60. Berchtold et al. 2001.

61. Hogervorst et al. 2000.

62. Zec and Trivedi 2002.

63. Lethaby et al. 2008.

64. Kronos Early Estrogen Prevention Study (http://www.keepstudy.org).

65. Harman et al. 2005.

66. http://clinicaltrials.gov/ct2/show/NCT00114517.

67. Greely et al. 2008.

68. DeSantis et al. 2008.

69. Mattay et al. 1996.

70. Mattay et al. 2000.

71. Elliott et al. 1997; Mehta et al. 2000.

72. Elliott et al. 1997.

73. Turner et al. 2003b.

74. Abi-Dargham et al. 2002; McNab et al. 2009.

75. Aalto et al. 2005.

76. Barch 2004.

77. Kimberg et al. 1997.

78. Arnsten et al. 1995, 1998.

79. Everitt and Robbins 1997.

80. Furey et al. 2000, 2008.

81. Mumenthaler et al. 2003.

82. Yesavage et al. 2002.

83. Turner et al. 2004.

84. Ibid.

85. Spence et al. 2005.

86. Minzenberg et al. 2008.

87. Greenwood 2007.

88. Minzenberg et al. 2008.

89. Rosso et al. 2008.

90. Ribeiro et al. 2002.

91. Deslandes et al. 2005.

92. Wentz and Magavi 2009.

93. Arendash et al. 2009.

94. Ritchie et al. 2007, 2010.

95. Santos et al. 2010.

96. Nehlig 2010.

Chapter 9

1. Rohwedder and Willis 2010.

2. Ibid.

3. Hultsch et al. 1999; Swaab et al. 2002.

4. Reviewed in Grossman et al. 2003.

5. Faherty et al. 2003; Greenough et al. 1985.

6. Arendash et al. 2004; Frick and Fernandez 2003.

7. Kempermann et al. 1997, 1998.

8. Kempermann and Gage 2002; Kempermann et al. 1997.

9. Wurm et al. 2007.

10. Drapeau et al. 2003; Kempermann et al. 1997; van Praag, Christie et al. 1999.

11. Hastings and Gould 1999.

12. See, e.g., Takasawa et al. 2002.

13. Leuner et al. 2004.

14. Chun et al. 2006.

15. Bruel-Jungerman et al. 2006.

16. Gould et al. 1999.

17. Kempermann 2008.

18. Kempermann and Gage 2002; Lemaire et al. 2000.

19. Goodrich-Hunsaker et al. 2008.

20. Garthe et al. 2009.

21. Goodrich-Hunsaker et al. 2008.

22. Diekelmann and Born 2010.

23. Bruel-Jungerman et al. 2006.

24. Abrous et al. 2005.

25. Curtis et al. 2007; Curtis et al. 2003; Eriksson et al. 1998.

26. Schooler et al. 2004.

27. Salthouse 2006.

28. Hertzog et al. 2009.

29. See, e.g., Willis et al. 2006.

30. Floyd and Scogin 1997; Verhaeghen et al. 1992.

31. Verhaeghen et al. 1992.

32. Bherer et al. 2005; Rasmusson et al. 1999.

33. See, e.g., Ball et al. 2002; Willis et al. 2006.

34. Greenwood 2007.

35. Gray et al. 2003.

36. Logan et al. 2002.

37. Cabeza 2002.

38. Grady et al. 2003.

39. Erickson, Colcombe, Wadhwa, et al. 2007.

40. Salat et al. 2002; Van Petten 2004; Van Petten et al. 2004.

41. DeRubeis et al. 2008.

42. Olesen et al. 2004.

43. Sawaguchi and Goldman-Rakic 1991.

44. Abi-Dargham et al. 2002.

45. Aalto et al. 2005.

46. McNab et al. 2009.

47. Demerens et al. 1996.

48. Peters and Sethares 2002.

49. Bartzokis 2004; Piguet et al. 2009.

50. Madden et al. 2004; Salat et al. 2005; Ziegler et al. 2010.

51. Takeuchi et al. 2010.

52. http://en.wikipedia.org/wiki/Cascade_(juggling).

53. Scholz et al. 2009.

54. Draganski et al. 2006.

55. Lövdén et al. 2010.

56. Demerens et al. 1996.

57. Draganski et al. 2006.

58. Walhovd et al. 2006.

59. Song et al. 2005.

60. Wheeler-Kingshott and Cercignani 2009.

61. Lövdén et al. 2010.

62. Demerens et al. 1996; Wake et al. 2011.

63. See, e.g., Yesavage and Rose 1984.

64. Ball et al. 2002.

65. Willis and Marsiske 1993.

66. Willis et al. 2006.

67. Craik 1983.

68. Bissig and Lustig 2007.

69. Lustig and Flegal 2008.

70. Derwinger, Stigsdotter Neely, and Bäckman 2005; Derwinger, Stigsdotter Neely, MacDonald, and Bäckman 2005.

71. Jonides 2004.

72. Dahlin, Nyberg, et al. 2008.

73. O'Reilly and Frank 2006.

74. Dahlin, Neely, et al. 2008.

75. Jonides 2004.

76. Dahlin, Nyberg, et al. 2008.

77. Dahlin, Neely, et al. 2008.

78. Kramer et al. 1995.

79. Bherer et al. 2005 2008.

80. Kramer et al. 1995.

81. Karbach and Kray 2009.

82. Tranter and Koutstaal 2008.

83. Ibid.

84. Basak et al. 2008.

85. Ibid.

86. Lövdén et al. 2010.

87. Schmiedek et al. 2010.

88. Ibid.

89. Stine-Morrow et al. 2008.

90. www.odysseyofthemind.org.

91. Carlson et al. 2009.

92. Stern et al. 1999; Tang et al. 2001; Zsembik and Peek 2001.

Chapter 10

1. Kattenstroth et al. 2010.

2. Wu, Ying, et al. 2008.

3. Joseph et al. 1998; Socci et al. 1995.

4. Cummings et al. 1996.

5. Head et al. 2002.

6. Milgram et al. 2002.

7. Pop et al. 2010.

8. Scarmeas et al. 2009.

9. Small et al. 2006.

10. Small 2004.

11. Fabre et al. 2002.

Chapter 11

1. Lee and Kirlik 2011; Wickens and Hollands 2000.

2. Fisk and Rogers 1997.

3. Charness and Schaie 2003; Fisk, Charness et al.; Pew and Hemel 2004.

4. Fozard and Gordon-Salant 2001; Kline and Scialfa 1995.

5. Schieber 2003.

6. Baldwin 2011.

7. Baldwin 2002.

8. Proctor and van Zandt 1984.

9. Fitts 1954.

10. Wickens and Hollands 2000.

11. Hancock et al. 2001.

12. Gardner et al. 1993.

13. Preiser and Ostroff 2001.

14. Card et al. 1980.

15. For a comprehensive survey of methods used to evaluate consumer products, see Stanton 1997.

16. Levin 1977.

17. Bhuyan 2004.

18. Majeed and Brown 2006.

19. Haigh et al. 2003.

20. Parasuraman and Miller 2004.

21. http://www.grandcare.com/page/home

22. http://dsc.discovery.com/videos/discovery-channel-cme-future-family-part-2 .html

23. http://beclose.com;   http://www.gehealthcare.com/usen/telehealth/quietcare/ proactive_eldercare_technology.html

24. Ho et al. 2005.

25. Parasuraman 2003; Parasuraman et al. 2008.

26. Parasuraman and Manzey 2010.

27. Parasuraman and Riley 1997.

28. NTSB 1997.

29. Parasuraman and Manzey 2010.

30. Hardy 1995; Vincenzi et al. 1996.

31. Sanchez et al. 2004.

32. Ezer et al. 2008.

33. Ho et al. 2005.

34. Parasuraman and Manzey 2010.

35. Bahner et al. 2008.

36. Convit et al. 2003; Wu, Brickman, et al. 2008.

37. Glasgow and Strycker 2000.

38. Rapoff 1999.

39. Blanson Henkemans et al. 2008.

40. Tran et al. 2005.

41. Pew Research Center 2009.

42. Aging In Place 2010.

Chapter 12

1. Reinvang et al. 2010.

2. Venter et al. 2001.

3. Plomin et al. 2001.

4. Davis et al. 2009.

5. McClearn et al. 1997.

6. Plomin and Crabbe 2000.

7. For reviews, see Goldberg and Weinberger 2004; Green et al. 2008; Greenwood and Parasuraman 2003.

8. Egan et al. 2001; Fan et al. 2003; Fossella et al. 2002; Fossella and Casey 2006; Greenwood et al. 2000; Parasuraman et al. 2005; Posner et al. 2007.

9. Espeseth et al. 2006; Greenwood, Lin, et al. 2009; Greenwood, Sundararajan, et al. 2009.

10. Hirschhorn and Daly 2005.

11. Davies et al. 2011.

12. Lambert et al. 2009.

13. Ibid.

14. Seshadri et al. 2010.

15. Purcell et al. 2009.

16. Pedersen 2010.

17. Gatz et al. 2006.

18. Parasuraman and Greenwood 2004.

19. Goldberg and Weinberger 2004.

20. Greenwood and Parasuraman 2003.

21. Saunders et al. 1993; Strittmatter et al. 1993.

22. Mahley 1988.

23. Arendt et al. 1997; White et al. 2001.

24. Corder et al. 1993.

25. Ibid.

26. Meyer et al. 1998.

27. Bondi et al. 1999; Espeseth et al., forthcoming; Parasuraman et al. 2002; Small et al. 2004.

28. Greenwood et al. 2000.

29. Goldberg and Weinberger 2004.

30. Payton 2009.

31. Green et al. 2008; Greenwood and Parasuraman 2003; Payton 2009.

32. Parasuraman and Greenwood 2004.

33. Corbetta et al. 2000; Yantis et al. 2002.

34. Davidson and Marrocco 2000.

35. Parasuraman et al. 2002.

36. Xiang et al. 1998.

37. Phillips et al. 2000.

38. Murphy and Klein 1998.

39. Levin and Simon 1998; Nordberg 2001.

40. Flores et al. 1996.

41. Parasuraman et al. 2005.

42. Ibid.

43. Posner 1980.

44. Espeseth et al. 2006, 2010.

45. Greenwood et al. 2005.

46. Parasuraman et al. 2005.

47. Cubells et al. 1998.

48. Cubells and Zabetian 2004.

49. Daly et al. 1999.

50. Parasuraman et al. 2005.

51. Cubells et al. 1998.

52. Cubells and Zabetian 2004.

53. Sundararajan et al. 2006.

54. Greenwood, Sundararajan, et al. 2009.

55. For a recent review, see Payton 2009.

56. Lin et al. 2008.

57. Caspi et al. 2002.

58. Meck et al. 2007.

59. Liu et al. 2009.

60. Witte et al. 2011.

61. Henderson et al. 1995; Meyer et al. 1998.

62. Kalaria et al. 2008.

63. Hendrie et al. 2001.

64. Palsdottir et al. 2008.

65. Rovio et al. 2005.

66. Podewils et al. 2005.

67. Zetterberg et al. 2002.

68. Hubacek et al. 2001.

69. Oria et al. 2005, 2007.

70. Mondadori et al. 2007.

71. Filippini et al. 2009.

72. Bondi et al. 2005.

73. Etnier et al. 2007.

74. Deeny et al. 2008.

75. Dipietro et al. 1993.

76. Smith et al. 2011.

77. Ibid.

78. Colcombe and Kramer 2003.

79. Lautenschlager et al. 2008.

80. Anstey and Christensen 2000; Breteler et al. 1994.

81. Lopez et al. 2003.

82. Bennet et al. 2007.

83. Cotman et al. 2007.

84. Williams et al. 2007.

85. Tran and Weltman 1985.

86. Hagberg et al. 1999.

87. Leon et al. 2004.

88. Obisesan et al. 2008.

89. Thompson et al. 2004.

90. Hagberg et al. 1999.

91. Yu et al. 1989.

92. Shadlen et al. 2005.

93. Tyas et al. 2007.

94. Friedland et al. 2001.

95. Wilson, Mendes De Leon, et al. 2002.

96. Gatz et al. 2005.

97. Brickell et al. 2007.

98. Crowe et al. 2003.

99. Carlson et al. 2008.

100. Niti et al. 2008.

Chapter 13

1. Drapeau et al. 2003; Willis and Schaie 1986; Wilson et al. 2003.

2. Burke and Barnes 2006.

3. McEwen et al. 2001; Cotman et al. 2007.

4. Small et al. 2002; Willis and Schaie 1986; Wilson et al. 2003.

5. Molteni et al. 2004.

6. Ibid.

7. Mozaffarian et al. 2010.

8. For a review, see Stillwell et al. 2005.

9. Sacks et al. 2009.

10. Fontana et al. 2004.

11. Witte et al. 2009.

12. Heilbronn et al. 2005.

13. Witte et al. 2009.

14. Wu, Brickman, et al. 2008.

15. Eriksson et al. 1998.

16. Gould et al. 1999.

17. Kempermann 2002.

18. Draganski et al. 2004.

19. Ibid.

20. Lövdén et al. 2010.

21. Willis et al. 2006.

22. Schmiedek et al. 2010.

23. Willis et al. 2006.

24. Schmiedek et al. 2010.

25. Basak et al. 2008.

26. Dahlin, Nyberg, et al. 2008.

27. Sherwin and Henry 2008; Simpkins and Singh 2008.

28. Greely et al. 2008.

29. Milgram et al. 2004.

30. Petersen 2004.

31. Rock et al. 2010.

32. Elderhostel Institute Network (http://www.roadscholar.org).

Bibliography

Aalto, S., Bruck, A., Laine, M., Nagren, K., and Rinne, J. O. 2005. Frontal and temporal dopamine release during working memory and attention tasks in healthy humans: A positron emission tomography study using the high-affinity dopamine D2 receptor ligand [11C]FLB 457. *Journal of Neuroscience* 25 (10): 2471–2477.

Abi-Dargham, A., Mawlawi, O., Lombardo, I., Gil, R., Martinez, D., Huang, Y., et al. 2002. Prefrontal dopamine D1 receptors and working memory in schizophrenia. *Journal of Neuroscience* 22 (9): 3708–3719.

Abrous, D. N., Koehl, M., and Le Moal, M. 2005. Adult neurogenesis: From precursors to network and physiology. *Physiological Reviews* 85 (2): 523–569.

Adam, S., Bay, C., Bonsang, E., Germain, S., and Perelman, S. 2007. Retirement and Cognitive Reserve: A Stochastic Frontier Approach Applied to Survey Data. Working paper 2007/04, Center of Research in Public Economics and Population Economics, University of Liège.

Aging In Place. 2010. Aging in Place Technology Watch (http://www.ageinplacetech .com).

Ahmet, I., Wan, R., Mattson, M. P., Lakatta, E. G., and Talan, M. I. 2010. Chronic alternate-day fasting results in reduced diastolic compliance and diminished systolic reserve in rats. *Journal of Cardiac Failure* 16 (10): 843–853.

Aisen, P. S., Schneider, L. S., Sano, M., Diaz-Arrastia, R., van Dyck, C. H., Weiner, M. F., et al. 2008. High-dose B vitamin supplementation and cognitive decline in Alzheimer disease: A randomized controlled trial. *Journal of the American Medical Association* 300 (15): 1774–1783.

Alkire, M. T., Haier, R. J., Fallon, J. H., and Cahill, L. 1998. Hippocampal, but not amygdala, activity at encoding correlates with long-term, free recall of nonemotional information. *Proceedings of the National Academy of Sciences* 95 (24): 14506–14510.

Allard, J. S., Heilbronn, L. K., Smith, C., Hunt, N. D. Ingram, D. K. Ravussin, E., et al. 2008. *In vitro* cellular adaptations of indicators of longevity in response to treatment with serum collected from humans on calorie restricted diets. *PLoS One* 3 (9): e3211.

Allen, J. S., Bruss, J., Brown, C. K., and Damasio, H. 2005. Normal neuroanatomical variation due to age: The major lobes and a parcellation of the temporal region. *Neurobiology of Aging* 26 (9): 1245–1260, discussion 1279–1282.

Alzheimer's Association. 2010. *Alzheimer's Disease Facts and Figures.*

Ambrogini, P., Cuppini, R., Lattanzi, D., Ciuffoli, S., Frontini, A., and Fanelli, M. 2010. Synaptogenesis in adult-generated hippocampal granule cells is affected by behavioral experiences. *Hippocampus* 20 (7): 799–810.

Amenta, F., Ferrante, F., Mancini, M., Sabbatini, M., Vega, J. A., and Zaccheo, D. 1995. Effect of long-term treatment with the dihydropyridine-type calcium channel blocker darodipine (PY 108–068): On the cerebral capillary network in aged rats. *Mechanisms of Ageing and Development* 78 (1): 27–37.

Ammons, A. R. 2005. "View of the fact." In *Bosh and Flapdoodle.* Norton.

Andersson, P. 2004. *Ageing and Employment Policies: Norway.* Organisation for Economic Cooperation and Development.

Ang, E. T., Wong, P. T., Moochhala, S., and Ng, Y. K. 2003. Neuroprotection associated with running: Is it a result of increased endogenous neurotrophic factors? *Neuroscience* 118 (2): 335–345.

Anson, R. M., Guo, Z., de Cabo, R., Iyun, T., Rios, M., Hagepanos, A., et al. 2003. Intermittent fasting dissociates beneficial effects of dietary restriction on glucose metabolism and neuronal resistance to injury from calorie intake. *Proceedings of the National Academy of Sciences* 100 (10): 6216–6220.

Anstey, K., and Christensen, H. 2000. Education, activity, health, blood pressure and apolipoprotein E as predictors of cognitive change in old age: A review. *Gerontology* 46 (3): 163–177.

Arendash, G. W., Garcia, M. F., Costa, D. A., Cracchiolo, J. R., Wefes, I. M., and Potter, H. 2004. Environmental enrichment improves cognition in aged Alzheimer's transgenic mice despite stable β-amyloid deposition. *Neuroreport* 15 (11): 1751–1754.

Arendash, G. W., Mori, T., Cao, C., Mamcarz, M., Runfeldt, M., Dickson, A., et al. 2009. Caffeine reverses cognitive impairment and decreases brain amyloid-β levels in aged Alzheimer's Disease mice. *Journal of Alzheimer's Disease* 17 (3): 661–680.

Arendt, T., Schindler, C., Bruckner, M. K., Eschrich, K., Bigl, V., Zedlick, D., et al. 1997. Plastic neuronal remodeling is impaired in patients with Alzheimer's disease carrying apolipoprotein ε4 allele. *Journal of Neuroscience* 17 (2): 516–529.

Arnsten, A. F., Cai, J. X., Steere, J. C., and Goldman-Rakic, P. S. 1995. Dopamine D2 receptor mechanisms contribute to age-related cognitive decline: The effects of quinpirole on memory and motor performance in monkeys. *Journal of Neuroscience* 15 (5): 3429–3439.

Arnsten, A. F., Steere, J. C., Jentsch, D. J., and Li, B. M. 1998. Noradrenergic influences on prefrontal cortical cognitive function: Opposing actions at postjunctional alpha 1 versus alpha 2-adrenergic receptors. *Advances in Pharmacology* 42: 764–767.

Arumugam, T. V., Phillips, T. M., Cheng, A., Morrell, C. H., Mattson, M. P., and Wan, R. 2010. Age and energy intake interact to modify cell stress pathways and stroke outcome. *Annals of Neurology* 67 (1): 41–52.

Arunachalam, G., Yao, H., Sundar, I. K., Caito, S., and Rahman, I. 2010. SIRT1 regulates oxidant- and cigarette smoke-induced eNOS acetylation in endothelial cells: Role of resveratrol. *Biochemical and Biophysical Research Communications* 393 (1): 66–72.

Awad, N., Gagnon, M., and Messier, C. 2004. The relationship between impaired glucose tolerance, type 2 diabetes, and cognitive function. *Journal of Clinical and Experimental Neuropsychology* 26 (8): 1044–1080.

Bäckman, L., Ginovart, N., Dixon, R. A., Wahlin, T. B., Wahlin, A., Halldin, C., et al. 2000. Age-related cognitive deficits mediated by changes in the striatal dopamine system. *American Journal of Psychiatry* 157 (4): 635–637.

Bäckman, L., Laukka, E. J., Wahlin, A., Small, B. J., and Fratiglioni, L. 2002. Influences of preclinical dementia and impending death on the magnitude of age-related cognitive deficits. *Psychology and Aging* 17 (3): 435–442.

Baddeley, A. D., Bressi, S., Della Sala, S., Logie, R., and Spinnler, H. 1991. The decline of working memory in Alzheimer's disease: A longitudinal study. *Brain* 114: 2521–2542.

Bagger, Y. Z., Tanko, L. B., Alexandersen, P., Qin, G., and Christiansen, C. 2005. Early postmenopausal hormone therapy may prevent cognitive impairment later in life. *Menopause* 12 (1): 12–17.

Bahner, E., Huper, A.-D., and Manzey, D. 2008. Misuse of automated decision aids: Complacency, automation bias and the impact of training experience. *International Journal of Human-Computer Studies* 66: 688–699.

Baldwin, C. L. 2002. Designing in-vehicle technologies for older drivers: Application of sensory-cognitive interaction theory. *Theoretical Issues in Ergonomics Science* 3 (4): 307–329.

Baldwin, C. L. 2011. *Auditory Cognition and Human Performance: Research and Applications*. CRC Press.

Ball, K., Berch, D. B., Helmers, K. F., Jobe, J. B., Leveck, M. D., Marsiske, M., et al. 2002. Effects of cognitive training interventions with older adults: A randomized controlled trial. *Journal of the American Medical Association* 288 (18): 2271–2281.

Ball, M. J. 1977. Neuronal loss, neurofibrillary tangles and granulovacuolar degeneration in the hippocampus with ageing and dementia: A quantitative study. *Acta Neuropathologica* 37 (2): 111–118.

Ball, M. J. 1978. Topographic distribution of neurofibrillary tangles and granulovacuolar degeneration in hippocampal cortex of aging and demented patients: A quantitative study. *Acta Neuropathologica* 42: 73–80.

Baltes, P. B., and Mayer, K. U., eds. 1999. *The Berlin Aging Study: Aging from 70 to 100*. Cambridge University Press.

Barberger-Gateau, P., Letenneur, L., Deschamps, V., Peres, K., Dartigues, J. F., and Renaud, S. 2002. Fish, meat, and risk of dementia: Cohort study. *British Medical Journal* 325 (7370): 932–933.

Barch, D. M. 2004. Pharmacological manipulation of human working memory. *Psychopharmacology* 174 (1): 126–135.

Barger, J. L., Kayo, T., Vann, J. M., Arias, E. B., Wang, J., Hacker, T. A., et al. 2008. A low dose of dietary resveratrol partially mimics caloric restriction and retards aging parameters in mice. *PLoS ONE* 3 (6): e2264.

Barnes, C. A. 1979. Memory deficits associated with senescence: A neurophysiological and behavioral study in the rat. *Journal of Comparative and Physiological Psychology* 93 (1): 74–104.

Barnes, C. A., and McNaughton, B. L. 1980. Physiological compensation for loss of afferent synapses in rat hippocampal granule cells during senescence. *Journal of Physiology* 309: 473–485.

Barnes, C. A., Suster, M. S., Shen, J., and McNaughton, B. L. 1997. Multistability of cognitive maps in the hippocampus of old rats. *Nature* 388 (6639): 272–275.

Bartzokis, G. 2004. Age-related myelin breakdown: A developmental model of cognitive decline and Alzheimer's disease. *Neurobiology of Aging* 25 (1): 5–18; author reply 49–62.

Basak, C., Boot, W. R., Voss, M. W., and Kramer, A. F. 2008. Can training in a real-time strategy video game attenuate cognitive decline in older adults? *Psychology and Aging* 23 (4): 765–777.

Baum, L., Lam, C. W., Cheung, S. K., Kwok, T., Lui, V., Tsoh, J., et al. 2008. Six-month randomized, placebo-controlled, double-blind, pilot clinical trial of curcumin in patients with Alzheimer disease. *Journal of Clinical Psychopharmacology* 28 (1): 110–113.

Baur, J. A., and Sinclair, D. A. 2006. Therapeutic potential of resveratrol: The in vivo evidence. *Nature Reviews Drug Discovery* 5 (6): 493–506.

Bayer-Carter, J., Green, P., Montine, T., VanFossen, B., Baker, L., Stennis Watson, G., et al. 2011. Diet intervention and cerebrospinal fluid biomarkers in amnestic mild cognitive impairment. *Archive of Neurology* 68 (6): 743–752.

Beaulieu, C. 2002. The basis of anisotropic water diffusion in the nervous system—a technical review. *NMR in Biomedicine* 15 (7–8): 435–455.

Benedetti, B., Charil, A., Rovaris, M., Judica, E., Valsasina, P., Sormani, M. P., et al. 2006. Influence of aging on brain gray and white matter changes assessed by conventional, MT, and DT MRI. *Neurology* 66 (4): 535–539.

Bennet, A. M., Di Angelantonio, E., Ye, Z., Wensley, F., Dahlin, A., Ahlbom, A., et al. 2007. Association of apolipoprotein E genotypes with lipid levels and coronary risk. *Journal of the American Medical Association* 298 (11): 1300–1311.

Berchtold, N. C., Chinn, G., Chou, M., Kesslak, J. P., and Cotman, C. W. 2005. Exercise primes a molecular memory for brain-derived neurotrophic factor protein induction in the rat hippocampus. *Neuroscience* 133 (3): 853–861.

Berchtold, N. C., Kesslak, J. P., Pike, C. J., Adlard, P. A., and Cotman, C. W. 2001. Estrogen and exercise interact to regulate brain-derived neurotrophic factor mRNA and protein expression in the hippocampus. *European Journal of Neuroscience* 14 (12): 1992–2002.

Bhardwaj, R. D., Curtis, M. A., Spalding, K. L., Buchholz, B. A., Fink, D., Bjork-Eriksson, T., et al. 2006. Neocortical neurogenesis in humans is restricted to development. *Proceedings of the National Academy of Sciences* 103 (33): 12564–12568.

Bherer, L., Kramer, A. F., Peterson, M. S., Colcombe, S. J., Erickson, K., and Becic, E. 2005. Training effects on dual-task performance: Are there age-related differences in plasticity of attentional control? *Psychology and Aging* 20 (4): 695–709.

Bherer, L., Kramer, A. F., Peterson, M. S., Colcombe, S. J., Erickson, K., and Becic, E. 2008. Transfer effects in task-set cost and dual-task cost after dual-task training in older and younger adults: Further evidence for cognitive plasticity in attentional control in late adulthood. *Experimental Aging Research* 34 (3): 188–219.

Bhuyan, K. K. 2004. Health promotion through self-care and community participation: Elements of a proposed programme in the developing countries. *BMC Public Health* 4: 11.

Binder, E. F., Schechtman, K. B., Birge, S. J., Williams, D. B., and Kohrt, W. M. 2001. Effects of hormone replacement therapy on cognitive performance in elderly women. *Maturitas* 38 (2): 137–146.

Bishop, J., and Simpkins, J. W. 1994. Estradiol treatment increases viability of glioma and neuroblastoma cells in vitro. *Molecular and Cellular Neurosciences* 5 (4): 303–308.

Bissig, D., and Lustig, C. 2007. Who benefits from memory training? *Psychological Science* 18 (8): 720–726.

Blanson Henkemans, O. A., Rogers, W. A., Fisk, A. D., Neerincx, M. A., Lindenberg, J., and van der Mast, C. A. 2008. Usability of an adaptive computer assistant that improves self-care and health literacy of older adults. *Methods of Information in Medicine* 47 (1): 82–88.

Bliss, T. V., and Collingridge, G. L. 1993. A synaptic model of memory: Long-term potentiation in the hippocampus. *Nature* 361 (6407): 31–39.

Bondi, M. W., Galasko, D., Salmon, D. P., and Thomas, R. G. 1999. Neuropsychological function and apolipoprotein E genotype in the preclinical detection of Alzheimer's disease. *Psychology and Aging* 14: 295–303.

Bondi, M. W., Houston, W. S., Eyler, L. T., and Brown, G. G. 2005. fMRI evidence of compensatory mechanisms in older adults at genetic risk for Alzheimer disease. *Neurology* 64 (3): 501–508.

Bookheimer, S. Y., Strojwas, M. H., Cohen, M. S., Saunders, A. M., Pericak-Vance, M. A., Mazziotta, J. C., et al. 2000. Patterns of brain activation in people at risk for Alzheimer's disease. *New England Journal of Medicine* 343 (7): 450–456.

Borghesani, P. R., Weaver, K. E., Aylward, E. H., Richards, A. L., Madhyastha, T. M., Kahn, A. R., et al. 2011. Midlife memory improvement predicts preservation of hippocampal volume in old age. *Neurobiology of Aging* (in press).

Bournival, J., Quessy, P., and Martinoli, M. G. 2009. Protective effects of resveratrol and quercetin against MPP+-induced oxidative stress act by modulating markers of apoptotic death in dopaminergic neurons. *Cellular and Molecular Neurobiology* 29 (8): 1169–1180.

Braak, H., and Braak, E. 1991. Neuropathological stageing of Alzheimer-related changes. *Acta Neuropathologica* 82: 239–259.

Braak, H., and Braak, E. 1995. Staging of Alzheimer's Disease-related neurofibrillary changes. *Neurobiology of Aging* 16: 271–284.

Bramham, C. R., and Messaoudi, E. 2005. BDNF function in adult synaptic plasticity: The synaptic consolidation hypothesis. *Progress in Neurobiology* 76 (2): 99–125.

Breitner, J. C., Wyse, B. W., Anthony, J. C., Welsh-Bohmer, K. A., Steffens, D. C., Norton, M. C., et al. 1999. APOE-ε4 count predicts age when prevalence of AD increases, then declines: The Cache County Study. *Neurology* 53 (2): 321–331.

Breteler, M. M., Claus, J. J., Grobbee, D. E., and Hofman, A. 1994. Cardiovascular disease and distribution of cognitive function in elderly people: The Rotterdam Study. *British Medical Journal* 308 (6944): 1604–1608.

Brickell, K. L., Leverenz, J. B., Steinbart, E. J., Rumbaugh, M., Schellenberg, G. D., Nochlin, D., et al. 2007. Clinicopathological concordance and discordance in three monozygotic twin pairs with familial Alzheimer's disease. *Journal of Neurology, Neurosurgery, and Psychiatry* 78 (10): 1050–1055.

Brickman, A. M., Stern, Y., and Small, S. A. 2011. Hippocampal subregions differentially associate with standardized memory tests. *Hippocampus.*

Brinton, R. D. 2009. Estrogen-induced plasticity from cells to circuits: Predictions for cognitive function. *Trends in Pharmacological Sciences* 30 (4): 212–222.

Brody, H. 1955. Organization of the cerebral cortex. III. A study of aging in the human cerebral cortex. *Journal of Comparative Neurology* 102 (2): 511–516.

Browner, W. S., Kahn, A. J., Ziv, E., Reiner, A. P., Oshima, J., Cawthon, R. M., et al. 2004. The genetics of human longevity. *American Journal of Medicine* 117 (11): 851–860.

Bruel-Jungerman, E., Davis, S., Rampon, C., and Laroche, S. 2006. Long-term potentiation enhances neurogenesis in the adult dentate gyrus. *Journal of Neuroscience* 26 (22): 5888–5893.

Bruel-Jungerman, E., Laroche, S., and Rampon, C. 2005. New neurons in the dentate gyrus are involved in the expression of enhanced long-term memory following environmental enrichment. *European Journal of Neuroscience* 21 (2): 513–521.

Bryan, J., and Tiggemann, M. 2001. The effect of weight-loss dieting on cognitive performance and psychological well-being in overweight women. *Appetite* 36 (2): 147–156.

Buckner, R. L. 2004. Memory and executive function in aging and AD: Multiple factors that cause decline and reserve factors that compensate. *Neuron* 44 (1): 195–208.

Bucur, B., Madden, D. J., Spaniol, J., Provenzale, J. M., Cabeza, R., White, L. E., et al. 2008. Age-related slowing of memory retrieval: Contributions of perceptual speed and cerebral white matter integrity. *Neurobiology of Aging* 29 (7): 1070–1079.

Buell, S. J., and Coleman, P. D. 1979. Dendritic growth in the aged human brain and failure of growth in senile dementia. *Science* 206: 854–856.

Bunce, D., Kivipelto, M., and Wahlin, A. 2004. Utilization of cognitive support in episodic free recall as a function of apolipoprotein E and vitamin B12 or folate among adults aged 75 years and older. *Neuropsychology* 18 (2): 362–370.

Burggren, A. C., Small, G. W., Sabb, F. W., and Bookheimer, S. Y. 2002. Specificity of brain activation patterns in people at genetic risk for Alzheimer disease. *American Journal of Geriatric Psychiatry* 10 (1): 44–51.

Burke, S. N., and Barnes, C. A. 2006. Neural plasticity in the ageing brain. *Nature Reviews Neuroscience* 7 (1): 30–40.

Burnett, C., Valentini, S., Cabreiro, F., Goss, M., Somogyvari, M., Piper, M., et al. 2011. Absence of effects of Sir2 overexpression on lifepan in *C. elegans* and *Drosophila. Nature* 477 (7365): 482–485.

Butcher, L. M., Davis, O. S., Craig, I. W., and Plomin, R. 2008. Genome-wide quantitative trait locus association scan of general cognitive ability using pooled DNA and 500K single nucleotide polymorphism microarrays. *Genes, Brain and Behavior* 7 (4): 435–446.

Bütefisch, C., Hummelsheim, H., Denzler, P., and Mauritz, K. H. 1995. Repetitive training of isolated movements improves the outcome of motor rehabilitation of the centrally paretic hand. *Journal of the Neurological Sciences* 130 (1): 59–68.

Cabeza, R. 2002. Hemispheric asymmetry reduction in older adults: The HAROLD model. *Psychology and Aging* 17 (1): 85–100.

Cabeza, R., Anderson, N. D., Locantore, J. K., and McIntosh, A. R. 2002. Aging gracefully: Compensatory brain activity in high-performing older adults. *NeuroImage* 17 (3): 1394–1402.

Cabeza, R., Grady, C. L., Nyberg, L., McIntosh, A. R., Tulving, E., Kapur, S., et al. 1997. Age-related differences in neural activity during memory encoding and retrieval: A positron emission tomography study. *Journal of Neuroscience* 17 (1): 391–400.

Calabrese, E. J., and Baldwin, L. A. 1998. Hormesis as a biological hypothesis. *Environmental Health Perspectives* 106 (suppl. 1): 357–362.

Campbell, K. L., Hasher, L., and Thomas, R. C. 2010. Hyper-binding: A unique age effect. *Psychological Science* 21 (3): 399–405.

Cao, Y., Vikingstad, E. M., George, K. P., Johnson, A. F., and Welch, K. M. 1999. Cortical language activation in stroke patients recovering from aphasia with functional MRI. *Stroke* 30 (11): 2331–2340.

Card, S. K., Moran, T. P., and Newell, A. 1980. Computer text-editing: An information-processing analysis of a routine cognitive skill. *Cognitive Psychology* 12: 32–74.

Carlson, M. C., Erickson, K. I., Kramer, A. F., Voss, M. W., Bolea, N., Mielke, M., et al. 2009. Evidence for neurocognitive plasticity in at-risk older adults: The Experience Corps program. *Journal of Gerontology A* 64 (12): 1275–1282.

Carlson, M. C., Hasher, L., Zacks, R. T., and Connelly, S. L. 1995. Aging, distraction, and the benefits of predictable location. *Psychology and Aging* 10 (3): 427–436.

Carlson, M. C., Helms, M. J., Steffens, D. C., Burke, J. R., Potter, G. G., and Plassman, B. L. 2008. Midlife activity predicts risk of dementia in older male twin pairs. *Alzheimer's & Dementia* 4 (5): 324–331.

Carlson, P. 1991. The strange case of the Silver Spring monkeys. *Washington Post Magazine*, February 24.

Carson, S. H., Peterson, J. B., and Higgins, D. M. 2003. Decreased latent inhibition is associated with increased creative achievement in high-functioning individuals. *Journal of Personality and Social Psychology* 85 (3): 499–506.

Carter, C. S., Leeuwenburgh, C., Daniels, M., and Foster, T. C. 2009. Influence of calorie restriction on measures of age-related cognitive decline: Role of increased physical activity. *Journal of Gerontology A* 64 (8): 850–859.

Caspi, A., McClay, J., Moffitt, T. E., Mill, J., Martin, J., Craig, I. W., et al. 2002. Role of genotype in the cycle of violence in maltreated children. *Science* 297: 851–854.

CDC (Centers for Disease Control and Prevention). 2008. Third National Health and Nutrition Examination Survey (http://www.cdc.gov).

Chang, Y. M., Rosene, D. L., Killiany, R. J., Mangiamele, L. A., and Luebke, J. I. 2005. Increased action potential firing rates of layer 2/3 pyramidal cells in the prefrontal cortex are significantly related to cognitive performance in aged monkeys. *Cerebral Cortex* 15 (4): 409–418.

Chantome, M., Perruchet, P., Hasboun, D., Dormont, D., Sahel, M., Sourour, N., et al. 1999. Is there a negative correlation between explicit memory and hippocampal volume? *NeuroImage* 10 (5): 589–595.

Chapillon, P., Manneche, C., Belzung, C., and Caston, J. 1999. Rearing environmental enrichment in two inbred strains of mice: 1. Effects on emotional reactivity. *Behavior Genetics* 29 (1): 41–46.

Charlton, R. A., Barrick, T. R., McIntyre, D. J., Shen, Y., O'Sullivan, M., Howe, F. A., et al. 2006. White matter damage on diffusion tensor imaging correlates with age-related cognitive decline. *Neurology* 66 (2): 217–222.

Charness, N., and Schaie, K. W. 2003. *Impact of Technology on Successful Aging*. Springer.

Chein, J. M., and Fiez, J. A. 2001. Dissociation of verbal working memory system components using a delayed serial recall task. *Cerebral Cortex* 11 (11): 1003–1014.

Cherubini, A., Andres-Lacueva, C., Martin, A., Lauretani, F., Iorio, A. D., Bartali, B., et al. 2007. Low plasma N-3 fatty acids and dementia in older persons: The InCHI-ANTI study. *Journal of Gerontology A* 62 (10): 1120–1126.

Chklovskii, D. B., Mel, B. W., and Svoboda, K. 2004. Cortical rewiring and information storage. *Nature* 431 (7010): 782–788.

Christensen, H., Korten, A. E., Jorm, A. F., Henderson, A. S., Jacomb, P. A., Rodgers, B., et al. 1997. Education and decline in cognitive performance: Compensatory but not protective. *International Journal of Geriatric Psychiatry* 12 (3): 323–330.

Chun, S. K., Sun, W., Park, J. J., and Jung, M. W. 2006. Enhanced proliferation of progenitor cells following long-term potentiation induction in the rat dentate gyrus. *Neurobiology of Learning and Memory* 86 (3): 322–329.

Cleophas, T. J. 1999. Wine, beer and spirits and the risk of myocardial infarction: A systematic review. *Biomedicine and Pharmacotherapy* 53 (9): 417–423.

Cockerham, W. C. 1991. *This Aging Society*. Prentice-Hall.

Colcombe, S. J., Erickson, K. I., Scalf, P. E., Kim, J. S., Prakash, R., McAuley, E., et al. 2006. Aerobic exercise training increases brain volume in aging humans. *Journal of Gerontology A* 61 (11): 1166–1170.

Colcombe, S. J., and Kramer, A. F. 2003. Fitness effects on the cognitive function of older adults: A meta-analytic study. *Psychological Science* 14 (2): 125–130.

Colcombe, S. J., Kramer, A. F., Erickson, K. I., and Scalf, P. 2005. The implications of cortical recruitment and brain morphology for individual differences in inhibitory function in aging humans. *Psychology and Aging* 20 (3): 363–375.

Colcombe, S. J., Kramer, A. F., Erickson, K. I., Scalf, P., McAuley, E., Cohen, N. J., et al. 2004. Cardiovascular fitness, cortical plasticity, and aging. *Proceedings of the National Academy of Sciences* 101 (9): 3316–3321.

Colditz, G. A. 1999. Economic costs of obesity and inactivity. *Medicine and Science in Sports and Exercise* 31 (11 suppl.): S663–S667.

Cole, B. F., Baron, J. A., Sandler, R. S., Haile, R. W., Ahnen, D. J., Bresalier, R. S., et al. 2007. Folic acid for the prevention of colorectal adenomas: A randomized clinical trial. *Journal of the American Medical Association* 297 (21): 2351–2359.

Cole, G. M., Ma, Q. L., and Frautschy, S. A. 2009. Omega-3 fatty acids and dementia. *Prostaglandins, Leukotrienes, and Essential Fatty Acids* 81 (2–3): 213–221.

Colman, R. J., Anderson, R. M., Johnson, S. C., Kastman, E. K., Kosmatka, K. J., Beasley, T. M., et al. 2009. Caloric restriction delays disease onset and mortality in rhesus monkeys. *Science* 325: 201–204.

Colsher, P. L., and Wallace, R. B. 1991. Longitudinal application of cognitive function measures in a defined population of community-dwelling elders. *Annals of Epidemiology* 1 (3): 215–230.

Convit, A., De Leon, M. J., Tarshish, C., De Santi, S., Tsui, W., Rusinek, H., et al. 1997. Specific hippocampal volume reductions in individuals at risk for Alzheimer's disease. *Neurobiology of Aging* 18 (2): 131–138.

Convit, A., Wolf, O. T., Tarshish, C., and de Leon, M. J. 2003. Reduced glucose tolerance is associated with poor memory performance and hippocampal atrophy among normal elderly. *Proceedings of the National Academy of Sciences* 100 (4): 2019–2022.

Cooke, A., Zurif, E. B., DeVita, C., Alsop, D., Koenig, P., Detre, J., et al. 2002. Neural basis for sentence comprehension: Grammatical and short-term memory components. *Human Brain Mapping* 15 (2): 80–94.

Corbetta, M., Kincade, J. M., Ollinger, J. M., McAvoy, M. P., and Shulman, G. L. 2000. Voluntary orienting is dissociated from target detection in human posterior parietal cortex. *Nature Neuroscience* 3 (3): 292–297.

Corder, E. H., Saunders, A. M., Strittmatter, W. J., Schmechel, D. E., Gaskell, P. C., Small, G. W., et al. 1993. Gene dose of Apolipoprotein E type 4 allele and the risk of Alzheimer's disease in late onset families. *Science* 261: 921–923.

Cotman, C. W., and Berchtold, N. C. 2002. Exercise: A behavioral intervention to enhance brain health and plasticity. *Trends in Neurosciences* 25 (6): 295–301.

Cotman, C. W., Berchtold, N. C., and Christie, L. A. 2007. Exercise builds brain health: Key roles of growth factor cascades and inflammation. *Trends in Neurosciences* 30 (9): 464–472.

Courchesne, E., Chisum, H. J., Townsend, J., Cowles, A., Covington, J., Egaas, B., et al. 2000. Normal brain development and aging: Quantitative analysis at in vivo MR imaging in healthy volunteers. *Radiology* 216 (3): 672–682.

Cracchiolo, J. R., Mori, T., Nazian, S. J., Tan, J., Potter, H., and Arendash, G. W. 2007. Enhanced cognitive activity—over and above social or physical activity—is required to protect Alzheimer's mice against cognitive impairment, reduce Aβ deposition, and increase synaptic immunoreactivity. *Neurobiology of Learning and Memory* 88 (3): 277–294.

Craik, F. I. M. 1983. On the transfer of information from temporary to permanent memory. *Philosophical Transactions of the Royal Society of London B* 301: 341–359.

Craik, F. I. M., and Salthouse, T. A., eds. 2007. *The Handbook of Aging and Cognition.* Erlbaum.

Crawford, M. A., Bloom, M., Broadhurst, C. L., Schmidt, W. F., Cunnane, S. C., Galli, C., et al. 1999. Evidence for the unique function of docosahexaenoic acid during the evolution of the modern hominid brain. *Lipids* 34 (suppl.): S39–S47.

Crowe, M., Andel, R., Pedersen, N. L., Johansson, B., and Gatz, M. 2003. Does participation in leisure activities lead to reduced risk of Alzheimer's disease? A prospective study of Swedish twins. *Journal of Gerontology B* 58 (5): 249–255.

Cubells, J. F., van Kammen, D. P., Kelley, M. E., Anderson, G. M., O'Connor, D. T., Price, L. H., et al. 1998. Dopamine β-hydroxylase: Two polymorphisms in linkage disequilibrium at the structural gene DBH associate with biochemical phenotypic variation. *Human Genetics* 102 (5): 533–540.

Cubells, J. F., and Zabetian, C. P. 2004. Human genetics of plasma dopamine β-hydroxylase activity: Applications to research in psychiatry and neurology. *Psychopharmacology* 174 (4): 463–476.

Cummings, B. J., Head, E., Afagh, A. J., Milgram, N. W., and Cotman, C. W. 1996. Beta-amyloid accumulation correlates with cognitive dysfunction in the aged canine. *Neurobiology of Learning and Memory* 66 (1): 11–23.

Curcio, C. A., and Hinds, J. W. 1983. Stability of synaptic density and spine volume in dentate gyrus of aged rats. *Neurobiology of Aging* 4 (1): 77–87.

Curtis, M. A., Eriksson, P. S., and Faull, R. L. 2007. Progenitor cells and adult neurogenesis in neurodegenerative diseases and injuries of the basal ganglia. *Clinical and Experimental Pharmacology and Physiology* 34 (5–6): 528–532.

Curtis, M. A., Penney, E. B., Pearson, A. G., van Roon-Mom, W. M., Butterworth, N. J., Dragunow, M., et al. 2003. Increased cell proliferation and neurogenesis in the adult human Huntington's disease brain. *Proceedings of the National Academy of Sciences* 100 (15): 9023–9027.

Dahlin, E., Neely, A. S., Larsson, A., Bäckman, L., and Nyberg, L. 2008. Transfer of learning after updating training mediated by the striatum. *Science* 320: 1510–1512.

Dahlin, E., Nyberg, L., Bäckman, L., and Neely, A. S. 2008. Plasticity of executive functioning in young and older adults: Immediate training gains, transfer, and long-term maintenance. *Psychology and Aging* 23 (4): 720–730.

Dalla, C., Papachristos, E. B., Whetstone, A. S., and Shors, T. J. 2009. Female rats learn trace memories better than male rats and consequently retain a greater proportion of new neurons in their hippocampi. *Proceedings of the National Academy of Sciences* 106 (8): 2927–2932.

Daly, G., Hawi, Z., Fitzgerald, M., and Gill, M. 1999. Mapping susceptibility loci in attention deficit hyperactivity disorder: Preferential transmission of parental alleles at DAT1, DBH and DRD5 to affected children. *Molecular Psychiatry* 4 (2): 192–196.

Dancause, N., Barbay, S., Frost, S. B., Plautz, E. J., Chen, D., Zoubina, E. V., et al. 2005. Extensive cortical rewiring after brain injury. *Journal of Neuroscience* 25 (44): 10167–10179.

Dangour, A. D., Allen, E., Elbourne, D., Fasey, N., Fletcher, A. E., Hardy, P., et al. 2010. Effect of 2-y n-3 long-chain polyunsaturated fatty acid supplementation on cognitive function in older people: A randomized, double-blind, controlled trial. *American Journal of Clinical Nutrition* 91 (6): 1725–1732.

Das, G. D., and Altman, J. 1971. Postnatal neurogenesis in the cerebellum of the cat and tritiated thymidine autoradiography. *Brain Research* 30 (2): 323–330.

Davidson, M. C., and Marrocco, R. T. 2000. Local infusion of scopolamine into intraparietal cortex slows covert orienting in rhesus monkeys. *Journal of Neurophysiology* 83 (3): 1536–1549.

Davies, G., Tenesa, A., Payton, A., Yang, J., Harris, S., Liewald, D., et al. 2011. Genome-wide association studies establish that human intelligence is highly heritable and polygenic. *Molecular Psychiatry* 16: 996–1005.

Davis, O. S., Haworth, C. M., and Plomin, R. 2009. Dramatic increase in heritability of cognitive development from early to middle childhood: An 8-year longitudinal study of 8,700 pairs of twins. *Psychological Science* 20 (10): 1301–1308.

Davis, S. W., Dennis, N. A., Daselaar, S. M., Fleck, M. S., and Cabeza, R. 2008. Que PASA? The posterior-anterior shift in aging. *Cerebral Cortex* 18 (5): 1201–1209.

de Brabander, J. M., Kramers, R. J., and Uylings, H. B. 1998. Layer-specific dendritic regression of pyramidal cells with ageing in the human prefrontal cortex. *European Journal of Neuroscience* 10 (4): 1261–1269.

de Moraes, S. A., Szklo, M., Knopman, D., and Park, E. 2001. Prospective assessment of estrogen replacement therapy and cognitive functioning: Atherosclerosis risk in communities study. *American Journal of Epidemiology* 154 (8): 733–739.

de Saint Hilaire, Z., Orosco, M., Rouch, C., Blanc, G., and Nicolaidis, S. 2001. Variations in extracellular monoamines in the prefrontal cortex and medial hypothalamus after modafinil administration: A microdialysis study in rats. *Neuroreport* 12 (16): 3533–3537.

Deary, I. J., Bastin, M. E., Pattie, A., Clayden, J. D., Whalley, L. J., Starr, J. M., et al. 2006. White matter integrity and cognition in childhood and old age. *Neurology* 66 (4): 505–512.

Deeny, S. P., Poeppel, D., Zimmerman, J. B., Roth, S. M., Brandauer, J., Witkowski, S., et al. 2008. Exercise, APOE, and working memory: MEG and behavioral evidence for benefit of exercise in ε4 carriers. *Biological Psychology* 78 (2): 179–187.

Demerens, C., Stankoff, B., Logak, M., Anglade, P., Allinquant, B., Couraud, F., et al. 1996. Induction of myelination in the central nervous system by electrical activity. *Proceedings of the National Academy of Sciences* 93 (18): 9887–9892.

Denk, W., Strickler, J. H., and Webb, W. W. 1990. Two-photon laser scanning fluorescence microscopy. *Science* 248: 73–76.

DeRubeis, R. J., Siegle, G. J., and Hollon, S. D. 2008. Cognitive therapy versus medication for depression: Treatment outcomes and neural mechanisms. *Nature Reviews Neuroscience* 9 (10): 788–796.

Derwinger, A., Stigsdotter Neely, A., and Bäckman, L. 2005. Design your own memory strategies! Self-generated strategy training versus mnemonic training in old age: An 8-month follow-up. *Neuropsychological Rehabilitation* 15 (1): 37–54.

Derwinger, A., Stigsdotter Neely, A., MacDonald, S., and Bäckman, L. 2005. Forgetting numbers in old age: Strategy and learning speed matter. *Gerontology* 51 (4): 277–284.

DeSantis, A. D., Webb, E. M., and Noar, S. M. 2008. Illicit use of prescription ADHD medications on a college campus: A multimethodological approach. *Journal of American College Health* 57 (3): 315–324.

Deslandes, A. C., Veiga, H., Cagy, M., Piedade, R., Pompeu, F., and Ribeiro, P. 2005. Effects of caffeine on the electrophysiological, cognitive and motor responses of the central nervous system. *Brazilian Journal of Medical and Biological Research* 38 (7): 1077–1086.

Diana, G., Scotti de Carolis, A., Frank, C., Domenici, M. R., and Sagratella, S. 1994. Selective reduction of hippocampal dentate frequency potentiation in aged rats with impaired place learning. *Brain Research Bulletin* 35 (2): 107–111.

Dieguez, D., Jr., and Barea-Rodriguez, E. J. 2004. Aging impairs the late phase of long-term potentiation at the medial perforant path-CA3 synapse in awake rats. *Synapse* 52 (1): 53–61.

Diekelmann, S., and Born, J. 2010. The memory function of sleep. *Nature Reviews Neuroscience* 11 (2): 114–126.

Dietrich, T., Krings, T., Neulen, J., Willmes, K., Erberich, S., Thron, A., et al. 2001. Effects of blood estrogen level on cortical activation patterns during cognitive activation as measured by functional MRI. *NeuroImage* 13 (3): 425–432.

Ding, Y. H., Ding, Y., Li, J., Bessert, D. A., and Rafols, J. A. 2006. Exercise preconditioning strengthens brain microvascular integrity in a rat stroke model. *Neurological Research* 28 (2): 184–189.

Dipietro, L., Caspersen, C. J., Ostfeld, A. M., and Nadel, E. R. 1993. A survey for assessing physical activity among older adults. *Medicine and Science in Sports and Exercise* 25 (5): 628–642.

Dobrossy, M. D., Drapeau, E., Aurousseau, C., Le Moal, M., Piazza, P. V., and Abrous, D. N. 2003. Differential effects of learning on neurogenesis: Learning increases or decreases the number of newly born cells depending on their birth date. *Molecular Psychiatry* 8 (12): 974–982.

Draganski, B., Gaser, C., Busch, V., Schuierer, G., Bogdahn, U., and May, A. 2004. Neuroplasticity: Changes in grey matter induced by training. *Nature* 427 (6972): 311–312.

Draganski, B., Gaser, C., Kempermann, G., Kuhn, H. G., Winkler, J., Buchel, C., et al. 2006. Temporal and spatial dynamics of brain structure changes during extensive learning. *Journal of Neuroscience* 26 (23): 6314–6317.

Drapeau, E., Mayo, W., Aurousseau, C., Le Moal, M., Piazza, P. V., and Abrous, D. N. 2003. Spatial memory performances of aged rats in the water maze predict levels of hippocampal neurogenesis. *Proceedings of the National Academy of Sciences* 100 (24): 14385–14390.

Duan, H., Wearne, S. L., Rocher, A. B., Macedo, A., Morrison, J. H., and Hof, P. R. 2003. Age-related dendritic and spine changes in corticocortically projecting neurons in macaque monkeys. *Cerebral Cortex* 13 (9): 950–961.

Dubos, R. 1968. *So Human an Animal*. Scribner.

Duckworth, A. L., and Seligman, M. E. 2005. Self-discipline outdoes IQ in predicting academic performance of adolescents. *Psychological Science* 16 (12): 939–944.

Duka, T., Tasker, R., and McGowan, J. F. 2000. The effects of 3-week estrogen hormone replacement on cognition in elderly healthy females. *Psychopharmacology* 149 (2): 129–139.

Duplus, E., Glorian, M., and Forest, C. 2000. Fatty acid regulation of gene transcription. *Journal of Biological Chemistry* 275 (40): 30749–30752.

Dupret, D., Revest, J. M., Koehl, M., Ichas, F., De Giorgi, F., Costet, P., et al. 2008. Spatial relational memory requires hippocampal adult neurogenesis. *PLoS One* 3 (4): e1959.

Eadie, B. D., Redila, V. A., and Christie, B. R. 2005. Voluntary exercise alters the cytoarchitecture of the adult dentate gyrus by increasing cellular proliferation, dendritic complexity, and spine density. *Journal of Comparative Neurology* 486 (1): 39–47.

Eckles-Smith, K., Clayton, D., Bickford, P., and Browning, M. D. 2000. Caloric restriction prevents age-related deficits in LTP and in NMDA receptor expression. *Brain Research. Molecular Brain Research* 78 (1–2): 154–162.

Egan, M. F., Goldberg, T. E., Kolachana, B. S., Callicott, J. H., Mazzanti, C. M., Straub, R. E., et al. 2001. Effect of COMT Val108/158 Met genotype on frontal lobe function and risk for schizophrenia. *Proceedings of the National Academy of Sciences* 98 (12): 6917–6922.

Elbaz, A., Clavel, J., Rathouz, P. J., Moisan, F., Galanaud, J. P., Delemotte, B., et al. 2009. Professional exposure to pesticides and Parkinson disease. *Annals of Neurology* 66 (4): 494–504.

Elbert, T., Pantev, C., Wienbruch, C., Rockstroh, B., and Taub, E. 1995. Increased cortical representation of the fingers of the left hand in string players. *Science* 270: 305–307.

Elliott, R., Sahakian, B. J., Matthews, K., Bannerjea, A., Rimmer, J., and Robbins, T. W. 1997. Effects of methylphenidate on spatial working memory and planning in healthy young adults. *Psychopharmacology* 131 (2): 196–206.

Englund, U., Fricker-Gates, R. A., Lundberg, C., Bjorklund, A., and Wictorin, K. 2002. Transplantation of human neural progenitor cells into the neonatal rat brain: Extensive migration and differentiation with long-distance axonal projections. *Experimental Neurology* 173 (1): 1–21.

Erickson, K. I., Colcombe, S. J., Elavsky, S., McAuley, E., Korol, D. L., Scalf, P. E., et al. 2007. Interactive effects of fitness and hormone treatment on brain health in postmenopausal women. *Neurobiology of Aging* 28 (2): 179–185.

Erickson, K. I., Colcombe, S. J., Raz, N., Korol, D. L., Scalf, P., Webb, A., et al. 2005. Selective sparing of brain tissue in postmenopausal women receiving hormone replacement therapy. *Neurobiology of Aging* 26 (8): 1205–1213.

Erickson, K. I., Colcombe, S. J., Wadhwa, R., Bherer, L., Peterson, M. S., Scalf, P. E., et al. 2007. Training-induced functional activation changes in dual-task processing: an FMRI study. *Cerebral Cortex* 17 (1): 192–204.

Erickson, K. I., Prakash, R. S., Voss, M. W., Chaddock, L., Heo, S., McLaren, M., et al. 2010. Brain-derived neurotrophic factor is associated with age-related decline in hippocampal volume. *Journal of Neuroscience* 30 (15): 5368–5375.

Erickson, K. I., Prakash, R. S., Voss, M. W., Chaddock, L., Hu, L., Morris, K. S., et al. 2009. Aerobic fitness is associated with hippocampal volume in elderly humans. *Hippocampus* 19 (10): 1030–1039.

Erickson, K. I., Raji, C. A., Lopez, O. L., Becker, J. T., Rosano, C., Newman, A. B., et al. 2010. Physical activity predicts gray matter volume in late adulthood: The Cardiovascular Health Study. *Neurology* 75 (16): 1415–1422.

Erickson, K. I., Voss, M. W., Prakash, R. S., Basak, C., Szabo, A., Chaddock, L., et al. 2011. Exercise training increases size of hippocampus and improves memory. *Proceedings of the National Academy of Sciences* 108: 3017–3022.

Erickson, K. I., Voss, M. W., Prakash, R. S., Chaddock, L., and Kramer, A. F. 2010. A cross-sectional study of hormone treatment and hippocampal volume in postmenopausal women: Evidence for a limited window of opportunity. *Neuropsychology* 24 (1): 68–76.

Eriksson, P. S., Perfilieva, E., Bjork-Eriksson, T., Alborn, A. M., Nordborg, C., Peterson, D. A., et al. 1998. Neurogenesis in the adult human hippocampus. *Nature Medicine* 4 (11): 1313–1317.

Espeland, M. A., Rapp, S. R., Shumaker, S. A., Brunner, R., Manson, J. E., Sherwin, B. B., et al. 2004. Conjugated equine estrogens and global cognitive function in postmenopausal women: Women's Health Initiative Memory Study. *Journal of the American Medical Association* 291 (24): 2959–2968.

Espeseth, T., Greenwood, P. M., Reinvang, I., Fjell, A. M., Walhovd, K. B., Westlye, L. T., et al. 2006. Interactive effects of APOE and CHRNA4 on attention and white matter volume in healthy middle-aged and older adults. *Cognitive, Affective & Behavioral Neuroscience* 6 (1): 31–43.

Espeseth, T., Sneve, M., Rootwelt, H., and Laeng, B. 2010. Nicotinic receptor gene CHRNA4 interacts with processing load in attention. *PLoS One* 5 (12): e14407.

Espeseth, T., Westlye, L. T., Fjell, A. M., Walhovd, K. B., Rootwelt, H., and Reinvang, I. 2008. Accelerated age-related cortical thinning in healthy carriers of apolipoprotein E ε4. *Neurobiology of Aging* 29 (3): 329–340.

Espeseth, T., Westlye, L. T., Walhovd, K. B., Fjell, A. M., Endestad, T., Rootwelt, H., et al. In press. Apolipoprotein E ε4-related thickening of the cerebral cortex modulates selective attention. *Neurobiology of Aging*.

Etnier, J. L., Caselli, R. J., Reiman, E. M., Alexander, G. E., Sibley, B. A., Tessier, D., et al. 2007. Cognitive performance in older women relative to ApoE-ε4 genotype and aerobic fitness. *Medicine and Science in Sports and Exercise* 39 (1): 199–207.

Etnier, J. L., Nowell, P. M., Landers, D. M., and Sibley, B. A. 2006. A meta-regression to examine the relationship between aerobic fitness and cognitive performance. *Brain Research Reviews* 52 (1): 119–130.

Everitt, B. J., and Robbins, T. W. 1997. Central cholinergic systems and cognition. *Annual Review of Psychology* 48: 649–684.

Ezer, N., Fisk, A. D., and Rogers, W. A. 2008. Age-related differences in reliance behavior attributable to costs within a human-decision aid system. *Human Factors* 50 (6): 853–863.

Fabel, K., Fabel, K., Tam, B., Kaufer, D., Baiker, A., Simmons, N., et al. 2003. VEGF is necessary for exercise-induced adult hippocampal neurogenesis. *European Journal of Neuroscience* 18 (10): 2803–2812.

Fabre, C., Chamari, K., Mucci, P., Masse-Biron, J., and Prefaut, C. 2002. Improvement of cognitive function by mental and/or individualized aerobic training in healthy elderly subjects. *International Journal of Sports Medicine* 23 (6): 415–421.

Faherty, C. J., Kerley, D., and Smeyne, R. J. 2003. A Golgi-Cox morphological analysis of neuronal changes induced by environmental enrichment. *Brain Research. Developmental Brain Research* 141 (1–2): 55–61.

Fan, J., Fossella, J., Sommer, T., Wu, Y., and Posner, M. I. 2003. Mapping the genetic variation of executive attention onto brain activity. *Proceedings of the National Academy of Sciences* 100: 7406–7411.

Farioli-Vecchioli, S., Saraulli, D., Costanzi, M., Pacioni, S., Cina, I., Aceti, M., et al. 2008. The timing of differentiation of adult hippocampal neurons is crucial for spatial memory. *PLoS Biology* 6 (10): e246.

Farmer, J., Zhao, X., van Praag, H., Wodtke, K., Gage, F. H., and Christie, B. R. 2004. Effects of voluntary exercise on synaptic plasticity and gene expression in the dentate gyrus of adult male Sprague-Dawley rats in vivo. *Neuroscience* 124 (1): 71–79.

Farquhar, C., Marjoribanks, J., Lethaby, A., Suckling, J. A., and Lamberts, Q. 2009. Long term hormone therapy for perimenopausal and postmenopausal women. *Cochrane Database of Systematic Reviews* 2: CD004143.

Farr, S. A., Yamada, K. A., Butterfield, D. A., Abdul, H. M., Xu, L., Miller, N. E., et al. 2008. Obesity and hypertriglyceridemia produce cognitive impairment. *Endocrinology* 149 (5): 2628–2636.

Farrell, D. 2008. *Talkin' 'bout My Generation: The Economic Impact of Aging U.S. Baby Boomers*. McKinsey Global Institute.

Feng, Y., Wang, X. P., Yang, S. G., Wang, Y. J., Zhang, X., Du, X. T., et al. 2009. Resveratrol inhibits β-amyloid oligomeric cytotoxicity but does not prevent oligomer formation. *Neurotoxicology* 30 (6): 986–995.

Fernandes, J. S., Mori, M. A., Ekuni, R., Oliveira, R. M., and Milani, H. 2008. Long-term treatment with fish oil prevents memory impairments but not hippocampal damage in rats subjected to transient, global cerebral ischemia. *Nutrition Research* 28 (11): 798–808.

Filippini, N., MacIntosh, B. J., Hough, M. G., Goodwin, G. M., Frisoni, G. B., Smith, S. M., et al. 2009. Distinct patterns of brain activity in young carriers of the APOE-ε4 allele. *Proceedings of the National Academy of Sciences* 106 (17): 7209–7214.

Fillmore, K. M., Stockwell, T., Chikritzhs, T., Bostrom, A., and Kerr, W. 2007. Moderate alcohol use and reduced mortality risk: Systematic error in prospective studies and new hypotheses. *Annals of Epidemiology* 17 (5 suppl.): S16–S23.

Fisk, A. D., Rogers, W. A., Charness, N., Czaja, S. J., and Sharit, J. 2009. *Designing for Older Adults: Principles and Creative Human Factors Approaches*, second edition. CRC Press.

Fisk, A. D., and Rogers, W. A. 1997. *Handbook of Human Factors and the Older Adult.* Academic Press.

Fitts, P. M. 1954. The information capacity of the human motor system in controlling the amplitude of movement. *Journal of Experimental Psychology* 47: 381–391.

Fjell, A. M., Walhovd, K. B., Fischl, B., and Reinvang, I. 2007. Cognitive function, P3a/P3b brain potentials, and cortical thickness in aging. *Human Brain Mapping* 28 (11): 1098–1116.

Fjell, A. M., Walhovd, K. B., Reinvang, I., Lundervold, A., Salat, D., Quinn, B. T., et al. 2006. Selective increase of cortical thickness in high-performing elderly—Structural indices of optimal cognitive aging. *NeuroImage* 29 (3): 984–994.

Fjell, A. M., Westlye, L. T., Greve, D. N., Fischl, B., Benner, T., van der Kouwe, A. J., et al. 2008. The relationship between diffusion tensor imaging and volumetry as measures of white matter properties. *NeuroImage* 42 (4): 1654–1668.

Flood, D. G. 1993. Critical issues in the analysis of dendritic extent in aging humans, primates, and rodents. *Neurobiology of Aging* 14 (6): 649–654.

Florence, S. L., Taub, H. B., and Kaas, J. H. 1998. Large-scale sprouting of cortical connections after peripheral injury in adult macaque monkeys. *Science* 282: 1117–1121.

Flores, C. M., DeCamp, R. M., Kilo, S., Rogers, S. W., and Hargreaves, K. M. 1996. Neuronal nicotinic receptor expression in sensory neurons of the rat trigeminal ganglion: Demonstration of $\alpha 3\beta 4$, a novel subtype in the mammalian nervous system. *Journal of Neuroscience* 16 (24): 7892–7901.

Floyd, M., and Scogin, F. 1997. Effects of memory training on the subjective memory functioning and mental health of older adults: A meta-analysis. *Psychology and Aging* 12 (1): 150–161.

Folstein, M. F., Folstein, S. E., and McHugh, P. R. 1975. "Mini-mental state"—a practical method for grading the cognitive state of patients for the clinician. *Journal of Current Psychiatric Research* 12: 189–198.

Fontan-Lozano, A., Saez-Cassanelli, J. L., Inda, M. C., de los Santos-Arteaga, M., Sierra-Dominguez, S. A., Lopez-Lluch, G., et al. 2007. Caloric restriction increases learning consolidation and facilitates synaptic plasticity through mechanisms dependent on NR2B subunits of the NMDA receptor. *Journal of Neuroscience* 27 (38): 10185–10195.

Fontana, L., and Klein, S. 2007. Aging, adiposity, and calorie restriction. *Journal of the American Medical Association* 297 (9): 986–994.

Fontana, L., Meyer, T. E., Klein, S., and Holloszy, J. O. 2004. Long-term calorie restriction is highly effective in reducing the risk for atherosclerosis in humans. *Proceedings of the National Academy of Sciences* 101 (17): 6659–6663.

Fossella, J., Sommer, T., Fan, J., Wu, Y., Swanson, J. M., Pfaff, D. W., et al. 2002. Assessing the molecular genetics of attention networks. *BMC Neuroscience* 3 (1): 14.

Fossella, J. A., and Casey, B. J. 2006. Genes, brain, and behavior: bridging disciplines. *Cognitive, Affective & Behavioral Neuroscience* 6 (1): 1–8.

Foster, J. K., Meikle, A., Goodson, G., Mayes, A. R., Howard, M., Sunram, S. I., et al. 1999. The hippocampus and delayed recall: Bigger is not necessarily better? *Memory* 7 (5–6): 715–732.

Fotenos, A. F., Mintun, M. A., Snyder, A. Z., Morris, J. C., and Buckner, R. L. 2008. Brain volume decline in aging: Evidence for a relation between socioeconomic status, preclinical Alzheimer disease, and reserve. *Archives of Neurology* 65 (1): 113–120.

Fotuhi, M., Hachinski, V., and Whitehouse, P. J. 2009. Changing perspectives regarding late-life dementia. *Nature Reviews Neurology* 5 (12): 649–658.

Foubert-Samier, A., Catheline, G., Amieva, H., Dilharreguy, B., Helmer, C., Allard, M., et al. In press. Education, occupation, leisure activities, and brain reserve: A population-based study. *Neurobiology of Aging*.

Freeman, E. W., Sammel, M. D., Lin, H., Gracia, C. R., Pien, G. W., Nelson, D. B., et al. 2007. Symptoms associated with menopausal transition and reproductive hormones in midlife women. *Obstetrics and Gynecology* 110 (2 pt. 1): 230–240.

Frick, K. M., and Fernandez, S. M. 2003. Enrichment enhances spatial memory and increases synaptophysin levels in aged female mice. *Neurobiology of Aging* 24 (4): 615–626.

Friedland, R. P., Fritsch, T., Smyth, K. A., Koss, E., Lerner, A. J., Chen, C. H., et al. 2001. Patients with Alzheimer's disease have reduced activities in midlife compared with healthy control-group members. *Proceedings of the National Academy of Sciences* 98 (6): 3440–3445.

Frost, R. 1936. "Provide, Provide." In *A Further Range*. Holt, Rinehart and Winston.

Frost, S. B., Barbay, S., Friel, K. M., Plautz, E. J., and Nudo, R. J. 2003. Reorganization of remote cortical regions after ischemic brain injury: A potential substrate for stroke recovery. *Journal of Neurophysiology* 89 (6): 3205–3214.

Furey, M. L., Pietrini, P., and Haxby, J. V. 2000. Cholinergic enhancement and increased selectivity of perceptual processing during working memory. *Science* 290: 2315–2319.

Furey, M. L., Pietrini, P., Haxby, J. V., and Drevets, W. C. 2008. Selective effects of cholinergic modulation on task performance during selective attention. *Neuropsychopharmacology* 33 (4): 913–923.

Galimberti, I., Gogolla, N., Alberi, S., Santos, A. F., Muller, D., and Caroni, P. 2006. Long-term rearrangements of hippocampal mossy fiber terminal connectivity in the adult regulated by experience. *Neuron* 50 (5): 749–763.

Gallagher, M., Bizon, J. L., Hoyt, E. C., Helm, K. A., and Lund, P. K. 2003. Effects of aging on the hippocampal formation in a naturally occurring animal model of mild cognitive impairment. *Experimental Gerontology* 38 (1–2): 71–77.

Gallagher, M., Burwell, R., and Burchinal, M. 1993. Severity of spatial learning impairment in aging: Development of a learning index for performance in the Morris water maze. *Behavioral Neuroscience* 107 (4): 618–626.

Gallagher, M., and Rapp, P. R. 1997. The use of animal models to study the effects of aging on cognition. *Annual Review of Psychology* 48: 339–370.

Gardner, L., Powell, L., and Page, M. 1993. An appraisal of a selection of products currently available to older consumers. *Applied Ergonomics* 24 (1): 35–39.

Garthe, A., Behr, J., and Kempermann, G. 2009. Adult-generated hippocampal neurons allow the flexible use of spatially precise learning strategies. *PLoS ONE* 4 (5): e5464.

Gatz, M., Fratiglioni, L., Johansson, B., Berg, S., Mortimer, J. A., Reynolds, C. A., et al. 2005. Complete ascertainment of dementia in the Swedish Twin Registry: The HARMONY study. *Neurobiology of Aging* 26 (4): 439–447.

Gatz, M., Reynolds, C. A., Fratiglioni, L., Johansson, B., Mortimer, J. A., Berg, S., et al. 2006. Role of genes and environments for explaining Alzheimer disease. *Archives of General Psychiatry* 63 (2): 168–174.

Ghebremedhin, E., Schultz, C., Braak, E., and Braak, H. 1998. High frequency of apolipoprotein E ε4 allele in young individuals with very mild Alzheimer's disease-related neurofibrillary changes. *Experimental Neurology* 153 (1): 152–155.

Gibbs, R. B., and Gabor, R. 2003. Estrogen and cognition: Applying preclinical findings to clinical perspectives. *Journal of Neuroscience Research* 74 (5): 637–643.

Giedd, J. N., Blumenthal, J., Jeffries, N. O., Castellanos, F. X., Liu, H., Zijdenbos, A., et al. 1999. Brain development during childhood and adolescence: A longitudinal MRI study. *Nature Neuroscience* 2 (10): 861–863.

Gigerenzer, G., and Gaissmaier, W. 2011. Heuristic decision making. *Annual Review of Psychology* 62: 451–482.

Gill, M., Haerich, P., Westcott, K., Godenick, K. L., and Tucker, J. A. 2006. Cognitive performance following modafinil versus placebo in sleep-deprived emergency physicians: A double-blind randomized crossover study. *Academic Emergency Medicine* 13 (2): 158–165.

Glasgow, R. E., and Strycker, L. A. 2000. Preventive care practices for diabetes management in two primary care samples. *American Journal of Preventive Medicine* 19 (1): 9–14.

Glisky, E. L., Rubin, S. R., and Davidson, P. S. 2001. Source memory in older adults: An encoding or retrieval problem? *Journal of Experimental Psychology: Learning, Memory, and Cognition* 27 (5): 1131–1146.

Goldberg, T. E., and Weinberger, D. R. 2004. Genes and the parsing of cognitive processes. *Trends in Cognitive Sciences* 8 (7): 325–335.

Gómez-Pinilla, F. 2008. Brain foods: The effects of nutrients on brain function. *Nature Reviews Neuroscience* 9 (7): 568–578.

Good, C. D., Johnsrude, I. S., Ashburner, J., Henson, R. N., Friston, K. J., and Frackowiak, R. S. 2001. A voxel-based morphometric study of ageing in 465 normal adult human brains. *NeuroImage* 14 (1 pt. 1): 21–36.

Goodrich-Hunsaker, N. J., Hunsaker, M. R., and Kesner, R. P. 2008. The interactions and dissociations of the dorsal hippocampus subregions: How the dentate gyrus, CA3, and CA1 process spatial information. *Behavioral Neuroscience* 122 (1): 16–26.

Gould, E., Beylin, A., Tanapat, P., Reeves, A., and Shors, T. J. 1999. Learning enhances adult neurogenesis in the hippocampal formation. *Nature Neuroscience* 2 (3): 260–265.

Gould, E., and Gross, C. G. 2002. Neurogenesis in adult mammals: Some progress and problems. *Journal of Neuroscience* 22 (3): 619–623.

Gould, E., McEwen, B. S., Tanapat, P., Galea, L. A., and Fuchs, E. 1997. Neurogenesis in the dentate gyrus of the adult tree shrew is regulated by psychosocial stress and NMDA receptor activation. *Journal of Neuroscience* 17 (7): 2492–2498.

Gould, E., Tanapat, P., McEwen, B. S., Flugge, G., and Fuchs, E. 1998. Proliferation of granule cell precursors in the dentate gyrus of adult monkeys is diminished by stress. *Proceedings of the National Academy of Sciences* 95 (6): 3168–3171.

Gould, E., Vail, N., Wagers, M., and Gross, C. G. 2001. Adult-generated hippocampal and neocortical neurons in macaques have a transient existence. *Proceedings of the National Academy of Sciences* 98 (19): 10910–10917.

Grady, C. L., Maisog, J. M., Horwitz, B., Ungerleider, L. G., Mentis, M. J., Salerna, J. A., et al. 1994. Age-related changes in cortical blood flow activation during visual processing of faces and location. *Journal of Neuroscience* 14: 1450–1462.

Grady, C. L., McIntosh, A. R., and Craik, F. I. 2003. Age-related differences in the functional connectivity of the hippocampus during memory encoding. *Hippocampus* 13 (5): 572–586.

Granholm, A. C., Bimonte-Nelson, H. A., Moore, A. B., Nelson, M. E., Freeman, L. R., and Sambamurti, K. 2008. Effects of a saturated fat and high cholesterol diet on memory and hippocampal morphology in the middle-aged rat. *Journal of Alzheimer's Disease* 14 (2): 133–145.

Gray, J. R., Chabris, C. F., and Braver, T. S. 2003. Neural mechanisms of general fluid intelligence. *Nature Neuroscience* 6 (3): 316–322.

Greely, H., Sahakian, B., Harris, J., Kessler, R. C., Gazzaniga, M., Campbell, P., et al. 2008. Towards responsible use of cognitive-enhancing drugs by the healthy. *Nature* 456 (7223): 702–705.

Green, A. E., Munafo, M. R., DeYoung, C. G., Fossella, J. A., Fan, J., and Gray, J. R. 2008. Using genetic data in cognitive neuroscience: From growing pains to genuine insights. *Nature Reviews Neuroscience* 9 (9): 710–720.

Green, M. W., and Rogers, P. J. 1998. Impairments in working memory associated with spontaneous dieting behaviour. *Psychological Medicine* 28 (5): 1063–1070.

Greenough, W. T., Hwang, H. M., and Gorman, C. 1985. Evidence for active synapse formation or altered postsynaptic metabolism in visual cortex of rats reared in complex environments. *Proceedings of the National Academy of Sciences* 82 (13): 4549–4552.

Greenwood, C. E., and Winocur, G. 1996. Cognitive impairment in rats fed high-fat diets: A specific effect of saturated fatty-acid intake. *Behavioral Neuroscience* 110 (3): 451–459.

Greenwood, C. E., and Winocur, G. 2005. High-fat diets, insulin resistance and declining cognitive function. *Neurobiology of Aging* 26 (suppl. 1): 42–45.

Greenwood, P. M. 2000. The frontal aging hypothesis evaluated. *Journal of the International Neuropsychological Society* 6 (6): 705–726.

Greenwood, P. M. 2007. Functional plasticity in cognitive aging: Review and hypothesis. *Neuropsychology* 21 (6): 657–673.

Greenwood, P. M., Fossella, J. A., and Parasuraman, R. 2005. Specificity of the effect of a nicotinic receptor polymorphism on individual differences in visuospatial attention. *Journal of Cognitive Neuroscience* 17 (10): 1611–1620.

Greenwood, P. M., Lin, M. K., Sundararajan, R., Fryxell, K. J., and Parasuraman, R. 2009. Synergistic effects of genetic variation in nicotinic and muscarinic receptors on visual attention but not working memory. *Proceedings of the National Academy of Sciences* 106 (9): 3633–3638.

Greenwood, P. M., and Parasuraman, R. 1994. Attentional disengagement deficit in nondemented elderly over 75 years of age. *Aging and Cognition* 1 (3): 188–202.

Greenwood, P. M., and Parasuraman, R. 2003. Normal genetic variation, cognition, and aging. *Behavioral and Cognitive Neuroscience Reviews* 2 (4): 278–306.

Greenwood, P. M., and Parasuraman, R. 2010. Neuronal and cognitive plasticity: A neurocognitive framework for ameliorating cognitive aging. *Frontiers in Aging Neuroscience* 2: 150.

Greenwood, P. M., Parasuraman, R., and Haxby, J. V. 1993. Changes in visuospatial attention over the adult lifespan. *Neuropsychologia* 31 (5): 471–485.

Greenwood, P. M., Sundararajan, R., Lin, M. K., Kumar, R., Fryxell, K. J., and Parasuraman, R. 2009. Both a nicotinic single nucleotide polymorphism (SNP) and a noradrenergic SNP modulate working memory performance when attention is manipulated. *Journal of Cognitive Neuroscience* 21 (11): 2139–2153.

Greenwood, P. M., Sunderland, T., Friz, J. L., and Parasuraman, R. 2000. Genetics and visual attention: Selective deficits in healthy adult carriers of the ε4 allele of the apolipoprotein E gene. *Proceedings of the National Academy of Sciences* 97 (21): 11661–11666.

Grieve, S. M., Williams, L. M., Paul, R. H., Clark, C. R., and Gordon, E. 2007. Cognitive aging, executive function, and fractional anisotropy: A diffusion tensor MR imaging study. *American Journal of Neuroradiology* 28 (2): 226–235.

Grodstein, F., Kang, J. H., Glynn, R. J., Cook, N. R., and Gaziano, J. M. 2007. A randomized trial of beta carotene supplementation and cognitive function in men: The Physicians' Health Study II. *Archives of Internal Medicine* 167 (20): 2184–2190.

Grossman, A. W., Churchill, J. D., Bates, K. E., Kleim, J. A., and Greenough, W. T. 2002. A brain adaptation view of plasticity: Is synaptic plasticity an overly limited concept? *Progress in Brain Research* 138: 91–108.

Grossman, A. W., Churchill, J. D., McKinney, B. C., Kodish, I. M., Otte, S. L., and Greenough, W. T. 2003. Experience effects on brain development: Possible contributions to psychopathology. *Journal of Child Psychology and Psychiatry* 44 (1): 33–63.

Guarente, L., and Picard, F. 2005. Calorie restriction—the SIR2 connection. *Cell* 120 (4): 473–482.

Gunstad, J., Paul, R. H., Cohen, R. A., Tate, D. F., Spitznagel, M. B., and Gordon, E. 2007. Elevated body mass index is associated with executive dysfunction in otherwise healthy adults. *Comprehensive Psychiatry* 48 (1): 57–61.

Gutchess, A. H., Welsh, R. C., Hedden, T., Bangert, A., Minear, M., Liu, L. L., et al. 2005. Aging and the neural correlates of successful picture encoding: Frontal activations compensate for decreased medial-temporal activity. *Journal of Cognitive Neuroscience* 17 (1): 84–96.

Hagberg, J. M., Ferrell, R. E., Katzel, L. I., Dengel, D. R., Sorkin, J. D., and Goldberg, A. P. 1999. Apolipoprotein E genotype and exercise training-induced increases in plasma high-density lipoprotein (HDL)- and HDL2-cholesterol levels in overweight men. *Metabolism: Clinical and Experimental* 48 (8): 943–945.

Haigh, K. Z., et al. 2003. *The Independent LifeStyle Assistant (I.L.S.A.): Lessons Learned.* Technical report ACS-PO3-023, Honeywell Laboratories (available at http://www .agingtech.org).

Hallett, M. 2000. Transcranial magnetic stimulation and the human brain. *Nature* 406 (6792): 147–150.

Hancock, H. E., Fisk, A. D., and Rogers, W. A. 2001. Everyday products: Easy to use . . . or not? *Ergonomics in Design* 9: 12–18.

Hanks, S. D., and Flood, D. G. 1991. Region-specific stability of dendritic extent in normal human aging and regression in Alzheimer's disease. I. CA1 of hippocampus. *Brain Research* 540 (1–2): 63–82.

Hao, J., Rapp, P. R., Leffler, A. E., Leffler, S. R., Janssen, W. G., Lou, W., et al. 2006. Estrogen alters spine number and morphology in prefrontal cortex of aged female rhesus monkeys. *Journal of Neuroscience* 26 (9): 2571–2578.

Hardy, D., Mouloua, M., Dwivedi, C., and Parasuraman, R. 1995. Monitoring of automation failures by young and older adults. In proceedings of International Symposium on Aviation Psychology, Columbus.

Harman, S. M., Brinton, E. A., Cedars, M., Lobo, R., Manson, J. E., Merriam, G. R., et al. 2005. KEEPS: The Kronos Early Estrogen Prevention Study. *Climacteric* 8 (1): 3–12.

Hartley, A. A., Kieley, J. M., and Slabach, E. H. 1990. Age differences and similarities in the effects of cues and prompts. *Journal of Experimental Psychology: Human Perception and Performance* 16: 523–537.

Hasher, L., Stoltzfus, E. R., Zacks, R. T., and Rypma, B. 1991. Age and inhibition. *Journal of Experimental Psychology: Learning, Memory, and Cognition* 17 (1): 163–169.

Hashimoto, M., Hossain, S., Tanabe, Y., Kawashima, A., Harada, T., Yano, T., et al. 2009. The protective effect of dietary eicosapentaenoic acid against impairment of spatial cognition learning ability in rats infused with amyloid $\beta(1–40)$. *Journal of Nutritional Biochemistry* 20 (12): 965–973.

Hastings, N. B., and Gould, E. 1999. Rapid extension of axons into the CA3 region by adult-generated granule cells. *Journal of Comparative Neurology* 413 (1): 146–154.

Hayes, K., Sprague, S., Guo, M., Davis, W., Friedman, A., Kumar, A., et al. 2008. Forced, not voluntary, exercise effectively induces neuroprotection in stroke. *Acta Neuropathologica* 115 (3): 289–296.

Head, E., Liu, J., Hagen, T. M., Muggenburg, B. A., Milgram, N. W., Ames, B. N., et al. 2002. Oxidative damage increases with age in a canine model of human brain aging. *Journal of Neurochemistry* 82 (2): 375–381.

Healey, M. K., Campbell, K. L., and Hasher, L. 2008. Cognitive aging and increased distractibility: Costs and potential benefits. *Progress in Brain Research* 169: 353–363.

Hebb, D. O. 1947. The effects of early experience on problem-solving at maturity. *American Psychologist* 2: 306–307.

Heilbronn, L. K., Civitarese, A. E., Bogacka, I., Smith, S. R., Hulver, M., and Ravussin, E. 2005. Glucose tolerance and skeletal muscle gene expression in response to alternate day fasting. *Obesity Research* 13 (3): 574–581.

Heilbronn, L. K., de Jonge, L., Frisard, M. I., DeLany, J. P., Larson-Meyer, D. E., Rood, J., et al. 2006. Effect of 6-month calorie restriction on biomarkers of longevity, metabolic adaptation, and oxidative stress in overweight individuals: A randomized controlled trial. *Journal of the American Medical Association* 295 (13): 1539–1548.

Heilbronn, L. K., and Ravussin, E. 2003. Calorie restriction and aging: review of the literature and implications for studies in humans. *American Journal of Clinical Nutrition* 78 (3): 361–369.

Henderson, A. S., Easteal, S., Jorm, A. F., Mackinnon, A. J., Korten, A. E., Christensen, H., et al. 1995. Apolipoprotein E allele ε4, dementia, and cognitive decline in a population sample. *Lancet* 346 (8987): 1387–1390.

Henderson, V. W., Paganini-Hill, A., Miller, B. L., Elble, R. J., Reyes, P. F., Shoupe, D., et al. 2000. Estrogen for Alzheimer's disease in women: Randomized, double-blind, placebo-controlled trial. *Neurology* 54 (2): 295–301.

Hendrie, H. C., Ogunniyi, A., Hall, K. S., Baiyewu, O., Unverzagt, F. W., Gureje, O., et al. 2001. Incidence of dementia and Alzheimer disease in 2 communities: Yoruba residing in Ibadan, Nigeria, and African Americans residing in Indianapolis, Indiana. *Journal of the American Medical Association* 285 (6): 739–747.

Hertzog, C., Kramer, A. F., Wilson, R. S., and Lindenberger, U. 2009. Enrichment effects on adult cognitive development: Can the functional capacity of older adults be preserved and enhanced? *Psychological Science in the Public Interest* 9 (1): 1–65.

Heyn, P., Abreu, B. C., and Ottenbacher, K. J. 2004. The effects of exercise training on elderly persons with cognitive impairment and dementia: A meta-analysis. *Archives of Physical Medicine and Rehabilitation* 85 (10): 1694–1704.

Hihara, S., Notoya, T., Tanaka, M., Ichinose, S., Ojima, H., Obayashi, S., et al. 2006. Extension of corticocortical afferents into the anterior bank of the intraparietal sulcus by tool-use training in adult monkeys. *Neuropsychologia* 44 (13): 2636–2646.

Hillman, C. H., Erickson, K. I., and Kramer, A. F. 2008. Be smart, exercise your heart: Exercise effects on brain and cognition. *Nature Reviews Neuroscience* 9 (1): 58–65.

Hirschhorn, J. N., and Daly, M. J. 2005. Genome-wide association studies for common diseases and complex traits. *Nature Reviews Genetics* 6 (2): 95–108.

Ho, G., Wheatley, D., and Scialfa, C. T. 2005. Age differences in trust and reliance of a medication management system. *Interacting with Computers* 17 (6): 690–710.

Hof, P. R., and Morrison, J. H. 2004. The aging brain: Morphomolecular senescence of cortical circuits. *Trends in Neurosciences* 27 (10): 607–613.

Hogervorst, E., Boshuisen, M., Riedel, W., Willeken, C., and Jolles, J. 1999. The effect of hormone replacement therapy on cognitive function in elderly women. *Psychoneuroendocrinology* 24 (1): 43–68.

Hogervorst, E., Williams, J., Budge, M., Riedel, W., and Jolles, J. 2000. The nature of the effect of female gonadal hormone replacement therapy on cognitive function in post-menopausal women: A meta-analysis. *Neuroscience* 101 (3): 485–512.

Holguin, S., Huang, Y., Liu, J., and Wurtman, R. 2008. Chronic administration of DHA and UMP improves the impaired memory of environmentally impoverished rats. *Behavioural Brain Research* 191 (1): 11–16.

Holtmaat, A., and Svoboda, K. 2009. Experience-dependent structural synaptic plasticity in the mammalian brain. *Nature Reviews Neuroscience* 10 (9): 647–658.

Honein, M. A., Paulozzi, L. J., Mathews, T. J., Erickson, J. D., and Wong, L. Y. 2001. Impact of folic acid fortification of the US food supply on the occurrence of neural tube defects. *Journal of the American Medical Association* 285 (23): 2981–2986.

Hu, P., Bretsky, P., Crimmins, E. M., Guralnik, J. M., Reuben, D. B., and Seeman, T. E. 2006. Association between serum beta-carotene levels and decline of cognitive function in high-functioning older persons with or without apolipoprotein E4 alleles: MacArthur studies of successful aging. *Journal of Gerontology A* 61 (6): 616–620.

Hubacek, J. A., Pitha, J., Skodova, Z., Adamkova, V., Lanska, V., and Poledne, R. 2001. A possible role of apolipoprotein E polymorphism in predisposition to higher education. *Neuropsychobiology* 43 (3): 200–203.

Hultsch, D. F., Hertzog, C., Small, B. J., and Dixon, R. A. 1999. Use it or lose it: Engaged lifestyle as a buffer of cognitive decline in aging? *Psychology and Aging* 14 (2): 245–263.

Huttenlocher, P. R. 1984. Synapse elimination and plasticity in developing human cerebral cortex. *American Journal of Mental Deficiency* 88 (5): 488–496.

Ingram, D. K., Weindruch, R., Spangler, E. L., Freeman, J. R., and Walford, R. L. 1987. Dietary restriction benefits learning and motor performance of aged mice. *Journal of Gerontology* 42 (1): 78–81.

Jacobs, B., Driscoll, L., and Schall, M. 1997. Life-span dendritic and spine changes in areas 10 and 18 of human cortex: A quantitative Golgi study. *Journal of Comparative Neurology* 386: 661–680.

Jacobs, B., Schall, M., and Scheibel, A. B. 1993. A quantitative dendritic analysis of Wernicke's area in humans. II. Gender, hemispheric, and environmental factors. *Journal of Comparative Neurology* 327 (1): 97–111.

Jacques, P. F., Selhub, J., Bostom, A. G., Wilson, P. W., and Rosenberg, I. H. 1999. The effect of folic acid fortification on plasma folate and total homocysteine concentrations. *New England Journal of Medicine* 340 (19): 1449–1454.

Jenkins, W. M., Merzenich, M. M., Ochs, M. T., Allard, T., and Guic-Robles, E. 1990. Functional reorganization of primary somatosensory cortex in adult owl monkeys after behaviorally controlled tactile stimulation. *Journal of Neurophysiology* 63 (1): 82–104.

Jernigan, T. L., Archibald, S. L., Fennema-Notestine, C., Gamst, A. C., Stout, J. C., Bonner, J., et al. 2001. Effects of age on tissues and regions of the cerebrum and cerebellum. *Neurobiology of Aging* 22 (4): 581–594.

Jiang, L. H., Shi, Y., Wang, L. S., and Yang, Z. R. 2009. The influence of orally administered docosahexaenoic acid on cognitive ability in aged mice. *Journal of Nutritional Biochemistry* 20 (9): 735–741.

Joffe, H., Hall, J. E., Gruber, S., Sarmiento, I. A., Cohen, L. S., Yurgelun-Todd, D., et al. 2006. Estrogen therapy selectively enhances prefrontal cognitive processes: A randomized, double-blind, placebo-controlled study with functional magnetic resonance imaging in perimenopausal and recently postmenopausal women. *Menopause* 13 (3): 411–422.

Johnson, J. B., Summer, W., Cutler, R. G., Martin, B., Hyun, D. H., Dixit, V. D., et al. 2007. Alternate day calorie restriction improves clinical findings and reduces markers of oxidative stress and inflammation in overweight adults with moderate asthma. *Free Radical Biology & Medicine* 42 (5): 665–674.

Jonides, J. 2004. How does practice makes perfect? *Nature Neuroscience* 7 (1): 10–11.

Joseph, J. A., Shukitt-Hale, B., Denisova, N. A., Prior, R. L., Cao, G., Martin, A., et al. 1998. Long-term dietary strawberry, spinach, or vitamin E supplementation retards the onset of age-related neuronal signal-transduction and cognitive behavioral deficits. *Journal of Neuroscience* 18 (19): 8047–8055.

Jucker, M., Battig, K., and Meier-Ruge, W. 1990. Effects of aging and vincamine derivatives on pericapillary microenvironment: Stereological characterization of the cerebral capillary network. *Neurobiology of Aging* 11 (1): 39–46.

Jump, D. B. 2002a. Dietary polyunsaturated fatty acids and regulation of gene transcription. *Current Opinion in Lipidology* 13 (2): 155–164.

Jump, D. B. 2002b. The biochemistry of n-3 polyunsaturated fatty acids. *Journal of Biological Chemistry* 277 (11): 8755–8758.

Kaas, J. H., Nelson, R. J., Sur, M., Dykes, R. W., and Merzenich, M. M. 1984. The somatotopic organization of the ventroposterior thalamus of the squirrel monkey, *Saimiri sciureus*. *Journal of Comparative Neurology* 226 (1): 111–140.

Kalaria, R. N., Maestre, G. E., Arizaga, R., Friedland, R. P., Galasko, D., Hall, K., et al. 2008. Alzheimer's disease and vascular dementia in developing countries: Prevalence, management, and risk factors. *Lancet Neurology* 7 (9): 812–826.

Kalisch, T., Ragert, P., Schwenkreis, P., Dinse, H. R., and Tegenthoff, M. 2009. Impaired tactile acuity in old age is accompanied by enlarged hand representations in somatosensory cortex. *Cerebral Cortex* 19 (7): 1530–1538.

Kang, J. H., Cook, N., Manson, J., Buring, J. E., Albert, C. M., and Grodstein, F. 2008. A trial of B vitamins and cognitive function among women at high risk of cardiovascular disease. *American Journal of Clinical Nutrition* 88 (6): 1602–1610.

Kaplan, M. S., and Hinds, J. W. 1977. Neurogenesis in the adult rat: Electron microscopic analysis of light radioautographs. *Science* 197: 1092–1094.

Karbach, J., and Kray, J. 2009. How useful is executive control training? Age differences in near and far transfer of task-switching training. *Developmental Science* 12 (6): 978–990.

Kattenstroth, J. C., Kolankowska, I., Kalisch, T., and Dinse, H. R. 2010. Superior sensory, motor, and cognitive performance in elderly individuals with multi-year dancing activities. *Frontiers in Aging Neuroscience* 2: 31.

Katzman, R. 1993. Education and the prevalence of dementia and Alzheimer's disease. *Neurology* 43 (1): 13–20.

Katzman, R., Zhang, M. Y., Chen, P. J., Gu, N., Jiang, S., Saitoh, T., et al. 1997. Effects of apolipoprotein E on dementia and aging in the Shanghai Survey of Dementia. *Neurology* 49 (3): 779–785.

Kempermann, G. 2002. Why new neurons? Possible functions for adult hippocampal neurogenesis. *Journal of Neuroscience* 22 (3): 635–638.

Kempermann, G. 2008. The neurogenic reserve hypothesis: What is adult hippocampal neurogenesis good for? *Trends in Neurosciences* 31 (4): 163–169.

Kempermann, G., and Gage, F. H. 1999. Experience-dependent regulation of adult hippocampal neurogenesis: Effects of long-term stimulation and stimulus withdrawal. *Hippocampus* 9 (3): 321–332.

Kempermann, G., and Gage, F. H. 2002. Genetic determinants of adult hippocampal neurogenesis correlate with acquisition, but not probe trial performance, in the water maze task. *European Journal of Neuroscience* 16 (1): 129–136.

Kempermann, G., Gast, D., and Gage, F. H. 2002. Neuroplasticity in old age: Sustained fivefold induction of hippocampal neurogenesis by long-term environmental enrichment. *Annals of Neurology* 52 (2): 135–143.

Kempermann, G., Kuhn, H. G., and Gage, F. H. 1997. More hippocampal neurons in adult mice living in an enriched environment. *Nature* 386 (6624): 493–495.

Kempermann, G., Kuhn, H. G., and Gage, F. H. 1998. Experience-induced neurogenesis in the senescent dentate gyrus. *Journal of Neuroscience* 18 (9): 3206–3212.

Kerr, D. S., Campbell, L. W., Hao, S. Y., and Landfield, P. W. 1989. Corticosteroid modulation of hippocampal potentials: Increased effect with aging. *Science* 245: 1505–1509.

Killiany, R. J., Gomez-Isla, T., Moss, M., Kikinis, R., Sandor, T., Jolesz, F., et al. 2000. Use of structural magnetic resonance imaging to predict who will get Alzheimer's disease. *Annals of Neurology* 47 (4): 430–439.

Kim, D., Nguyen, M. D., Dobbin, M. M., Fischer, A., Sananbenesi, F., Rodgers, J. T., et al. 2007. SIRT1 deacetylase protects against neurodegeneration in models for Alzheimer's disease and amyotrophic lateral sclerosis. *EMBO Journal* 26 (13): 3169–3179.

Kim, H., Lee, S. H., Kim, S. S., Yoo, J. H., and Kim, C. J. 2007. The influence of maternal treadmill running during pregnancy on short-term memory and hippocampal cell survival in rat pups. *International Journal of Developmental Neuroscience* 25: 243–249.

Kim, S., Hasher, L., and Zacks, R. T. 2007. Aging and a benefit of distractibility. *Psychonomic Bulletin & Review* 14 (2): 301–305.

Kimberg, D. Y., D'Esposito, M., and Farah, M. J. 1997. Effects of bromocriptine on human subjects depend on working memory capacity. *Neuroreport* 8 (16): 3581–3585.

Kinsella, K., and Phillips, D. R. 2005. Global aging: The challenge of success. *Population Bulletin* 60 (1): 1–40.

Kleim, J. A., Jones, T. A., and Schallert, T. 2003. Motor enrichment and the induction of plasticity before or after brain injury. *Neurochemical Research* 28 (11): 1757–1769.

Knaepen, K., Goekint, M., Heyman, E. M., and Meeusen, R. 2010. Neuroplasticity—exercise-induced response of peripheral brain-derived neurotrophic factor: A systematic review of experimental studies in human subjects. *Sports Medicine* 40 (9): 765–801.

Knoth, R., Singec, I., Ditter, M., Pantazis, G., Capetian, P., Meyer, R. P., et al. 2010. Murine features of neurogenesis in the human hippocampus across the lifespan from 0 to 100 years. *PLoS ONE* 5 (1): e8809.

Kokoeva, M. V., Yin, H., and Flier, J. S. 2005. Neurogenesis in the hypothalamus of adult mice: Potential role in energy balance. *Science* 310: 679–683.

Kramer, A. F., Larish, J. F., and Strayer, D. L. 1995. Training for attentional control in dual task settings: A comparison of young and older adults. *Journal of Experimental Psychology: Applied* 1: 50–76.

Kronenberg, G., Bick-Sander, A., Bunk, E., Wolf, C., Ehninger, D., and Kempermann, G. 2006. Physical exercise prevents age-related decline in precursor cell activity in the mouse dentate gyrus. *Neurobiology of Aging* 27 (10): 1505–1513.

Krug, R., Born, J., and Rasch, B. 2006. A 3-day estrogen treatment improves prefrontal cortex-dependent cognitive function in postmenopausal women. *Psychoneuroendocrinology* 31 (8): 965–975.

Kuhn, H. G., Dickinson-Anson, H., and Gage, F. H. 1996. Neurogenesis in the dentate gyrus of the adult rat: Age-related decrease of neuronal progenitor proliferation. *Journal of Neuroscience* 16 (6): 2027–2033.

Kuiper, G. G., Carlsson, B., Grandien, K., Enmark, E., Haggblad, J., Nilsson, S., et al. 1997. Comparison of the ligand binding specificity and transcript tissue distribution of estrogen receptors α and β. *Endocrinology* 138 (3): 863–870.

Kunkel, A., Kopp, B., Muller, G., Villringer, K., Villringer, A., Taub, E., et al. 1999. Constraint-induced movement therapy for motor recovery in chronic stroke patients. *Archives of Physical Medicine and Rehabilitation* 80 (6): 624–628.

Kuo, H. K., Jones, R. N., Milberg, W. P., Tennstedt, S., Talbot, L., Morris, J. N., et al. 2006. Cognitive function in normal-weight, overweight, and obese older adults: An analysis of the Advanced Cognitive Training for Independent and Vital Elderly cohort. *Journal of the American Geriatrics Society* 54 (1): 97–103.

Lafenetre, P., Leske, O., Ma-Hogemeie, Z., Haghikia, A., Bichler, Z., Wahle, P., et al. 2010. Exercise can rescue recognition memory impairment in a model with reduced adult hippocampal neurogenesis. *Frontiers in Behavioral Neuroscience* 3: 34.

Laitinen, M. H., Ngandu, T., Rovio, S., Helkala, E. L., Uusitalo, U., Viitanen, M., et al. 2006. Fat intake at midlife and risk of dementia and Alzheimer's disease: A population-based study. *Dementia and Geriatric Cognitive Disorders* 22 (1): 99–107.

Lakshman, R., Garige, M., Gong, M., Leckey, L., Varatharajalu, R., and Zakhari, S. 2010. Is alcohol beneficial or harmful for cardioprotection? *Genes & Nutrition* 5 (2): 111–120.

Lambert, J. C., Heath, S., Even, G., Campion, D., Sleegers, K., Hiltunen, M., et al. 2009. Genome-wide association study identifies variants at CLU and CR1 associated with Alzheimer's disease. *Nature Genetics* 41 (10): 1094–1099.

Landfield, P. W., McGaugh, J. L., and Lynch, G. 1978. Impaired synaptic potentiation processes in the hippocampus of aged, memory-deficient rats. *Brain Research* 150 (1): 85–101.

Lane, M. A. 2000. Nonhuman primate models in biogerontology. *Experimental Gerontology* 35 (5): 533–541.

Larson, E. B., Wang, L., Bowen, J. D., McCormick, W. C., Teri, L., Crane, P., et al. 2006. Exercise is associated with reduced risk for incident dementia among persons 65 years of age and older. *Annals of Internal Medicine* 144 (2): 73–81.

Lautenschlager, N. T., Cox, K. L., Flicker, L., Foster, J. K., van Bockxmeer, F. M., Xiao, J., et al. 2008. Effect of physical activity on cognitive function in older adults at risk for Alzheimer disease: A randomized trial. *Journal of the American Medical Association* 300 (9): 1027–1037.

Leary, W. E. 1991. Renewal of brain is found in disputed monkey tests. *New York Times*, June 28.

Lee, J. D., and Kirlik, A. 2011. *The Oxford Handbook of Cognitive Engineering*. Oxford University Press.

Lee, J. M., Ross, E. R., Gower, A., Paris, J. M., Martensson, R., and Lorens, S. A. 1994. Spatial learning deficits in the aged rat: Neuroanatomical and neurochemical correlates. *Brain Research Bulletin* 33 (5): 489–500.

Lee, W. C., Huang, H., Feng, G., Sanes, J. R., Brown, E. N., So, P. T., et al. 2006. Dynamic remodeling of dendritic arbors in GABAergic interneurons of adult visual cortex. *PLoS Biology* 4 (2): e29.

Lee, Y. J., and Shacter, E. 1999. Oxidative stress inhibits apoptosis in human lymphoma cells. *Journal of Biological Chemistry* 274 (28): 19792–19798.

Lemaire, V., Koehl, M., Le Moal, M., and Abrous, D. N. 2000. Prenatal stress produces learning deficits associated with an inhibition of neurogenesis in the hippocampus. *Proceedings of the National Academy of Sciences* 97 (20): 11032–11037.

Leon, A. S., Togashi, K., Rankinen, T., Despres, J. P., Rao, D. C., Skinner, J. S., et al. 2004. Association of apolipoprotein E polymorphism with blood lipids and maximal oxygen uptake in the sedentary state and after exercise training in the HERITAGE family study. *Metabolism: Clinical and Experimental* 53 (1): 108–116.

Leonard, B. W., Mastroeni, D., Grover, A., Liu, Q., Yang, K., Gao, M., et al. 2009. Subventricular zone neural progenitors from rapid brain autopsies of elderly subjects with and without neurodegenerative disease. *Journal of Comparative Neurology* 515 (4): 269–294.

Lethaby, A., Hogervorst, E., Richards, M., Yesufu, A., and Yaffe, K. 2008. Hormone replacement therapy for cognitive function in postmenopausal women. *Cochrane Database of Systematic Reviews* 1: CD003122.

Leuner, B., Mendolia-Loffredo, S., Kozorovitskiy, Y., Samburg, D., Gould, E., and Shors, T. J. 2004. Learning enhances the survival of new neurons beyond the time when the hippocampus is required for memory. *Journal of Neuroscience* 24 (34): 7477–7481.

Levin, E. D., and Simon, B. B. 1998. Nicotinic acetylcholine involvement in cognitive function in animals. *Psychopharmacology* 138 (3–4): 217–230.

Levin, L. S. 1977. Self-care and health planning. *Social Policy* 8 (3): 47–54.

Lewington, S., Whitlock, G., Clarke, R., Sherliker, P., Emberson, J., Halsey, J., et al. 2007. Blood cholesterol and vascular mortality by age, sex, and blood pressure: A meta-analysis of individual data from 61 prospective studies with 55,000 vascular deaths. *Lancet* 370 (9602): 1829–1839.

Li, F., and Tsien, J. Z. 2009. Memory and the NMDA receptors. *New England Journal of Medicine* 361 (3): 302–303.

Li, G., Peskind, E. R., Millard, S. P., Chi, P., Sokal, I., Yu, C. E., et al. 2009. Cerebrospinal fluid concentration of brain-derived neurotrophic factor and cognitive function in non-demented subjects. *PLoS ONE* 4 (5): e5424.

Liepert, J., Miltner, W. H., Bauder, H., Sommer, M., Dettmers, C., Taub, E., et al. 1998. Motor cortex plasticity during constraint-induced movement therapy in stroke patients. *Neuroscience Letters* 250 (1): 5–8.

Lim, W. S., Gammack, J. K., Van Niekerk, J., and Dangour, A. D. 2006. Omega 3 fatty acid for the prevention of dementia. *Cochrane Database of Systematic Reviews* 1: CD005379.

Lin, M.-K., Greenwood, P. M., Sundararajan, R., Fryxell, K. J., and Parasuraman, R. 2008. The effect of COMT Val158Met on human working memory depends on age, task difficulty, and memory load. Paper presented at annual meeting of Society for Neuroscience, Washington.

Lindqvist, A., Mohapel, P., Bouter, B., Frielingsdorf, H., Pizzo, D., Brundin, P., et al. 2006. High-fat diet impairs hippocampal neurogenesis in male rats. *European Journal of Neurology* 13 (12): 1385–1388.

Lindsay, J., Laurin, D., Verreault, R., Hebert, R., Helliwell, B., Hill, G. B., et al. 2002. Risk factors for Alzheimer's disease: A prospective analysis from the Canadian Study of Health and Aging. *American Journal of Epidemiology* 156 (5): 445–453.

Liu, F., Day, M., Muniz, L. C., Bitran, D., Arias, R., Revilla-Sanchez, R., et al. 2008. Activation of estrogen receptor-β regulates hippocampal synaptic plasticity and improves memory. *Nature Neuroscience* 11 (3): 334–343.

Liu, L., van Groen, T., Kadish, I., and Tollefsbol, T. O. 2009. DNA methylation impacts on learning and memory in aging. *Neurobiology of Aging* 30 (4): 549–560.

Logan, J. M., Sanders, A. L., Snyder, A. Z., Morris, J. C., and Buckner, R. L. 2002. Under-recruitment and nonselective recruitment: Dissociable neural mechanisms associated with aging. *Neuron* 33 (5): 827–840.

Lombard, D., Pletcher, S., Canto, C., and Auwerx, J. Ageing: Longevity hits a roadblock. *Nature* 477 (7365): 410–411.

Lonn, E., Yusuf, S., Arnold, M. J., Sheridan, P., Pogue, J., Micks, M., et al. 2006. Homocysteine lowering with folic acid and B vitamins in vascular disease. *New England Journal of Medicine* 354 (15): 1567–1577.

Lopez, O. L., Jagust, W. J., Dulberg, C., Becker, J. T., DeKosky, S. T., Fitzpatrick, A., et al. 2003. Risk factors for mild cognitive impairment in the Cardiovascular Health Study Cognition Study: Part 2. *Archives of Neurology* 60 (10): 1394–1399.

Lord, C., Buss, C., Lupien, S. J., and Pruessner, J. C. 2008. Hippocampal volumes are larger in postmenopausal women using estrogen therapy compared to past users, never users and men: A possible window of opportunity effect. *Neurobiology of Aging* 29 (1): 95–101.

Lou, S. J., Liu, J. Y., Chang, H., and Chen, P. J. 2008. Hippocampal neurogenesis and gene expression depend on exercise intensity in juvenile rats. *Brain Research* 1210: 48–55.

Lövdén, M., Bodammer, N. C., Kuhn, S., Kaufmann, J., Schutze, H., Tempelmann, C., et al. 2010. Experience-dependent plasticity of white-matter microstructure extends into old age. *Neuropsychologia* 48 (13): 3878–3883.

Luine, V., Park, D., Joh, T., Reis, D., and McEwen, B. 1980. Immunochemical demonstration of increased choline acetyltransferase concentration in rat preoptic area after estradiol administration. *Brain Research* 191 (1): 273–277.

Luine, V. N. 1985. Estradiol increases choline acetyltransferase activity in specific basal forebrain nuclei and projection areas of female rats. *Experimental Neurology* 89 (2): 484–490.

Lukiw, W. J., Cui, J. G., Marcheselli, V. L., Bodker, M., Botkjaer, A., Gotlinger, K., et al. 2005. A role for docosahexaenoic acid-derived neuroprotectin D1 in neural cell survival and Alzheimer disease. *Journal of Clinical Investigation* 115 (10): 2774–2783.

Luo, C. X., Jiang, J., Zhou, Q. G., Zhu, X. J., Wang, W., Zhang, Z. J., et al. 2007. Voluntary exercise-induced neurogenesis in the postischemic dentate gyrus is associated with spatial memory recovery from stroke. *Journal of Neuroscience Research* 85 (8): 1637–1646.

Lustig, C., and Flegal, K. E. 2008. Targeting latent function: Encouraging effective encoding for successful memory training and transfer. *Psychology and Aging* 23 (4): 754–764.

Ma, Y., Pagoto, S. L., Griffith, J. A., Merriam, P. A., Ockene, I. S., Hafner, A. R., et al. 2007. A dietary quality comparison of popular weight-loss plans. *Journal of the American Dietetic Association* 107 (10): 1786–1791.

MacLennan, A. H., Henderson, V. W., Paine, B. J., Mathias, J., Ramsay, E. N., Ryan, P., et al. 2006. Hormone therapy, timing of initiation, and cognition in women aged older than 60 years: The REMEMBER pilot study. *Menopause* 13 (1): 28–36.

Madden, D. J., Whiting, W. L., Huettel, S. A., White, L. E., MacFall, J. R., and Provenzale, J. M. 2004. Diffusion tensor imaging of adult age differences in cerebral white matter: Relation to response time. *NeuroImage* 21 (3): 1174–1181.

Maguire, E. A., Gadian, D. G., Johnsrude, I. S., Good, C. D., Ashburner, J., Frackowiak, R. S., et al. 2000. Navigation-related structural change in the hippocampi of taxi drivers. *Proceedings of the National Academy of Sciences* 97 (8): 4398–4403.

Maguire, E. A., Woollett, K., and Spiers, H. J. 2006. London taxi drivers and bus drivers: A structural MRI and neuropsychological analysis. *Hippocampus* 16 (12): 1091–1101.

Maher, P., Akaishi, T., Schubert, D., and Abe, K. 2010. A pyrazole derivative of curcumin enhances memory. *Neurobiology of Aging* 31 (4): 706–709.

Mahley, R. W. 1988. Apolipoprotein E: Cholesterol transport protein with expanding role in cell biology. *Science* 240: 622–630.

Majeed, B. A., and Brown, S. J. 2006. Developing a well-being monitoring system—Modeling and data analysis techniques. *Applied Soft Computing* 6: 384–393.

Malouf, R., and Grimley Evans, J. 2008. Folic acid with or without vitamin B12 for the prevention and treatment of healthy elderly and demented people. *Cochrane Database of Systematic Reviews* 4: CD004514.

Markham, J. A., McKian, K. P., Stroup, T. S., and Juraska, J. M. 2005. Sexually dimorphic aging of dendritic morphology in CA1 of hippocampus. *Hippocampus* 15 (1): 97–103.

Martin, C. K., Anton, S. D., Han, H., York-Crowe, E., Redman, L. M., Ravussin, E., et al. 2007. Examination of cognitive function during six months of calorie restriction: Results of a randomized controlled trial. *Rejuvenation Research* 10 (2): 179–190.

Masliah, E., Mallory, M., Hansen, L., DeTeresa, R., and Terry, R. D. 1993. Quantitative synaptic alterations in the human neocortex during normal aging. *Neurology* 43 (1): 192–197.

Maswood, N., Young, J., Tilmont, E., Zhang, Z., Gash, D. M., Gerhardt, G. A., et al. 2004. Caloric restriction increases neurotrophic factor levels and attenuates neurochemical and behavioral deficits in a primate model of Parkinson's disease. *Proceedings of the National Academy of Sciences* 101 (52): 18171–18176.

Mattay, V. S., Berman, K. F., Ostrem, J. L., Esposito, G., Van Horn, J. D., Bigelow, L. B., et al. 1996. Dextroamphetamine enhances "neural network-specific" physiological signals: A positron-emission tomography rCBF study. *Journal of Neuroscience* 16 (15): 4816–4822.

Mattay, V. S., Callicott, J. H., Bertolino, A., Heaton, I., Frank, J. A., Coppola, R., et al. 2000. Effects of dextroamphetamine on cognitive performance and cortical activation. *NeuroImage* 12 (3): 268–275.

Matthews, K., Cauley, J., Yaffe, K., and Zmuda, J. M. 1999. Estrogen replacement therapy and cognitive decline in older community women. *Journal of the American Geriatrics Society* 47 (5): 518–523.

Matthews, K. A., Kuller, L. H., Wing, R. R., Meilahn, E. N., and Plantinga, P. 1996. Prior to use of estrogen replacement therapy, are users healthier than nonusers? *American Journal of Epidemiology* 143 (10): 971–978.

Mattson, M. P. 2000. Neuroprotective signaling and the aging brain: Take away my food and let me run. *Brain Research* 886 (1–2): 47–53.

Mattson, M. P., Duan, W., Chan, S. L., Cheng, A., Haughey, N., Gary, D. S., et al. 2002. Neuroprotective and neurorestorative signal transduction mechanisms in brain aging: Modification by genes, diet and behavior. *Neurobiology of Aging* 23 (5): 695–705.

Mattson, M. P., Duan, W., and Guo, Z. 2003. Meal size and frequency affect neuronal plasticity and vulnerability to disease: Cellular and molecular mechanisms. *Journal of Neurochemistry* 84 (3): 417–431.

Mattson, M. P., Maudsley, S., and Martin, B. 2004. BDNF and 5-HT: A dynamic duo in age-related neuronal plasticity and neurodegenerative disorders. *Trends in Neurosciences* 27 (10): 589–594.

McClearn, G. E., Johansson, B., Berg, S., Pedersen, N. L., Ahern, F., Petrill, S. A., et al. 1997. Substantial genetic influence on cognitive abilities in twins 80 or more years old. *Science* 276: 1560–1563.

McEwen, B., Akama, K., Alves, S., Brake, W. G., Bulloch, K., Lee, S., et al. 2001. Tracking the estrogen receptor in neurons: Implications for estrogen-induced synapse formation. *Proceedings of the National Academy of Sciences* 98 (13): 7093–7100.

McGahon, B. M., Martin, D. S., Horrobin, D. F., and Lynch, M. A. 1999. Age-related changes in LTP and antioxidant defenses are reversed by an alpha-lipoic acid-enriched diet. *Neurobiology of Aging* 20 (6): 655–664.

McKhann, G., Drachman, D., Folstein, M., Katzman, R., Price, D., and Stadlan, E. M. 1984. Clinical diagnosis of Alzheimer's disease: Report on the NINCDS-ADRDA Work Group under the auspices of the Department of Health and Human Services Task Force on Alzheimer's Disease. *Neurology* 34: 939–944.

McNab, F., Varrone, A., Farde, L., Jucaite, A., Bystritsky, P., Forssberg, H., et al. 2009. Changes in cortical dopamine D1 receptor binding associated with cognitive training. *Science* 323: 800–802.

McNamara, R. K., Hahn, C. G., Jandacek, R., Rider, T., Tso, P., Stanford, K. E., et al. 2007. Selective deficits in the omega-3 fatty acid docosahexaenoic acid in the postmortem orbitofrontal cortex of patients with major depressive disorder. *Biological Psychiatry* 62 (1): 17–24.

McNay, E. C., Fries, T. M., and Gold, P. E. 2000. Decreases in rat extracellular hippocampal glucose concentration associated with cognitive demand during a spatial task. *Proceedings of the National Academy of Sciences* 97 (6): 2881–2885.

McNay, E. C., and Gold, P. E. 2001. Age-related differences in hippocampal extracellular fluid glucose concentration during behavioral testing and following systemic glucose administration. *Journals of Gerontology A* 56 (2): B66–B71.

Meck, W. H., Williams, C. L., Cermak, J. M., and Blusztajn, J. K. 2007. Developmental periods of choline sensitivity provide an ontogenetic mechanism for regulating memory capacity and age-related dementia. *Frontiers in Integrative Neuroscience* 1: 7.

Mehta, M. A., Owen, A. M., Sahakian, B. J., Mavaddat, N., Pickard, J. D., and Robbins, T. W. 2000. Methylphenidate enhances working memory by modulating discrete frontal and parietal lobe regions in the human brain. *Journal of Neuroscience* 20 (6), RC65.

Mensink, R. P., Zock, P. L., Kester, A. D., and Katan, M. B. 2003. Effects of dietary fatty acids and carbohydrates on the ratio of serum total to HDL cholesterol and on serum lipids and apolipoproteins: A meta-analysis of 60 controlled trials. *American Journal of Clinical Nutrition* 77 (5): 1146–1155.

Merzenich, M. M., Nelson, R. J., Stryker, M. P., Cynader, M. S., Schoppmann, A., and Zook, J. M. 1984. Somatosensory cortical map changes following digit amputation in adult monkeys. *Journal of Comparative Neurology* 224 (4): 591–605.

Meshi, D., Drew, M. R., Saxe, M., Ansorge, M. S., David, D., Santarelli, L., et al. 2006. Hippocampal neurogenesis is not required for behavioral effects of environmental enrichment. *Nature Neuroscience* 9 (6): 729–731.

Meyer, M. R., Tschanz, J. T., Norton, M. C., Welsh-Bohmer, K. A., Steffens, D. C., Wyse, B. W., et al. 1998. APOE genotype predicts when—not whether—one is predisposed to develop Alzheimer disease. *Nature Genetics* 19: 321–322.

Milgram, N. W., Head, E., Muggenburg, B., Holowachuk, D., Murphey, H., Estrada, J., et al. 2002. Landmark discrimination learning in the dog: Effects of age, an antioxidant fortified food, and cognitive strategy. *Neuroscience and Biobehavioral Reviews* 26 (6): 679–695.

Milgram, N. W., Head, E., Zicker, S. C., Ikeda-Douglas, C., Murphey, H., Muggenberg, B. A., et al. 2004. Long-term treatment with antioxidants and a program of behavioral enrichment reduces age-dependent impairment in discrimination and reversal learning in beagle dogs. *Experimental Gerontology* 39 (5): 753–765.

Milner, B. 1958. Psychological defects produced by temporal lobe excision. *Research Publications—Association for Research in Nervous and Mental Disease* 36: 244–257.

Mine, M., Okumura, Y., Ichimaru, M., Nakamura, T., and Kondo, S. 1990. Apparently beneficial effect of low to intermediate doses of A-bomb radiation on human lifespan. *International Journal of Radiation Biology* 58 (6): 1035–1043.

Minzenberg, M. J., Watrous, A. J., Yoon, J. H., Ursu, S., and Carter, C. S. 2008. Modafinil shifts human locus coeruleus to low-tonic, high-phasic activity during functional MRI. *Science* 322: 1700–1702.

Molteni, R., Wu, A., Vaynman, S., Ying, Z., Barnard, R. J., and Gómez-Pinilla, F. 2004. Exercise reverses the harmful effects of consumption of a high-fat diet on synaptic and behavioral plasticity associated to the action of brain-derived neurotrophic factor. *Neuroscience* 123 (2): 429–440.

Mondadori, C. R., de Quervain, D. J., Buchmann, A., Mustovic, H., Wollmer, M. A., Schmidt, C. F., et al. 2007. Better memory and neural efficiency in young apolipoprotein E ε4 carriers. *Cerebral Cortex* 17 (8): 1934–1947.

Morcom, A. M., Good, C. D., Frackowiak, R. S., and Rugg, M. D. 2003. Age effects on the neural correlates of successful memory encoding. *Brain* 126: 213–229.

Morris, M. C., Evans, D. A., Bienias, J. L., Tangney, C. C., Bennett, D. A., Aggarwal, N., et al. 2003. Dietary fats and the risk of incident Alzheimer disease. *Archives of Neurology* 60 (2): 194–200.

Morris, M. C., Evans, D. A., Bienias, J. L., Tangney, C. C., and Wilson, R. S. 2004. Dietary fat intake and 6-year cognitive change in an older biracial community population. *Neurology* 62 (9): 1573–1579.

Morris, R. 1984. Developments of a water-maze procedure for studying spatial learning in the rat. *Journal of Neuroscience Methods* 11 (1): 47–60.

Morris, R. G., and Becker, J. T., eds. 2004. *Cognitive Neuropsychology of Alzheimer's Disease*, second edition. Oxford University Press.

Morris, R. G., and Kopelman, M. D. 1986. The memory deficits in Alzheimer-type dementia: A review. *Quarterly Journal of Experimental Psychology A* 38 (4): 575–602.

Morrison, J. H., Brinton, R. D., Schmidt, P. J., and Gore, A. C. 2006. Estrogen, menopause, and the aging brain: How basic neuroscience can inform hormone therapy in women. *Journal of Neuroscience* 26 (41): 10332–10348.

Morrison, J. H., and Hof, P. R. 1997. Life and death of neurons in the aging brain. *Science* 278: 412–419.

Morse, C. K. 1993. Does variability increase with age? An archival study of cognitive measures. *Psychology and Aging* 8 (2): 156–164.

Moser, E. I., Kropff, E., and Moser, M. B. 2008. Place cells, grid cells, and the brain's spatial representation system. *Annual Review of Neuroscience* 31: 69–89.

Mouret, A., Gheusi, G., Gabellec, M. M., de Chaumont, F., Olivo-Marin, J. C., and Lledo, P. M. 2008. Learning and survival of newly generated neurons: When time matters. *Journal of Neuroscience* 28 (45): 11511–11516.

Moyer, J. R., Jr., Thompson, L. T., Black, J. P., and Disterhoft, J. F. 1992. Nimodipine increases excitability of rabbit CA1 pyramidal neurons in an age- and concentration-dependent manner. *Journal of Neurophysiology* 68 (6): 2100–2109.

Mozaffarian, D., Micha, R., and Wallace, S. 2010. Effects on coronary heart disease of increasing polyunsaturated fat in place of saturated fat: A systematic review and meta-analysis of randomized controlled trials. *PLoS Medicine* 7 (3): e1000252.

Muldoon, M. F., Ryan, C. M., Sheu, L., Yao, J. K., Conklin, S. M., and Manuck, S. B. 2010. Serum phospholipid docosahexaenonic acid is associated with cognitive functioning during middle adulthood. *Journal of Nutrition* 140 (4): 848–853.

Mumenthaler, M. S., Yesavage, J. A., Taylor, J. L., O'Hara, R., Friedman, L., Lee, H., et al. 2003. Psychoactive drugs and pilot performance: A comparison of nicotine, donepezil, and alcohol effects. *Neuropsychopharmacology* 28 (7): 1366–1373.

Murphy, D. D., Cole, N. B., Greenberger, V., and Segal, M. 1998. Estradiol increases dendritic spine density by reducing GABA neurotransmission in hippocampal neurons. *Journal of Neuroscience* 18 (7): 2550–2559.

Murphy, F. C., and Klein, R. M. 1998. The effects of nicotine on spatial and non-spatial expectancies in a covert orienting task. *Neuropsychologia* 36 (11): 1103–1114.

Murray, A. J., Knight, N. S., Cochlin, L. E., McAleese, S., Deacon, R. M., Rawlins, J. N., et al. 2009. Deterioration of physical performance and cognitive function in rats with short-term high-fat feeding. *FASEB Journal* 23 (12): 4353–4360.

Naimi, T. S., Brown, D. W., Brewer, R. D., Giles, W. H., Mensah, G., Serdula, M. K., et al. 2005. Cardiovascular risk factors and confounders among nondrinking and moderate-drinking U.S. adults. *American Journal of Preventive Medicine* 28 (4): 369–373.

Naylor, A. S., Bull, C., Nilsson, M. K., Zhu, C., Bjork-Eriksson, T., Eriksson, P. S., et al. 2008. Voluntary running rescues adult hippocampal neurogenesis after irradiation of the young mouse brain. *Proceedings of the National Academy of Sciences* 105 (38): 14632–14637.

Nehlig, A. 2010. Is caffeine a cognitive enhancer? *Journal of Alzheimer's Disease* 20 (suppl. 1): S85–S94.

Newman, M. C., and Kaszniak, A. W. 2000. Spatial memory and aging: Performance on a human analog of the Morris water maze task. *Aging, Neuropsychology, and Cognition* 7: 86–93.

Nichol, K. E., Parachikova, A. I., and Cotman, C. W. 2007. Three weeks of running wheel exposure improves cognitive performance in the aged Tg2576 mouse. *Behavioural Brain Research* 184 (2): 124–132.

Nicholson, D. A., Yoshida, R., Berry, R. W., Gallagher, M., and Geinisman, Y. 2004. Reduction in size of perforated postsynaptic densities in hippocampal axospinous synapses and age-related spatial learning impairments. *Journal of Neuroscience* 24 (35): 7648–7653.

Nilsen, J., and Brinton, R. D. 2003. Mechanism of estrogen-mediated neuroprotection: Regulation of mitochondrial calcium and Bcl-2 expression. *Proceedings of the National Academy of Sciences* 100 (5): 2842–2847.

Niti, M., Yap, K. B., Kua, E. H., Tan, C. H., and Ng, T. P. 2008. Physical, social and productive leisure activities, cognitive decline and interaction with APOE-ε4 genotype in Chinese older adults. *International Psychogeriatrics* 20 (2): 237–251.

Nordberg, A. 2001. Nicotinic receptor abnormalities of Alzheimer's disease: Therapeutic implications. *Biological Psychiatry* 49 (3): 200–210.

Nottebohm, F. 2002. Neuronal replacement in adult brain. *Brain Research Bulletin* 57 (6): 737–749.

NTSB (National Transportation Safety Board). 1997. Grounding of the Panamanian Passenger Ship Royal Majesty on Rose and Crown Shoal Near Nantucket, Massachusetts, June 10, 1995. Marine Accident Report.

Nudo, R. J., and Friel, K. M. 1999. Cortical plasticity after stroke: Implications for rehabilitation. *Revue Neurologique* 155 (9): 713–717.

Nudo, R. J., Plautz, E. J., and Frost, S. B. 2001. Role of adaptive plasticity in recovery of function after damage to motor cortex. *Muscle & Nerve* 24 (8): 1000–1019.

Nudo, R. J., Wise, B. M., SiFuentes, F., and Milliken, G. W. 1996. Neural substrates for the effects of rehabilitative training on motor recovery after ischemic infarct. *Science* 272: 1791–1794.

O'Reilly, R. C., and Frank, M. J. 2006. Making working memory work: A computational model of learning in the prefrontal cortex and basal ganglia. *Neural Computation* 18 (2): 283–328.

O'Sullivan, M., Summers, P. E., Jones, D. K., Jarosz, J. M., Williams, S. C., and Markus, H. S. 2001. Normal-appearing white matter in ischemic leukoaraiosis: A diffusion tensor MRI study. *Neurology* 57 (12): 2307–2310.

Obisesan, T. O., Ferrell, R. E., Goldberg, A. P., Phares, D. A., Ellis, T. J., and Hagberg, J. M. 2008. APOE genotype affects black-white responses of high-density lipoprotein cholesterol subspecies to aerobic exercise training. *Metabolism: Clinical and Experimental* 57 (12): 1669–1676.

Olesen, P. J., Westerberg, H., and Klingberg, T. 2004. Increased prefrontal and parietal activity after training of working memory. *Nature Neuroscience* 7 (1): 75–79.

Oomen, C. A., Farkas, E., Roman, V., van der Beek, E. M., Luiten, P. G., and Meerlo, P. 2009. Resveratrol preserves cerebrovascular density and cognitive function in aging mice. *Frontiers in Aging Neuroscience* 1 (4).

Oria, R. B., Patrick, P. D., Blackman, J. A., Lima, A. A., and Guerrant, R. L. 2007. Role of apolipoprotein E4 in protecting children against early childhood diarrhea outcomes and implications for later development. *Medical Hypotheses* 68 (5): 1099–1107.

Oria, R. B., Patrick, P. D., Zhang, H., Lorntz, B., de Castro Costa, C. M., Brito, G. A., et al. 2005. APOE4 protects the cognitive development in children with heavy diarrhea burdens in Northeast Brazil. *Pediatric Research* 57 (2): 310–316.

Oster, T., and Pillot, T. 2010. Docosahexaenoic acid and synaptic protection in Alzheimer's disease mice. *Biochimica et Biophysica Acta* 1801 (8): 791–798.

Palsdottir, A., Helgason, A., Palsson, S., Bjornsson, H. T., Bragason, B. T., Gretarsdottir, S., et al. 2008. A drastic reduction in the life span of cystatin C L68Q carriers due to life-style changes during the last two centuries. *PLoS Genetics* 4 (6): e1000099.

Pang, P. T., and Lu, B. 2004. Regulation of late-phase LTP and long-term memory in normal and aging hippocampus: Role of secreted proteins tPA and BDNF. *Ageing Research Reviews* 3 (4): 407–430.

Parasuraman, R. 2003. Neuroergonomics: Research and practice. *Theoretical Issues in Ergonomics Science* 4: 5–20.

Parasuraman, R., and Greenwood, P. M. 2004. Visual attention, genetics, and Alzheimer's disease. In *Interdisciplinary Topics in Gerontology*, ed. A. Cronin-Golomb and P. Hof. Karger.

Parasuraman, R., Greenwood, P. M., Haxby, J. V., and Grady, C. L. 1992. Visuospatial attention in dementia of the Alzheimer type. *Brain* 115: 711–733.

Parasuraman, R., Greenwood, P. M., Kumar, R., and Fossella, J. 2005. Beyond heritability: Neurotransmitter genes differentially modulate visuospatial attention and working memory. *Psychological Science* 16 (3): 200–207.

Parasuraman, R., Greenwood, P. M., and Sunderland, T. 2002. The apolipoprotein E gene, attention, and brain function. *Neuropsychology* 16 (2): 254–274.

Parasuraman, R., and Haxby, J. 1993. Attention and brain function in Alzheimer's disease. *Revue de Neuropsychologie* 7: 242–272.

Parasuraman, R., and Manzey, D. 2010. Complacency and bias in human use of automation: An attentional integration. *Human Factors* 52 (3): 381–410.

Parasuraman, R., and Martin, A. 1994. Cognition in Alzheimer's disease: Disorders of attention and semantic knowledge. *Neurobiology* 4: 237–244.

Parasuraman, R., and Miller, C. 2004. Trust and etiquette in high-criticality automated systems. *Communications of the ACM* 47: 51–55.

Parasuraman, R., and Riley, V. 1997. Humans and automation: Use, misuse, disuse, abuse. *Human Factors* 39: 230–253.

Parasuraman, R., Sheridan, T. B., and Wickens, C. D. 2008. Situation awareness, mental workload, and trust in automation: Viable, empirically supported cognitive engineering constructs. *Journal of Cognitive Engineering and Decision Making* 2: 141–161.

Park, D. C., Lautenschlager, G., Hedden, T., Davidson, N. S., Smith, A. D., and Smith, P. K. 2002. Models of visuospatial and verbal memory across the adult life span. *Psychology and Aging* 17 (2): 299–320.

Park, D. C., and Schwarz, N., eds. 2000. *Cognitive Aging: A Primer*. Psychology Press.

Park, H. R., Park, M., Choi, J., Park, K. Y., Chung, H. Y., and Lee, J. 2010. A high-fat diet impairs neurogenesis: Involvement of lipid peroxidation and brain-derived neurotrophic factor. *Neuroscience Letters* 482 (3): 235–239.

Paulus, M. P., Hozack, N., Frank, L., and Brown, G. G. 2002. Error rate and outcome predictability affect neural activation in prefrontal cortex and anterior cingulate during decision-making. *NeuroImage* 15 (4): 836–846.

Pawlowski, T. L., Bellush, L. L., Wright, A. W., Walker, J. P., Colvin, R. A., and Huentelman, M. J. 2009. Hippocampal gene expression changes during age-related cognitive decline. *Brain Research* 1256: 101–110.

Payton, A. 2009. The impact of genetic research on our understanding of normal cognitive ageing: 1995 to 2009. *Neuropsychology Review* 19 (4): 451–477.

Pearson, K. J., Baur, J. A., Lewis, K. N., Peshkin, L., Price, N. L., Labinskyy, N., et al. 2008. Resveratrol delays age-related deterioration and mimics transcriptional aspects of dietary restriction without extending life span. *Cell Metabolism* 8 (2): 157–168.

Pedersen, N. L. 2010. Reaching the limits of genome-wide significance in Alzheimer disease: Back to the environment. *Journal of the American Medical Association* 303 (18): 1864–1865.

Pereira, A. C., Huddleston, D. E., Brickman, A. M., Sosunov, A. A., Hen, R., McKhann, G. M., et al. 2007. An in vivo correlate of exercise-induced neurogenesis in the adult dentate gyrus. *Proceedings of the National Academy of Sciences* 104 (13): 5638–5643.

Persson, J., Nyberg, L., Lind, J., Larsson, A., Nilsson, L. G., Ingvar, M., et al. 2006. Structure-function correlates of cognitive decline in aging. *Cerebral Cortex* 16 (7): 907–915.

Peters, A., Moss, M. B., and Sethares, C. 2000. Effects of aging on myelinated nerve fibers in monkey primary visual cortex. *Journal of Comparative Neurology* 419 (3): 364–376.

Peters, A., and Sethares, C. 2002. Aging and the myelinated fibers in prefrontal cortex and corpus callosum of the monkey. *Journal of Comparative Neurology* 442 (3): 277–291.

Petersen, R. C. 2004. Mild cognitive impairment as a diagnostic entity. *Journal of Internal Medicine* 256 (3): 183–194.

Pew Research Center. 2009. *Generations Online in 2009* (http://pewresearch.org).

Pew, R. W., and Hemel, V. 2004. *Technology for Adaptive Aging*. National Academies Press.

Pfluger, P. T., Herranz, D., Velasco-Miguel, S., Serrano, M., and Tschop, M. H. 2008. Sirt1 protects against high-fat diet-induced metabolic damage. *Proceedings of the National Academy of Sciences* 105 (28): 9793–9798.

Phillips, J. M., McAlonan, K., Robb, W. G., and Brown, V. J. 2000. Cholinergic neurotransmission influences covert orientation of visuospatial attention in the rat. *Psychopharmacology* 150 (1): 112–116.

Picciotto, M. R., and Zoli, M. 2002. Nicotinic receptors in aging and dementia. *Journal of Neurobiology* 53 (4): 641–655.

Pieper, A. A., Xie, S., Capota, E., Estill, S. J., Zhong, J., Long, J. M., et al. 2010. Discovery of a proneurogenic, neuroprotective chemical. *Cell* 142 (1): 39–51.

Piguet, O., Double, K. L., Kril, J. J., Harasty, J., Macdonald, V., McRitchie, D. A., et al. 2009. White matter loss in healthy ageing: A postmortem analysis. *Neurobiology of Aging* 30 (8): 1288–1295.

Pistell, P. J., Morrison, C. D., Gupta, S., Knight, A. G., Keller, J. N., Ingram, D. K., et al. 2010. Cognitive impairment following high fat diet consumption is associated with brain inflammation. *Journal of Neuroimmunology* 219 (1–2): 25–32.

Plomin, R., and Crabbe, J. 2000. DNA. *Psychological Bulletin* 126 (6): 806–828.

Plomin, R., DeFries, J. C., McClearn, G. E., and McGuffin, P. 2001. *Behavioral Genetics*, fourth edition. Worth.

Podewils, L. J., Guallar, E., Kuller, L. H., Fried, L. P., Lopez, O. L., Carlson, M., et al. 2005. Physical activity, APOE genotype, and dementia risk: Findings from the Cardiovascular Health Cognition Study. *American Journal of Epidemiology* 161 (7): 639–651.

Pons, T. P., Garraghty, P. E., Ommaya, A. K., Kaas, J. H., Taub, E., and Mishkin, M. 1991. Massive cortical reorganization after sensory deafferentation in adult macaques. *Science* 252: 1857–1860.

Pontifex, M. B., Hillman, C. H., Fernhall, B., Thompson, K. M., and Valentini, T. A. 2009. The effect of acute aerobic and resistance exercise on working memory. *Medicine and Science in Sports and Exercise* 41 (4): 927–934.

Poo, M. M. 2001. Neurotrophins as synaptic modulators. *Nature Reviews Neuroscience* 2 (1): 24–32.

Pop, V., Head, E., Hill, M. A., Gillen, D., Berchtold, N. C., Muggenburg, B. A., et al. 2010. Synergistic effects of long-term antioxidant diet and behavioral enrichment on β-amyloid load and non-amyloidogenic processing in aged canines. *Journal of Neuroscience* 30 (29): 9831–9839.

Posner, M. I. 1980. Orienting of attention. *Quarterly Journal of Experimental Psychology* 32: 3–25.

Posner, M. I., Rothbart, M. K., and Sheese, B. E. 2007. Attention genes. *Developmental Science* 10 (1): 24–29.

Potier, B., Lamour, Y., and Dutar, P. 1993. Age-related alterations in the properties of hippocampal pyramidal neurons among rat strains. *Neurobiology of Aging* 14 (1): 17–25.

Preiser, W. F. E., and Ostroff, E. 2001. *Universal Design Handbook*. McGraw-Hill.

Proctor, R., and van Zandt, T. 1984. *Human Factors in Simple and Complex Systems*. Allyn and Bacon.

Purcell, S. M., Wray, N. R., Stone, J. L., Visscher, P. M., O'Donovan, M. C., Sullivan, P. F., et al. 2009. Common polygenic variation contributes to risk of schizophrenia and bipolar disorder. *Nature* 460 (7256): 748–752.

Puskas, L. G., Kitajka, K., Nyakas, C., Barcelo-Coblijn, G., and Farkas, T. 2003. Short-term administration of omega 3 fatty acids from fish oil results in increased trans-thyretin transcription in old rat hippocampus. *Proceedings of the National Academy of Sciences* 100 (4): 1580–1585.

Pyapali, G. K., and Turner, D. A. 1996. Increased dendritic extent in hippocampal CA1 neurons from aged F344 rats. *Neurobiology of Aging* 17 (4): 601–611.

Quallo, M. M., Price, C. J., Ueno, K., Asamizuya, T., Cheng, K., Lemon, R. N., et al. 2009. Gray and white matter changes associated with tool-use learning in macaque monkeys. *Proceedings of the National Academy of Sciences* 106 (43): 18379–18384.

Ragaz, J., Wilson, K., Muraca, G., Budlovsky, J., and Froehlich, J. 2010. Dual estrogen effects on breast cancer: Endogenous estrogen stimulates, exogenous estrogen protects. Further investigation of estrogen chemoprevention is warranted. Presented at San Antonio Breast Cancer Symposium.

Rainer, G., and Miller, E. K. 2000. Effects of visual experience on the representation of objects in the prefrontal cortex. *Neuron* 27 (1): 179–189.

Rakic, P. 1985. Limits of neurogenesis in primates. *Science* 227: 1054–1056.

Rakic, P. 2002. Adult neurogenesis in mammals: An identity crisis. *Journal of Neuroscience* 22 (3): 614–618.

Ramachandran, V. S. 2005. Plasticity and functional recovery in neurology. *Clinical Medicine* 5 (4): 368–373.

Ramachandran, V. S., and Hirstein, W. 1998. The perception of phantom limbs. The D. O. Hebb lecture. *Brain* 121: 1603–1630.

Rampon, C., Tang, Y. P., Goodhouse, J., Shimizu, E., Kyin, M., and Tsien, J. Z. 2000. Enrichment induces structural changes and recovery from nonspatial memory deficits in CA1 NMDAR1-knockout mice. *Nature Neuroscience* 3 (3): 238–244.

Randall, D. C., Fleck, N. L., Shneerson, J. M., and File, S. E. 2004. The cognitive-enhancing properties of modafinil are limited in non-sleep-deprived middle-aged volunteers. *Pharmacology, Biochemistry, and Behavior* 77 (3): 547–555.

Rapoff, M. A. 1999. *Adherence to Pediatric Medical Regimens*. Kluwer/Plenum.

Rapp, P. R., and Amaral, D. G. 1991. Recognition memory deficits in a subpopulation of aged monkeys resemble the effects of medial temporal lobe damage. *Neurobiology of Aging* 12: 481–486.

Rapp, P. R., and Gallagher, M. 1996. Preserved neuron number in the hippocampus of aged rats with spatial learning deficits. *Proceedings of the National Academy of Sciences* 93: 9926–9930.

Rapp, P. R., Kansky, M. T., and Roberts, J. A. 1997. Impaired spatial information processing in aged monkeys with preserved recogntion memory. *Neuroreport* 8: 1923–1928.

Rapp, P. R., Morrison, J. H., and Roberts, J. A. 2003. Cyclic estrogen replacement improves cognitive function in aged ovariectomized rhesus monkeys. *Journal of Neuroscience* 23 (13): 5708–5714.

Rasmussen, P., Brassard, P., Adser, H., Pedersen, M. V., Leick, L., Hart, E., et al. 2009. Evidence for a release of brain-derived neurotrophic factor from the brain during exercise. *Experimental Physiology* 94 (10): 1062–1069.

Rasmusson, D. X., Rebok, G. W., and Brandt, J. 1999. Effects of three types of memory training in normal elderly. *Aging, Neuropsychology, and Cognition* 6: 56–66.

Raz, N. 2000. Aging of the brain and its impact on cognitive performance: Integration of structural and functional findings. In *Handbook of Aging and Cognition*, second edition, ed. F. Craik and T. Salthouse. Erlbaum.

Raz, N., Gunning-Dixon, F., Head, D., Rodrigue, K. M., Williamson, A., and Acker, J. D. 2004. Aging, sexual dimorphism, and hemispheric asymmetry of the cerebral cortex: Replicability of regional differences in volume. *Neurobiology of Aging* 25 (3): 377–396.

Raz, N., Lindenberger, U., Rodrigue, K. M., Kennedy, K. M., Head, D., Williamson, A., et al. 2005. Regional brain changes in aging healthy adults: General trends, individual differences and modifiers. *Cerebral Cortex* 15 (11): 1676–1689.

Raz, N., Rodrigue, K. M., Kennedy, K. M., and Acker, J. D. 2007. Vascular health and longitudinal changes in brain and cognition in middle-aged and older adults. *Neuropsychology* 21 (2): 149–157.

Rebeck, G. W., Perls, T. T., West, H. L., Sodhi, P., Lipsitz, L. A., and Hyman, B. T. 1994. Reduced apolipoprotein ε4 allele frequency in the oldest old Alzheimer's patients and cognitively normal individuals. *Neurology* 44 (8): 1513–1516.

Reinvang, I., Deary, I. J., Fjell, A. M., Steen, V. M., Espeseth, T., and Parasuraman, R. 2010. Neurogenetic effects on cognition in aging brains: A window of opportunity for intervention? *Frontiers in Aging Neuroscience* 2: 143.

Renaud, S., and de Lorgeril, M. 1992. Wine, alcohol, platelets, and the French paradox for coronary heart disease. *Lancet* 339 (8808): 1523–1526.

Resnick, S. M., Maki, P. M., Rapp, S. R., Espeland, M. A., Brunner, R., Coker, L. H., et al. 2006. Effects of combination estrogen plus progestin hormone treatment on cognition and affect. *Journal of Clinical Endocrinology and Metabolism* 91 (5): 1802–1810.

Resnick, S. M., Metter, E. J., and Zonderman, A. B. 1997. Estrogen replacement therapy and longitudinal decline in visual memory: A possible protective effect? *Neurology* 49 (6): 1491–1497.

Restrepo, H. E., and Rozental, M. 1994. The social impact of aging populations: Some major issues. *Social Science & Medicine* 39: 1323–1338.

Reuter-Lorenz, P. A., Jonides, J., Smith, E. E., Hartley, A., Miller, A., Marshuetz, C., et al. 2000. Age differences in the frontal lateralization of verbal and spatial working memory revealed by PET. *Journal of Cognitive Neuroscience* 12 (1): 174–187.

Rhyu, I. J., Bytheway, J. A., Kohler, S. J., Lange, H., Lee, K. J., Boklewski, J., et al. 2010. Effects of aerobic exercise training on cognitive function and cortical vascularity in monkeys. *Neuroscience* 167 (4): 1239–1248.

Ribeiro, J. A., Sebastiao, A. M., and de Mendonca, A. 2002. Adenosine receptors in the nervous system: Pathophysiological implications. *Progress in Neurobiology* 68 (6): 377–392.

Ritchie, K., Artero, S., Portet, F., Brickman, A., Muraskin, J., Beanino, E., et al. 2010. Caffeine, cognitive functioning, and white matter lesions in the elderly: Establishing causality from epidemiological evidence. *Journal of Alzheimer's Disease* 20 (suppl. 1): S161–S166.

Ritchie, K., Carriere, I., de Mendonca, A., Portet, F., Dartigues, J. F., Rouaud, O., et al. 2007. The neuroprotective effects of caffeine: A prospective population study (the Three City Study). *Neurology* 69 (6): 536–545.

Ritz, B., and Costello, S. 2006. Geographic model and biomarker-derived measures of pesticide exposure and Parkinson's disease. *Annals of the New York Academy of Sciences* 1076: 378–387.

Ro, T., Noser, E., Boake, C., Johnson, R., Gaber, M., Speroni, A., et al. 2006. Functional reorganization and recovery after constraint-induced movement therapy in subacute stroke: Case reports. *Neurocase* 12 (1): 50–60.

Roberge, M. C., Hotte-Bernard, J., Messier, C., and Plamondon, H. 2008. Food restriction attenuates ischemia-induced spatial learning and memory deficits despite extensive CA1 ischemic injury. *Behavioural Brain Research* 187 (1): 123–132.

Rock, C. L., Flatt, S. W., Sherwood, N. E., Karanja, N., Pakiz, B., and Thomson, C. A. 2010. Effect of a free prepared meal and incentivized weight loss program on weight loss and weight loss maintenance in obese and overweight women: A randomized controlled trial. *Journal of the American Medical Association* 304 (16): 1803–1810.

Rodrigue, K. M., and Raz, N. 2004. Shrinkage of the entorhinal cortex over five years predicts memory performance in healthy adults. *Journal of Neuroscience* 24 (4): 956–963.

Rogers, S. W., Gahring, L. C., Collins, A. C., and Marks, M. 1998. Age-related changes in neuronal nicotinic acetylcholine receptor subunit α4 expression are modified by long-term nicotine administration. *Journal of Neuroscience* 18 (13): 4825–4832.

Rohwedder, S., and Willis, R. J. 2010. Mental retirement. *Journal of Economic Perspectives* 24 (1): 119–138.

Rolland, Y., Pillard, F., Klapouszczak, A., Reynish, E., Thomas, D., Andrieu, S., et al. 2007. Exercise program for nursing home residents with Alzheimer's disease: A 1-year randomized, controlled trial. *Journal of the American Geriatrics Society* 55 (2): 158–165.

Rosenzweig, M. R., and Bennett, E. L. 1972. Cerebral changes in rats exposed individually to an enriched environment. *Journal of Comparative and Physiological Psychology* 80 (2): 304–313.

Rosenzweig, M. R., and Bennett, E. L. 1996. Psychobiology of plasticity: Effects of training and experience on brain and behavior. *Behavioural Brain Research* 78 (1): 57–65.

Rosso, A., Mossey, J., and Lippa, C. F. 2008. Caffeine: Neuroprotective functions in cognition and Alzheimer's disease. *American Journal of Alzheimer's Disease and Other Dementias* 23 (5): 417–422.

Roth, G. S., Mattison, J. A., Ottinger, M. A., Chachich, M. E., Lane, M. A., and Ingram, D. K. 2004. Aging in rhesus monkeys: Relevance to human health interventions. *Science* 305: 1423–1426.

Rovio, S., Kareholt, I., Helkala, E. L., Viitanen, M., Winblad, B., Tuomilehto, J., et al. 2005. Leisure-time physical activity at midlife and the risk of dementia and Alzheimer's disease. *Lancet Neurology* 4 (11): 705–711.

Rugg, M. D., Otten, L. J., and Henson, R. N. 2002. The neural basis of episodic memory: Evidence from functional neuroimaging. *Philosophical Transactions of the Royal Society of London B, Biological Sciences* 357 (1424): 1097–1110.

Rypma, B., Berger, J. S., Genova, H. M., Rebbechi, D., and D'Esposito, M. 2005. Dissociating age-related changes in cognitive strategy and neural efficiency using event-related fMRI. *Cortex* 41 (4): 582–594.

Rypma, B., and D'Esposito, M. 2000. Isolating the neural mechanisms of age-related changes in human working memory. *Nature Neuroscience* 3 (5): 509–515.

Sabia, S., Kivimaki, M., Shipley, M. J., Marmot, M. G., and Singh-Manoux, A. 2009. Body mass index over the adult life course and cognition in late midlife:

The Whitehall II Cohort Study. *American Journal of Clinical Nutrition* 89 (2): 601–607.

Sacks, F. M., Bray, G. A., Carey, V. J., Smith, S. R., Ryan, D. H., Anton, S. D., et al. 2009. Comparison of weight-loss diets with different compositions of fat, protein, and carbohydrates. *New England Journal of Medicine* 360 (9): 859–873.

Salat, D. H., Kaye, J. A., and Janowsky, J. S. 2002. Greater orbital prefrontal volume selectively predicts worse working memory performance in older adults. *Cerebral Cortex* 12 (5): 494–505.

Salat, D. H., Tuch, D. S., Hevelone, N. D., Fischl, B., Corkin, S., Rosas, H. D., et al. 2005. Age-related changes in prefrontal white matter measured by diffusion tensor imaging. *Annals of the New York Academy of Sciences* 1064: 37–49.

Salthouse, T. A. 2006. Mental exercise and mental aging. *Perspectives on Psychological Science* 1: 68–87.

Salthouse, T. A. 2009. When does age-related cognitive decline begin? *Neurobiology of Aging* 30 (4): 507–514.

Sanchez, J., Fisk, A. D., and A., R. W. 2004. Reliability and age-related effects on trust and reliance of a decision support aid. In proceedings of 48th Annual Meeting of the Human Factors and Ergonomics Society, Santa Monica.

Sandstrom, N. J., and Williams, C. L. 2001. Memory retention is modulated by acute estradiol and progesterone replacement. *Behavioral Neuroscience* 115 (2): 384–393.

Santos, C., Lunet, N., Azevedo, A., de Mendonca, A., Ritchie, K., and Barros, H. 2010. Caffeine intake is associated with a lower risk of cognitive decline: A cohort study from Portugal. *Journal of Alzheimer's Disease* 20 (suppl. 1): S175–S185.

Saunders, A. M., Strittmatter, W. J., Schmechel, D., St. George-Hyslop, P. H., Pericak-Vance, M. A., Joo, S. H., et al. 1993. Association of apolipoprotein E allele e4 with late-onset familial and sporadic Alzheimer's disease. *Neurology* 43: 1467–1472.

Sawaguchi, T., and Goldman-Rakic, P. S. 1991. D1 dopamine receptors in prefrontal cortex: Involvement in working memory. *Science* 251: 947–950.

Scarmeas, N., Luchsinger, J. A., Schupf, N., Brickman, A. M., Cosentino, S., Tang, M. X., et al. 2009. Physical activity, diet, and risk of Alzheimer disease. *Journal of the American Medical Association* 302 (6): 627–637.

Schaie, K. W. 1989. Perceptual speed in adulthood: Cross-sectional and longitudinal studies. *Psychology and Aging* 4: 443–453.

Schaie, K. W. 1994. The course of adult intellectual development. *American Psychologist* 49 (4): 304–313.

Schaie, K. W. 2005. What can we learn from longitudinal studies of adult development? *Research in Human Development* 2 (3): 133–158.

Schaie, K. W. 2009. "When does age-related cognitive decline begin?" Salthouse again reifies the "cross-sectional fallacy." *Neurobiology of Aging* 30 (4): 528–529, discussion 530–533.

Scheff, S. W., Price, D. A., Schmitt, F. A., DeKosky, S. T., and Mufson, E. J. 2007. Synaptic alterations in CA1 in mild Alzheimer disease and mild cognitive impairment. *Neurology* 68 (18): 1501–1508.

Schieber, F., ed. 2003. *Human Factors and Aging: Identifying and Compensating for Age-Related Deficits in Sensory and Cognitive Function.* Springer.

Schmiedek, F., Lövdén, M., and Lindenberger, U. 2010. Hundred days of cognitive training enhance broad cognitive abilities in adulthood: Findings from the COGITO Study. *Frontiers of Aging Neuroscience* 2: 27.

Schmierer, K., Wheeler-Kingshott, C. A., Boulby, P. A., Scaravilli, F., Altmann, D. R., Barker, G. J., et al. 2007. Diffusion tensor imaging of post mortem multiple sclerosis brain. *NeuroImage* 35 (2): 467–477.

Schoenhofen, E. A., Wyszynski, D. F., Andersen, S., Pennington, J., Young, R., Terry, D. F., et al. 2006. Characteristics of 32 supercentenarians. *Journal of the American Geriatrics Society* 54 (8): 1237–1240.

Scholz, J., Klein, M. C., Behrens, T. E., and Johansen-Berg, H. 2009. Training induces changes in white-matter architecture. *Nature Neuroscience* 12 (11): 1370–1371.

Schooler, C., and Mulatu, M., and Oates, G. 2004. Occupational self-direction, intellectual functioning, and self-directed orientation in older workers: Findings and implications for individuals and societies. *American Journal of Sociology* 110 (1): 161–197.

Schupf, N., Costa, R., Tang, M. X., Andrews, H., Tycko, B., Lee, J. H., et al. 2004. Preservation of cognitive and functional ability as markers of longevity. *Neurobiology of Aging* 25 (9): 1231–1240.

Seshadri, S., Beiser, A., Selhub, J., Jacques, P. F., Rosenberg, I. H., D'Agostino, R. B., et al. 2002. Plasma homocysteine as a risk factor for dementia and Alzheimer's disease. *New England Journal of Medicine* 346 (7): 476–483.

Seshadri, S., Fitzpatrick, A. L., Ikram, M. A., DeStefano, A. L., Gudnason, V., Boada, M., et al. 2010. Genome-wide analysis of genetic loci associated with Alzheimer disease. *Journal of the American Medical Association* 303 (18): 1832–1840.

Shadlen, M. F., Larson, E. B., Wang, L., Phelan, E. A., McCormick, W. C., Jolley, L., et al. 2005. Education modifies the effect of apolipoprotein ε4 on cognitive decline. *Neurobiology of Aging* 26 (1): 17–24.

Shamosh, N. A., Deyoung, C. G., Green, A. E., Reis, D. L., Johnson, M. R., Conway, A. R., et al. 2008. Individual differences in delay discounting: Relation to intelligence, working memory, and anterior prefrontal cortex. *Psychological Science* 19 (9): 904–911.

Shaper, A. G., Wannamethee, G., and Whincup, P. 1988. Alcohol and blood pressure in middle-aged British men. *Journal of Human Hypertension* 2 (2): 71–78.

Shaw, P., Greenstein, D., Lerch, J., Clasen, L., Lenroot, R., Gogtay, N., et al. 2006. Intellectual ability and cortical development in children and adolescents. *Nature* 440 (7084): 676–679.

Shaywitz, S. E., Shaywitz, B. A., Pugh, K. R., Fulbright, R. K., Skudlarski, P., Mencl, W. E., et al. 1999. Effect of estrogen on brain activation patterns in postmenopausal women during working memory tasks. *Journal of the American Medical Association* 281 (13): 1197–1202.

Sherwin, B. B. 1988. Estrogen and/or androgen replacement therapy and cognitive functioning in surgically menopausal women. *Psychoneuroendocrinology* 13 (4): 345–357.

Sherwin, B. B. 2002. Estrogen and cognitive aging in women. *Trends in Pharmacological Sciences* 23 (11): 527–534.

Sherwin, B. B., and Henry, J. F. 2008. Brain aging modulates the neuroprotective effects of estrogen on selective aspects of cognition in women: A critical review. *Frontiers in Neuroendocrinology* 29 (1): 88–113.

Shi, L., Adams, M. M., Linville, M. C., Newton, I. G., Forbes, M. E., Long, A. B., et al. 2007. Caloric restriction eliminates the aging-related decline in NMDA and AMPA receptor subunits in the rat hippocampus and induces homeostasis. *Experimental Neurology* 206 (1): 70–79.

Shors, T. J., Miesegaes, G., Beylin, A., Zhao, M., Rydel, T., and Gould, E. 2001. Neurogenesis in the adult is involved in the formation of trace memories. *Nature* 410 (6826): 372–376.

Shumaker, S. A., Legault, C., Thal, L., Wallace, R. B., Ockene, J. K., Hendrix, S. L., et al. 2003. Estrogen plus progestin and the incidence of dementia and mild cognitive impairment in postmenopausal women: The Women's Health Initiative Memory Study: A randomized controlled trial. *Journal of the American Medical Association* 289 (20): 2651–2662.

Silva, A. J., Kogan, J. H., Frankland, P. W., and Kida, S. 1998. CREB and memory. *Annual Review of Neuroscience* 21: 127–148.

Simopoulos, A. P. 2002. Omega-3 fatty acids in inflammation and autoimmune diseases. *Journal of the American College of Nutrition* 21 (6): 495–505.

Simpkins, J. W., and Singh, M. 2008. More than a decade of estrogen neuroprotection. *Alzheimer's & Dementia* 4 (1 suppl. 1): S131–S136.

Singh, M., Meyer, E. M., Millard, W. J., and Simpkins, J. W. 1994. Ovarian steroid
deprivation results in a reversible learning impairment and compromised cholinergic function in female Sprague-Dawley rats. *Brain Research* 644 (2): 305–312.

Sliwinski, M., Lipton, R. B., Buschke, H., and Stewart, W. 1996. The effects of pre-
clinical dementia on estimates of normal cognitive functioning in aging. *Journal of
Gerontology B* 51: 217–225.

Small, B. J., Rosnick, C. B., Fratiglioni, L., and Bäckman, L. 2004. Apolipoprotein
E and cognitive performance: A meta-analysis. *Psychology and Aging* 19 (4): 592–
600.

Small, G. W., Silverman, D. H., Siddarth, P., Ercoli, L. M., Miller, K. J., Lavretsky, H.,
et al. 2006. Effects of a 14-day healthy longevity lifestyle program on cognition and
brain function. *American Journal of Geriatric Psychiatry* 14 (6): 538–545.

Small, G. W. 2004. *The Memory Prescription: Dr. Gary Small's 14-Day Plan to Keep Your
Brain and Body Young.* Hyperion.

Small, S. A., Tsai, W. Y., DeLaPaz, R., Mayeux, R., and Stern, Y. 2002. Imaging hippocampal function across the human life span: Is memory decline normal or not?
Annals of Neurology 51 (3): 290–295.

Smiley-Oyen, A. L., Lowry, K. A., Francois, S. J., Kohut, M. L., and Ekkekakis, P. 2008.
Exercise, fitness, and neurocognitive function in older adults: The "selective improvement" and "cardiovascular fitness" hypotheses. *Annals of Behavioral Medicine* 36 (3):
280–291.

Smith, A. D., Smith, S. M., de Jager, C. A., Whitbread, P., Johnston, C., Agacinski,
G., et al. 2010. Homocysteine-lowering by B vitamins slows the rate of accelerated
brain atrophy in mild cognitive impairment: A randomized controlled trial. *PLoS
ONE* 5 (9): e12244.

Smith, D. E., Rapp, P. R., McKay, H. M., Roberts, J. A., and Tuszynski, M. H. 2004.
Memory impairment in aged primates is associated with focal death of cortical
neurons and atrophy of subcortical neurons. *Journal of Neuroscience* 24 (18):
4373–4381.

Smith, D. E., Roberts, J., Gage, F. H., and Tuszynski, M. H. 1999. Age-
associated neuronal atrophy occurs in the primate brain and is reversible by growth
factor gene therapy. *Proceedings of the National Academy of Sciences* 96 (19):
10893–10898.

Smith, E. E., Geva, A., Jonides, J., Miller, A., Reuter-Lorenz, P., and Koeppe, R. A.
2001. The neural basis of task-switching in working memory: Effects of performance
and aging. *Proceedings of the National Academy of Sciences* 98 (4): 2095–2100.

Smith, J. C., Nielson, K. A., Woodard, J. L., Seidenberg, M., Durgerian, S., Antuono, P., et al. 2011. Interactive effects of physical activity and APOE-ε4 on BOLD semantic memory activation in healthy elders. *NeuroImage* 54 (1): 635–644.

Smith, J. P., and Kington, R. 1997. Demographic and economic correlates of health in old age. *Demography* 34 (1): 159–170.

Snyder, J. S., Hong, N. S., McDonald, R. J., and Wojtowicz, J. M. 2005. A role for adult neurogenesis in spatial long-term memory. *Neuroscience* 130 (4): 843–852.

Socci, D. J., Crandall, B. M., and Arendash, G. W. 1995. Chronic antioxidant treatment improves the cognitive performance of aged rats. *Brain Research* 693 (1–2): 88–94.

Solfrizzi, V., Colacicco, A. M., D'Introno, A., Capurso, C., Torres, F., Rizzo, C., et al. 2006. Dietary intake of unsaturated fatty acids and age-related cognitive decline: A 8.5-year follow-up of the Italian Longitudinal Study on Aging. *Neurobiology of Aging* 27 (11): 1694–1704.

Song, H. J., Stevens, C. F., and Gage, F. H. 2002. Neural stem cells from adult hippocampus develop essential properties of functional CNS neurons. *Nature Neuroscience* 5 (5): 438–445.

Song, S. K., Yoshino, J., Le, T. Q., Lin, S. J., Sun, S. W., Cross, A. H., et al. 2005. Demyelination increases radial diffusivity in corpus callosum of mouse brain. *NeuroImage* 26 (1): 132–140.

Sowell, E. R., Thompson, P. M., Tessner, K. D., and Toga, A. W. 2001. Mapping continued brain growth and gray matter density reduction in dorsal frontal cortex: Inverse relationships during postadolescent brain maturation. *Journal of Neuroscience* 21 (22): 8819–8829.

Sowell, E. R., Thompson, P. M., and Toga, A. W. 2004. Mapping changes in the human cortex throughout the span of life. *Neuroscientist* 10 (4): 372–392.

Specter, M. 2001. Rethinking the brain: How the songs of canaries upset a fundamental principle of science. *The New Yorker*, July 23: 42–53.

Spence, S. A., Green, R. D., Wilkinson, I. D., and Hunter, M. D. 2005. Modafinil modulates anterior cingulate function in chronic schizophrenia. *British Journal of Psychiatry* 187: 55–61.

Spencer, J., Waters, E., Romeo, R., Wood, G., Milner, T., and McEwen, B. 2008. Uncovering the mechanisms of estrogen effects on hippocampal function. *Frontiers in Neuroendocrinology* 29 (2): 219–237.

Springer, M. V., McIntosh, A. R., Winocur, G., and Grady, C. L. 2005. The relation between brain activity during memory tasks and years of education in young and older adults. *Neuropsychology* 19 (2): 181–192.

Srivareerat, M., Tran, T. T., Salim, S., Aleisa, A. M., and Alkadhi, K. A. 2011. Chronic nicotine restores normal Aβ levels and prevents short-term memory and E-LTP impairment in Aβ rat model of Alzheimer's disease. *Neurobiology of Aging* 32 (5): 834–844.

Stanton, N. 1997. *Human Factors in Consumer Products*. Taylor and Francis.

Starr, J. M., Deary, I. J., Inch, S., Cross, S., and MacLennan, W. J. 1997. Age-associated cognitive decline in healthy old people. *Age and Ageing* 26 (4): 295–300.

Stepanyants, A., Tamas, G., and Chklovskii, D. B. 2004. Class-specific features of neuronal wiring. *Neuron* 43 (2): 251–259.

Stern, Y. 2002. What is cognitive reserve? Theory and research application of the reserve concept. *Journal of the International Neuropsychological Society* 8 (3): 448–460.

Stern, Y., Albert, S., Tang, M. X., and Tsai, W. Y. 1999. Rate of memory decline in AD is related to education and occupation: Cognitive reserve? *Neurology* 53 (9): 1942–1947.

Sterr, A., Muller, M. M., Elbert, T., Rockstroh, B., Pantev, C., and Taub, E. 1998. Changed perceptions in Braille readers. *Nature* 391 (6663): 134–135.

Stillwell, W., Shaikh, S. R., Zerouga, M., Siddiqui, R., and Wassall, S. R. 2005. Docosahexaenoic acid affects cell signaling by altering lipid rafts. *Reproduction, Nutrition, Development* 45 (5): 559–579.

Stine-Morrow, E. A., Parisi, J. M., Morrow, D. G., and Park, D. C. 2008. The effects of an engaged lifestyle on cognitive vitality: A field experiment. *Psychology and Aging* 23 (4): 778–786.

Stote, K. S., Baer, D. J., Spears, K., Paul, D. R., Harris, G. K., Rumpler, W. V., et al. 2007. A controlled trial of reduced meal frequency without caloric restriction in healthy, normal-weight, middle-aged adults. *American Journal of Clinical Nutrition* 85 (4): 981–988.

Stranahan, A. M., Arumugam, T. V., Cutler, R. G., Lee, K., Egan, J. M., and Mattson, M. P. 2008. Diabetes impairs hippocampal function through glucocorticoid-mediated effects on new and mature neurons. *Nature Neuroscience* 11 (3): 309–317.

Strandberg, J., Kondziella, D., Thorlin, T., and Asztely, F. 2008. Ketogenic diet does not disturb neurogenesis in the dentate gyrus in rats. *Neuroreport* 19 (12): 1235–1237.

Straube, T., Korz, V., and Frey, J. U. 2003. Bidirectional modulation of long-term potentiation by novelty-exploration in rat dentate gyrus. *Neuroscience Letters* 344 (1): 5–8.

Strittmatter, W. J., Saunders, A. M., Schmechel, D., Pericak-Vance, M., Enghild, J., Salvesen, G. S., et al. 1993. Apolipoprotein-E: High-avidity binding to β-amyloid and increased frequency of type-4 allele in late-onset familial Alzheimer disease. *Proceedings of the National Academy of Sciences* 90: 1977–1981.

Stroemer, R. P., Kent, T. A., and Hulsebosch, C. E. 1995. Neocortical neural sprouting, synaptogenesis, and behavioral recovery after neocortical infarction in rats. *Stroke* 26 (11): 2135–2144.

Sullivan, E. V., Adalsteinsson, E., and Pfefferbaum, A. 2006. Selective age-related degradation of anterior callosal fiber bundles quantified in vivo with fiber tracking. *Cerebral Cortex* 16 (7): 1030–1039.

Sundararajan, R., Fryxell, K. J., Lin, M.-K., Greenwood, P. M., and Parasuraman, R. 2006. Comparison of the effect of two SNPs in the DBH gene on working memory. Paper presented to Society for Neuroscience, Atlanta.

Swaab, D. F., Dubelaar, E. J., Hofman, M. A., Scherder, E. J., van Someren, E. J., and Verwer, R. W. 2002. Brain aging and Alzheimer's disease; use it or lose it. *Progress in Brain Research* 138: 343–373.

Takasawa, K., Kitagawa, K., Yagita, Y., Sasaki, T., Tanaka, S., Matsushita, K., et al. 2002. Increased proliferation of neural progenitor cells but reduced survival of newborn cells in the contralateral hippocampus after focal cerebral ischemia in rats. *Journal of Cerebral Blood Flow and Metabolism* 22 (3): 299–307.

Takeuchi, H., Sekiguchi, A., Taki, Y., Yokoyama, S., Yomogida, Y., Komuro, N., et al. 2010. Training of working memory impacts structural connectivity. *Journal of Neuroscience* 30 (9): 3297–3303.

Tanapat, P., Hastings, N. B., Reeves, A. J., and Gould, E. 1999. Estrogen stimulates a transient increase in the number of new neurons in the dentate gyrus of the adult female rat. *Journal of Neuroscience* 19 (14): 5792–5801.

Tang, M. X., Cross, P., Andrews, H., Jacobs, D. M., Small, S., Bell, K., et al. 2001. Incidence of AD in African-Americans, Caribbean Hispanics, and Caucasians in northern Manhattan. *Neurology* 56 (1): 49–56.

Tangney, J. P., Baumeister, R. F., and Boone, A. L. 2004. High self-control predicts good adjustment, less pathology, better grades, and interpersonal success. *Journal of Personality* 72 (2): 271–324.

Tanner, C. M., Ross, G. W., Jewell, S. A., Hauser, R. A., Jankovic, J., Factor, S. A., et al. 2009. Occupation and risk of Parkinsonism: A multicenter case-control study. *Archives of Neurology* 66 (9): 1106–1113.

Taub, E., Miller, N. E., Novack, T. A., Cook, E. W., III, Fleming, W. C., Nepomuceno, C. S., et al. 1993. Technique to improve chronic motor deficit after stroke. *Archives of Physical Medicine and Rehabilitation* 74 (4): 347–354.

Taub, E., Uswatte, G., and Elbert, T. 2002. New treatments in neurorehabilitation founded on basic research. *Nature Reviews Neuroscience* 3 (3): 228–236.

Taylor, J. L., Kennedy, Q., Noda, A., and Yesavage, J. A. 2007. Pilot age and expertise predict flight simulator performance: A 3-year longitudinal study. *Neurology* 68 (9): 648–654.

Teipel, S. J., Meindl, T., Wagner, M., Kohl, T., Burger, K., Reiser, M. F., et al. 2009. White Matter Microstructure in Relation to Education in Aging and Alzheimer's Disease. *Journal of Alzheimers Disease* 17 (3): 457–467.

Teng, E. L., and Chui, H. C. 1987. The Modified Mini-Mental State (3MS) examination. *Journal of Clinical Psychiatry* 48 (8): 314–318.

Teri, L., Gibbons, L. E., McCurry, S. M., Logsdon, R. G., Buchner, D. M., Barlow, W. E., et al. 2003. Exercise plus behavioral management in patients with Alzheimer disease: A randomized controlled trial. *Journal of the American Medical Association* 290 (15): 2015–2022.

Thompson, P. D., Tsongalis, G. J., Seip, R. L., Bilbie, C., Miles, M., Zoeller, R., et al. 2004. Apolipoprotein E genotype and changes in serum lipids and maximal oxygen uptake with exercise training. *Metabolism: Clinical and Experimental* 53 (2): 193–202.

Tierney, M. C., Oh, P., Moineddin, R., Greenblatt, E. M., Snow, W. G., Fisher, R. H., et al. 2009. A randomized double-blind trial of the effects of hormone therapy on delayed verbal recall in older women. *Psychoneuroendocrinology* 34 (7): 1065–1074.

Tisserand, D. J., Pruessner, J. C., Sanz Arigita, E. J., van Boxtel, M. P., Evans, A. C., Jolles, J., et al. 2002. Regional frontal cortical volumes decrease differentially in aging: An MRI study to compare volumetric approaches and voxel-based morphometry. *NeuroImage* 17 (2): 657–669.

Tisserand, D. J., van Boxtel, M. P., Pruessner, J. C., Hofman, P., Evans, A. C., and Jolles, J. 2004. A voxel-based morphometric study to determine individual differences in gray matter density associated with age and cognitive change over time. *Cerebral Cortex* 14 (9): 966–973.

Tombaugh, G. C., Rowe, W. B., Chow, A. R., Michael, T. H., and Rose, G. M. 2002. Theta-frequency synaptic potentiation in CA1 in vitro distinguishes cognitively impaired from unimpaired aged Fischer 344 rats. *Journal of Neuroscience* 22 (22): 9932–9940.

Toni, N., Laplagne, D. A., Zhao, C., Lombardi, G., Ribak, C. E., Gage, F. H., et al. 2008. Neurons born in the adult dentate gyrus form functional synapses with target cells. *Nature Neuroscience* 11 (8): 901–907.

Tozuka, Y., Wada, E., and Wada, K. 2009. Diet-induced obesity in female mice leads to peroxidized lipid accumulations and impairment of hippocampal neurogenesis during the early life of their offspring. *FASEB Journal* 23 (6): 1920–1934.

Tran, Q. T., Calcaterra, G., and Mynatt, E. D. 2005. Cook's collage: Memory aid display for cooking. In proceedings of HOIT 2005, York.

Tran, Z. V., and Weltman, A. 1985. Differential effects of exercise on serum lipid and lipoprotein levels seen with changes in body weight: A meta-analysis. *Journal of the American Medical Association* 254 (7): 919–924.

Tranter, L. J., and Koutstaal, W. 2008. Age and flexible thinking: An experimental demonstration of the beneficial effects of increased cognitively stimulating activity on fluid intelligence in healthy older adults. *Neuropsychology, Development, and Cognition B* 15 (2): 184–207.

Trejo, J. L., Carro, E., and Torres-Aleman, I. 2001. Circulating insulin-like growth factor I mediates exercise-induced increases in the number of new neurons in the adult hippocampus. *Journal of Neuroscience* 21 (5): 1628–1634.

Troen, A., and Rosenberg, I. 2005. Homocysteine and cognitive function. *Seminars in Vascular Medicine* 5 (2): 209–214.

Tschop, M., and Thomas, G. 2006. Fat fuels insulin resistance through Toll-like receptors. *Nature Medicine* 12 (12): 1359–1361.

Turner, D. A., and Deupree, D. L. 1991. Functional elongation of CA1 hippo-campal neurons with aging in Fischer 344 rats. *Neurobiology of Aging* 12 (3): 201–210.

Turner, D. C., Clark, L., Dowson, J., Robbins, T. W., and Sahakian, B. J. 2004. Modafinil improves cognition and response inhibition in adult attention-deficit/hyperactivity disorder. *Biological Psychiatry* 55 (10): 1031–1040.

Turner, D. C., Robbins, T. W., Clark, L., Aron, A. R., Dowson, J., and Sahakian, B. J. 2003a. Cognitive enhancing effects of modafinil in healthy volunteers. *Psychopharmacology* 165 (3): 260–269.

Turner, D. C., Robbins, T. W., Clark, L., Aron, A. R., Dowson, J., and Sahakian, B. J. 2003b. Relative lack of cognitive effects of methylphenidate in elderly male volunteers. *Psychopharmacology* 168 (4): 455–464.

Tyas, S. L., Salazar, J. C., Snowdon, D. A., Desrosiers, M. F., Riley, K. P., Mendiondo, M. S., et al. 2007. Transitions to mild cognitive impairments, dementia, and death: Findings from the Nun Study. *American Journal of Epidemiology* 165 (11): 1231–1238.

Urakawa, S., Hida, H., Masuda, T., Misumi, S., Kim, T. S., and Nishino, H. 2007. Environmental enrichment brings a beneficial effect on beam walking and enhances the migration of doublecortin-positive cells following striatal lesions in rats. *Neuroscience* 144 (3): 920–933.

Uylings, H. B., and de Brabander, J. M. 2002. Neuronal changes in normal human aging and Alzheimer's disease. *Brain and Cognition* 49 (3): 268–276.

Valdez, G., Tapia, J. C., Kang, H., Clemenson, G. D., Jr., Gage, F. H., Lichtman, J. W., et al. 2010. Attenuation of age-related changes in mouse neuromuscular synapses by caloric restriction and exercise. *Proceedings of the National Academy of Sciences* 107 (33): 14863–14868.

van de Rest, O., Geleijnse, J. M., Kok, F. J., van Staveren, W. A., Dullemeijer, C., Olderikkert, M. G., et al. 2008. Effect of fish oil on cognitive performance in older subjects: A randomized, controlled trial. *Neurology* 71 (6): 430–438.

van de Rest, O., Geleijnse, J. M., Kok, F. J., van Staveren, W. A., Hoefnagels, W. H., Beekman, A. T., et al. 2008. Effect of fish-oil supplementation on mental well-being in older subjects: A randomized, double-blind, placebo-controlled trial. *American Journal of Clinical Nutrition* 88 (3): 706–713.

van de Rest, O., Spiro, A., III, Krall-Kaye, E., Geleijnse, J. M., de Groot, L. C., and Tucker, K. L. 2009. Intakes of (n-3): fatty acids and fatty fish are not associated with cognitive performance and 6-year cognitive change in men participating in the Veterans Affairs Normative Aging Study. *Journal of Nutrition* 139 (12): 2329–2336.

van Gelder, B. M., Tijhuis, M., Kalmijn, S., and Kromhout, D. 2007. Fish consumption, n-3 fatty acids, and subsequent 5-y cognitive decline in elderly men: The Zutphen Elderly Study. *American Journal of Clinical Nutrition* 85 (4): 1142–1147.

Van Petten, C. 2004. Relationship between hippocampal volume and memory ability in healthy individuals across the lifespan: Review and meta-analysis. *Neuropsychologia* 42 (10): 1394–1413.

Van Petten, C., Plante, E., Davidson, P. S., Kuo, T. Y., Bajuscak, L., and Glisky, E. L. 2004. Memory and executive function in older adults: Relationships with temporal and prefrontal gray matter volumes and white matter hyperintensities. *Neuropsychologia* 42 (10): 1313–1335.

van Praag, H., Christie, B. R., Sejnowski, T. J., and Gage, F. H. 1999. Running enhances neurogenesis, learning, and long-term potentiation in mice. *Proceedings of the National Academy of Sciences* 96 (23): 13427–13431.

van Praag, H., Kempermann, G., and Gage, F. H. 1999. Running increases cell proliferation and neurogenesis in the adult mouse dentate gyrus. *Nature Neuroscience* 2 (3): 266–270.

van Praag, H., Kempermann, G., and Gage, F. H. 2000. Neural consequences of environmental enrichment. *Nature Reviews Neuroscience* 1 (3): 191–198.

van Praag, H., Schinder, A. F., Christie, B. R., Toni, N., Palmer, T. D., and Gage, F. H. 2002. Functional neurogenesis in the adult hippocampus. *Nature* 415 (6875): 1030–1034.

van Praag, H., Shubert, T., Zhao, C., and Gage, F. H. 2005. Exercise enhances learning and hippocampal neurogenesis in aged mice. *Journal of Neuroscience* 25 (38): 8680–8685.

Varady, K. A., and Hellerstein, M. K. 2007. Alternate-day fasting and chronic disease prevention: A review of human and animal trials. *American Journal of Clinical Nutrition* 86 (1): 7–13.

Vaynman, S., Ying, Z., and Gómez-Pinilla, F. 2004. Exercise induces BDNF and synapsin I to specific hippocampal subfields. *Journal of Neuroscience Research* 76 (3): 356–362.

Venter, J. C., Adams, M. D., Myers, E. W., Li, P. W., Mural, R. J., Sutton, G. G., et al. 2001. The sequence of the human genome. *Science* 291: 1304–1351.

Vercambre, M. N., Grodstein, F., and Kang, J. H. 2010. Dietary fat intake in relation to cognitive change in high-risk women with cardiovascular disease or vascular factors. *European Journal of Clinical Nutrition* 64 (10): 1134–1140.

Verhaeghen, P., Marcoen, A., and Goossens, L. 1992. Improving memory performance in the aged through mnemonic training: A meta-analytic study. *Psychology and Aging* 7 (2): 242–251.

Vernooij, M. W., Ikram, M. A., Vrooman, H. A., Wielopolski, P. A., Krestin, G. P., Hofman, A., et al. 2009. White matter microstructural integrity and cognitive function in a general elderly population. *Archives of General Psychiatry* 66 (5): 545–553.

Vincenzi, D. A., Muldoon, R., Mouloua, M., Parasuraman, R., and Molloy, R. 1996. Effects of aging and workload on monitoring of automation failures. In proceedings of 40th Annual Meeting of the Human Factors and Ergonomics Society, Santa Monica.

Virtanen, J. K., Siscovick, D. S., Longstreth, W. T., Jr., Kuller, L. H., and Mozaffarian, D. 2008. Fish consumption and risk of subclinical brain abnormalities on MRI in older adults. *Neurology* 71 (6): 439–446.

Viswanathan, M., and Guarente, L., 2011. Regulation of *Caenorhabditis elegans* lifespan by sir-2.1 transgenes. *Nature* 477 (7365): E1–E2.

Volkow, N. D., Gur, R. C., Wang, G. J., Fowler, J. S., Moberg, P. J., Ding, Y. S., et al. 1998. Association between decline in brain dopamine activity with age and cognitive and motor impairment in healthy individuals. *American Journal of Psychiatry* 155 (3): 344–349.

Voss, H. U., Uluc, A. M., Dyke, J. P., Watts, R., Kobylarz, E. J., McCandliss, B. D., et al. 2006. Possible axonal regrowth in late recovery from the minimally conscious state. *Journal of Clinical Investigation* 116 (7): 2005–2011.

Voytko, M. L., Mach, R. H., Gage, H. D., Ehrenkaufer, R. L., Efange, S. M., and Tobin, J. R. 2001. Cholinergic activity of aged rhesus monkeys revealed by positron emission tomography. *Synapse* 39 (1): 95–100.

Waddell, J., and Shors, T. J. 2008. Neurogenesis, learning and associative strength. *European Journal of Neuroscience* 27 (11): 3020–3028.

Wake, H., Lee, P., and Fields, R., 2011. Control of local protein synthesis and initial events in myelination by action potentials. *Science* 333: 1647–1651.

Walford, R. L., Mock, D., MacCallum, T., and Laseter, J. L. 1999. Physiologic changes in humans subjected to severe, selective calorie restriction for two years in biosphere 2: Health, aging, and toxicological perspectives. *Toxicological Sciences* 52 (2 suppl.): 61–65.

Walford, R. L., Mock, D., Verdery, R., and MacCallum, T. 2002. Calorie restriction in biosphere 2: Alterations in physiologic, hematologic, hormonal, and biochemical parameters in humans restricted for a 2-year period. *Journal of Gerontology A* 57 (6): B211–B224.

Walhovd, K. B., Fjell, A. M., Reinvang, I., Lundervold, A., Fischl, B., Quinn, B. T., et al. 2004. Size does matter in the long run: Hippocampal and cortical volume predict recall across weeks. *Neurology* 63 (7): 1193–1197.

Walhovd, K. B., Fjell, A. M., Reinvang, I., Lundervold, A., Dale, A. M., Eilertsen, D. E., et al. 2005. Effects of age on volumes of cortex, white matter and subcortical structures. *Neurobiology of Aging* 26 (9): 1261–1270; discussion 1275–1268.

Walhovd, K. B., Fjell, A. M., Dale, A. M., Fischl, B., Quinn, B. T., Makris, N., et al. 2006. Regional cortical thickness matters in recall after months more than minutes. *NeuroImage* 31 (3): 1343–1351.

Walker, C. D., Naef, L., d'Asti, E., Long, H., Xu, Z., Moreau, A., et al. 2008. Perinatal maternal fat intake affects metabolism and hippocampal function in the offspring: A potential role for leptin. *Annals of the New York Academy of Sciences* 1144: 189–202.

Walsh, V., and Cowey, A. 2000. Transcranial magnetic stimulation and cognitive neuroscience. *Nature Reviews Neuroscience* 1 (1): 73–79.

Wang, L., Goldstein, F. C., Veledar, E., Levey, A. I., Lah, J. J., Meltzer, C. C., et al. 2009. Alterations in cortical thickness and white matter integrity in mild cognitive impairment measured by whole-brain cortical thickness mapping and diffusion tensor imaging. *American Journal of Neuroradiology* 30 (5): 893–899.

Wentz, C. T., and Magavi, S. S. 2009. Caffeine alters proliferation of neuronal precursors in the adult hippocampus. *Neuropharmacology* 56 (6–7): 994–1000.

Westerberg, H., and Klingberg, T. 2007. Changes in cortical activity after training of working memory—a single-subject analysis. *Physiology & Behavior* 92 (1–2): 186–192.

Wheeler-Kingshott, C. A., and Cercignani, M. 2009. About "axial" and "radial" diffusivities. *Magnetic Resonance in Medicine* 61 (5): 1255–1260.

White, C. L., Pistell, P. J., Purpera, M. N., Gupta, S., Fernandez-Kim, S. O., Hise, T. L., et al. 2009. Effects of high fat diet on Morris maze performance, oxidative stress, and inflammation in rats: Contributions of maternal diet. *Neurobiology of Disease* 35 (1): 3–13.

White, F., Nicoll, J. A., Roses, A. D., and Horsburgh, K. 2001. Impaired neuronal plasticity in transgenic mice expressing human apolipoprotein E4 compared to E3 in a model of entorhinal cortex lesion. *Neurobiology of Disease* 8 (4): 611–625.

Whitlock, J. R., Heynen, A. J., Shuler, M. G., and Bear, M. F. 2006. Learning induces long-term potentiation in the hippocampus. *Science* 313: 1093–1097.

Wickens, C. D., and Hollands, J. G. 2000. *Engineering Psychology and Human Performance*. Prentice-Hall.

Williams, M. A., Haskell, W. L., Ades, P. A., Amsterdam, E. A., Bittner, V., Franklin, B. A., et al. 2007. Resistance exercise in individuals with and without cardiovascular disease: 2007 update: A scientific statement from the American Heart Association Council on Clinical Cardiology and Council on Nutrition, Physical Activity, and Metabolism. *Circulation* 116 (5): 572–584.

Williams, R. S., and Matthysse, S. 1986. Age-related changes in Down syndrome brain and the cellular pathology of Alzheimer disease. *Progress in Brain Research* 70: 49–67.

Willis, S. L., and Marsiske, M. 1993. *Manual for the Everyday Problems Test*. Pennsylvania State University.

Willis, S. L., and Schaie, K. W. 1986. Training the elderly on the ability factors of spatial orientation and inductive reasoning. *Psychology and Aging* 1 (3): 239–247.

Willis, S. L., Tennstedt, S. L., Marsiske, M., Ball, K., Elias, J., Koepke, K. M., et al. 2006. Long-term effects of cognitive training on everyday functional outcomes in older adults. *Journal of the American Medical Association* 296 (23): 2805–2814.

Wilson, I. A., Ikonen, S., Gallagher, M., Eichenbaum, H., and Tanila, H. 2005. Age-associated alterations of hippocampal place cells are subregion specific. *Journal of Neuroscience* 25 (29): 6877–6886.

Wilson, I. A., Ikonen, S., McMahan, R. W., Gallagher, M., Eichenbaum, H., and Tanila, H. 2003. Place cell rigidity correlates with impaired spatial learning in aged rats. *Neurobiology of Aging* 24 (2): 297–305.

Wilson, M. A., and McNaughton, B. L. 1993. Dynamics of the hippocampal ensemble code for space. *Science* 261: 1055–1058.

Wilson, R. S., Beckett, L. A., Barnes, L. L., Schneider, J. A., Bach, J., Evans, D. A., et al. 2002. Individual differences in rates of change in cognitive abilities of older persons. *Psychology and Aging* 17 (2): 179–193.

Wilson, R. S., Mendes De Leon, C. F., Barnes, L. L., Schneider, J. A., Bienias, J. L., Evans, D. A., et al. 2002. Participation in cognitively stimulating activities and risk of incident Alzheimer disease. *Journal of the American Medical Association* 287 (6): 742–748.

Witte, A. V., Fobker, M., Gellner, R., Knecht, S., and Floel, A. 2009. Caloric restriction improves memory in elderly humans. *Proceedings of the National Academy of Sciences* 106 (4): 1255–1260.

Witte, A. V., Jansen, S., Schirmacher, A., Young, P., and Floel, A. 2011. COMT Val-158Met polymorphism modulates cognitive effects of dietary intervention. *Frontiers in Aging Neuroscience* 2: 146.

Wittenberg, G. F., Chen, R., Ishii, K., Bushara, K. O., Eckloff, S., Croarkin, E., et al. 2003. Constraint-induced therapy in stroke: Magnetic-stimulation motor maps and cerebral activation. *Neurorehabilitation and Neural Repair* 17 (1): 48–57.

Wolf, S. L., Thompson, P. A., Winstein, C. J., Miller, J. P., Blanton, S. R., Nichols-Larsen, D. S., et al. 2010. The EXCITE stroke trial: Comparing early and delayed constraint-induced movement therapy. *Stroke* 41 (10): 2309–2315.

Woolley, C. S. 1999. Electrophysiological and cellular effects of estrogen on neuronal function. *Critical Reviews in Neurobiology* 13 (1): 1–20.

Woolley, C. S., and McEwen, B. S. 1994. Estradiol regulates hippocampal dendritic spine density via an N-methyl-D-aspartate receptor-dependent mechanism. *Journal of Neuroscience* 14 (12): 7680–7687.

Wozniak, J. R., and Lim, K. O. 2006. Advances in white matter imaging: A review of in vivo magnetic resonance methodologies and their applicability to the study of development and aging. *Neuroscience and Biobehavioral Reviews* 30 (6): 762–774.

Wu, A., Ying, Z., and Gómez-Pinilla, F. 2006. Dietary curcumin counteracts the outcome of traumatic brain injury on oxidative stress, synaptic plasticity, and cognition. *Experimental Neurology* 197 (2): 309–317.

Wu, A., Ying, Z., and Gómez-Pinilla, F. 2008. Docosahexaenoic acid dietary supplementation enhances the effects of exercise on synaptic plasticity and cognition. *Neuroscience* 155 (3): 751–759.

Wu, W., Brickman, A. M., Luchsinger, J., Ferrazzano, P., Pichiule, P., Yoshita, M., et al. 2008. The brain in the age of old: The hippocampal formation is targeted differentially by diseases of late life. *Annals of Neurology* 64 (6): 698–706.

Wurm, F., Keiner, S., Kunze, A., Witte, O. W., and Redecker, C. 2007. Effects of skilled forelimb training on hippocampal neurogenesis and spatial learning after focal cortical infarcts in the adult rat brain. *Stroke* 38 (10): 2833–2840.

Xi, M. C., Liu, R. H., Engelhardt, J. K., Morales, F. R., and Chase, M. H. 1999. Changes in the axonal conduction velocity of pyramidal tract neurons in the aged cat. *Neuroscience* 92 (1): 219–225.

Xiang, Z., Huguenard, J. R., and Prince, D. A. 1998. Cholinergic switching within neocortical inhibitory networks. *Science* 281: 985–988.

Xu, T., Yu, X., Perlik, A. J., Tobin, W. F., Zweig, J. A., Tennant, K., et al. 2009. Rapid formation and selective stabilization of synapses for enduring motor memories. *Nature* 462 (7275): 915–919.

Yaffe, K., Barnes, D., Lindquist, K., Cauley, J., Simonsick, E. M., Penninx, B., et al. 2007. Endogenous sex hormone levels and risk of cognitive decline in an older biracial cohort. *Neurobiology of Aging* 28 (2): 171–178.

Yaffe, K., Sawaya, G., Lieburg, I., and Grady, D. 1998. Estrogen therapy in postmenopausal women: Effects on cognitive function and dementia. *Journal of the American Medical Association* 279 (9): 688–695.

Yamamoto, H., Schoonjans, K., and Auwerx, J. 2007. Sirtuin functions in health and disease. *Molecular Endocrinology* 21 (8): 1745–1755.

Yanai, S., Okaichi, Y., and Okaichi, H. 2004. Long-term dietary restriction causes negative effects on cognitive functions in rats. *Neurobiology of Aging* 25 (3): 325–332.

Yantis, S., Schwarzbach, J., Serences, J. T., Carlson, R. L., Steinmetz, M. A., Pekar, J. J., et al. 2002. Transient neural activity in human parietal cortex during spatial attention shifts. *Nature Neuroscience* 5 (10): 995–1002.

Yesavage, J. A., Mumenthaler, M. S., Taylor, J. L., Friedman, L., O'Hara, R., Sheikh, J., et al. 2002. Donepezil and flight simulator performance: Effects on retention of complex skills. *Neurology* 59 (1): 123–125.

Yesavage, J. A., and Rose, T. L. 1984. Semantic elaboration and the method of loci: A new trip for older learners. *Experimental Aging Research* 10 (3): 155–159.

Yeung, E. H., Zhang, C., Louis, G. M., Willett, W. C., and Hu, F. B. 2010. Childhood size and life course weight characteristics in association with the risk of incident type 2 diabetes. *Diabetes Care* 33 (6): 1364–1369.

Yu, E. S., Liu, W. T., Levy, P., Zhang, M. Y., Katzman, R., Lung, C. T., et al. 1989. Cognitive impairment among elderly adults in Shanghai, China. *Journal of Gerontology* 44 (3): S97–S106.

Yurko-Mauro, K., McCarthy, D., Rom, D., Nelson, E. B., Ryan, A. S., Blackwell, A., et al. 2010. Beneficial effects of docosahexaenoic acid on cognition in age-related cognitive decline. *Alzheimers & Dementia* 6 (6): 456–464.

Zec, R. F., and Trivedi, M. A. 2002. Effects of hormone replacement therapy on cognitive aging and dementia risk in postmenopausal women: A review of ongoing large-scale, long-term clinical trials. *Climacteric* 5 (2): 122–134.

Zetterberg, H., Palmer, M., Ricksten, A., Poirier, J., Palmqvist, L., Rymo, L., et al. 2002. Influence of the apolipoprotein E ε4 allele on human embryonic development. *Neuroscience Letters* 324 (3): 189–192.

Ziegler, D. A., Piguet, O., Salat, D. H., Prince, K., Connally, E., and Corkin, S. 2010. Cognition in healthy aging is related to regional white matter integrity, but not cortical thickness. *Neurobiology of Aging* 31 (11): 1912–1926.

Zsembik, B. A., and Peek, M. K. 2001. Race differences in cognitive functioning among older adults. *Journal of Gerontology B* 56 (5): S266–S274.

Index

Obesity, 67, 81, 83, 105, 108, 110, 125,
    211

Parkinson's disease, 5, 99, 112
Pathology, 24, 25, 35, 43, 73, 77–79,
    86, 104, 112, 118, 141
Plasticity, 5, 8, 10, 13, 16–18, 28–36, 39,
    42, 47–59, 62–72, 76, 89, 91, 96,
    105–110, 123–126, 129, 137, 145,
    147, 151–154, 187, 209, 210, 213, 216

Reasoning, 14, 26, 31, 59, 117, 136,
    156–164, 212
Retirement, 3, 5, 143, 144, 186, 215,
    216

Synapses, 10, 13, 20, 33–44, 47–57,
    65–68, 72, 75, 76, 79, 89–93, 96, 107,
    109, 110, 116, 123, 128, 129, 141,
    145, 147, 194, 196

Training, cognitive, 17, 29, 30–36, 43,
    47, 50–58, 63, 65, 68, 69, 75–78, 84,
    87–89, 139, 143–165, 169–176, 183,
    198, 203, 204, 211–216

Visuospatial attention, 6, 117, 162,
    192–197
VO_2 max, 88, 94, 95, 200, 204–207

Walking, 43, 81, 86–89, 94, 171, 199,
    204, 210, 215
Wechsler Memory Scale, 6, 46, 171
White matter, 16, 20–22, 35, 44–47,
    54, 66, 76, 77, 129, 135, 151–156,
    212